WITHDRAWN

D1326313

EEG SIGNAL PROCESSING

EEG SIGNAL PROCESSING

Saeid Sanei and J.A. Chambers

Centre of Digital Signal Processing
Cardiff University, UK

John Wiley & Sons, Ltd

Telephone (+44) 1243 779777

Email (for orders and customer service enquiries): cs-books@wiley.co.uk
Visit our Home Page on www.wileyeurope.com or www.wiley.com

Other Wiley Editorial Offices

John Wiley & Sons Inc., 111 River Street, Hoboken, NJ 07030, USA

Jossey-Bass, 989 Market Street, San Francisco, CA 94103-1741, USA

Wiley-VCH Verlag GmbH, Boschstr. 12, D-69469 Weinheim, Germany

John Wiley & Sons Australia Ltd, 42 McDougall Street, Milton, Queensland 4064, Australia

John Wiley & Sons (Asia) Pte Ltd, 2 Clementi Loop #02-01, Jin Xing Distripark, Singapore 129809

John Wiley & Sons Canada Ltd, 6045 Freemont Blvd, Mississauga, Ontario, L5R 4J3, Canada

Wiley also publishes its books in a variety of electronic formats. Some content that appears in print may not be available in electronic books.

Anniversary Logo Design: Richard J. Pacifico

Library of Congress Cataloging-in-Publication Data

Sanei, Saeid.
EEG signal processing / Saeid Sanei and Jonathon Chambers.
p. ; cm.
Includes bibliographical references.
ISBN 978-0-470-02581-9 (alk. paper)
1. Electroencephalography. 2. Signal processing. I. Chambers, Jonathon. II. Title.
[DNLM: 1. Electroencephalography – methods. 2. Evoked Potentials. 3. Signal Processing, Computer-Assisted. WL 150 S223e 2007]
RC386.6.E43S252 2007
616.8′047547 – dc22

2007019900

British Library Cataloguing in Publication Data

A catalogue record for this book is available from the British Library

ISBN-13 978-0-470-02581-9

Typeset in 10/12 Times by Laserwords Private Limited, Chennai, India
Printed and bound in Great Britain by Antony Rowe Ltd, Chippenham, Wiltshire
This book is printed on acid-free paper responsibly manufactured from sustainable forestry in which at least two trees are planted for each one used for paper production.

Contents

Preface

There is ever-increasing global demand for more affordable and effective clinical and healthcare services. New techniques and equipment must therefore be developed to aid in the diagnosis, monitoring, and treatment of abnormalities and diseases of the human body. Biomedical signals (biosignals) in their manifold forms are rich information sources, which when appropriately processed have the potential to facilitate such advancements. In today's technology, such processing is very likely to be digital, as confirmed by the inclusion of digital signal processing concepts as core training in biomedical engineering degrees. Recent advancements in digital signal processing are expected to underpin key aspects of the future progress in biomedical research and technology, and it is the purpose of this research monograph to highlight this trend for the processing of measurements of brain activity, primarily electroencephalograms (EEGs).

Most of the concepts in multichannel EEG digital signal processing have their origin in distinct application areas such as communications engineering, seismics, speech and music signal processing, together with the processing of other physiological signals, such as electrocardiograms (ECGs). The particular topics in digital signal processing first explained in this research monograph include definitions; illustrations; time-domain, frequency-domain, and time-frequency domain processing; signal conditioning; signal transforms; linear and nonlinear filtering; chaos definition, evaluation, and measurement; certain classification algorithms; adaptive systems; independent component analysis; and multivariate autoregressive modelling. In addition, motivated by research in the field over the last two decades, techniques specifically related to EEG processing such as brain source localization, detection and classification of event related potentials, sleep signal analysis, seizure detection and prediction, together with brain–computer interfacing are comprehensively explained and, with the help of suitable graphs and (topographic) images, simulation results are provided to assess the efficacy of the methods.

Chapter 1 of this research monograph is a comprehensive biography of the history and generation of EEG signals, together with a discussion of their significance and diagnostic capability. Chapter 2 provides an in-depth introduction to the mathematical algorithms and tools commonly used in the processing of EEG signals. Most of these algorithms have only been recently developed by experts in the signal processing community and then applied to the analysis of EEG signals for various purposes. In Chapter 3, event-related potentials are explained and the schemes for their detection and classification are explored. Many neurological and psychiatric brain disorders are diagnosed and monitored using these techniques. Chapter 4 complements the previous chapter by specifically looking at the behaviour of EEG signals in patients suffering from epilepsy. Some very recent methods in seizure prediction are demonstrated. This chapter concludes by opening up a new methodology in joint, or bimodal, EEG–fMRI analysis of epileptic seizure signals. Localization of brain source signals is next covered in Chapter 5. Traditional dipole methods are described and some very recent processing techniques such as blind source separation are briefly reviewed. In Chapter 6, the concepts developed for the analysis and description of EEG sleep recordings are summarized and the important parameters and terminologies are explained. Finally, in Chapter 7, one of the most important applications of the developed mathematical tools for processing of EEG signals, namely brain–computer interfacing, is explored and recent advancements are briefly explained. Results of the application of these algorithms are described.

In the treatment of various topics covered within this research monograph it is assumed that the reader has a background in the fundamentals of digital signal processing and wishes to focus on processing of EEGs. It is hoped that the concepts covered in each chapter provide a foundation for future research and development in the field.

In conclusion, we do wish to stress that in this book there is no attempt to challenge previous clinical or diagnostic knowledge. Instead, the tools and algorithms described in this book can, we believe, potentially enhance the significant clinically related information within EEG signals and thereby aid physicians and ultimately provide more cost-effective and efficient diagnostic tools.

Both authors wish to thank most sincerely our previous and current PhD students who have contributed so much to the material in this work and our understanding of the field. Special thanks to Min Jing, Tracey Lee, Kianoush Nazarpour, Leor Shoker, Loukianous Spyrou, and Wenwu Wang, who contributed to providing some of the illustrations. Finally, this book became truly possible due to spiritual support and encouragement of Maryam Zahabsaniei, Erfan Sanei, and Ideen Sanei.

Saeid Sanei
Jonathon Chambers
January 2007

List of Abbreviations

3D	Three-dimensional
ACT	Adaptive chirplet transform
AD	Alzheimer's disease
ADC	Analogue-to-digital converter
ADD	Attention deficit disorder
ADHD	Attention deficit hyperactivity disorder
AE	Approximate entropy
AEP	Audio evoked potential
Ag–AgCl	Silver–silver chloride
AIC	Akaike information criterion
ALF	Adaptive standardized LORETA/FOCUSS
ALS	Alternating least squares
AMDF	Average magnitude difference function
AMI	Average mutual information
ANN	Artificial neural network
AP	Action potential
APGARCH	Asymmetric power GARCH
AR	Autoregressive modelling
ARMA	Autoregressive moving average
ASDA	American Sleep Disorders Association
BCI	Brain–computer interfacing/interaction
BDS	Brock, Dechert, and Scheinkman
BEM	Boundary element method
BMI	Brain–machine interfacing
BOLD	Blood oxygenation level dependence
BSS	Blind source separation
Ca	Calcium
CANDECOMP	Canonical decomposition
CDR	Current distributed-source reconstruction
CF	Characteristic function
CJD	Creutzfeldt–Jakob disease
Cl	Chloride
CNS	Central nervous system
CSD	Current source density
CT	Computerized tomography

DC	Direct current
DCT	Discrete cosine transform
DLE	Digitally linked ears
DSM	*Diagnostic and Statistical Manual*
DTF	Directed transfer function
DWT	Discrete wavelet transform
ECD	Electric current dipole
ECG	Electrocardiogram/electrocardiography
ECoG	Electrocorticogram
ED	Error distance
EEG	Electroencephalogram/electroencephalography
EGARCH	Exponential GARCH
EGG	Electrogastrography
EKG	Electrocardiogram/electrocardiography
EM	Expectation maximization
EMG	Electromyogram/electromyography
EOG	Electrooculogram
EP	Evoked potential
EPSP	Excitatory postsynaptic potential
ERD	Event-related desynchronization
ERP	Event-related potential
ERS	Event-related synchronization
FA	Factor analysis
FEM	Finite element model
FFNN	Feedforward neural network
FHWA	First half-wave amplitude
FHWD	First half-wave duration
FHWS	First half-wave slope
fICA	Fast independent component analysis
FIR	Finite impulse response
fMRI	Functional magnetic resonance imaging
FOCUSS	Focal underdetermined system solver
FSP	Falsely detected source number percentage
GA	Genetic algorithm
GARCH	Generalized autoregressive conditional heteroskedasticity
GARCH-M	GARCH-in-mean
GFNN	Global false nearest neighbours
GJR-GARCH	Glosten, Jagannathan, and Runkle GARCH
HCI	Human–computer interfacing/interaction
HMM	Hidden Markov model
HOS	Higher-order statistics
IBE	International Bureau for Epilepsy
ICA	Independent component analysis
IIR	Infinite impulse response
ILAE	International League Against Epilepsy

IPSP	Inhibitory postsynaptic potential
IR	Impulse response
ISODATA	Iterative self-organizing data analysis technique algorithm
JADE	Joint approximate diagonalization of eigenmatrices
K	Potassium
KL	Kullback–Laibler
KLT	Karhunen–Loéve transform
KT	Kuhn–Tucker
LD	Linear discriminants
LDA	Linear discriminant analysis
LDA	Long-delta activity
LE	Lyapunov exponent
LEM	Local EEG model
LLE	Largest Lyapunov exponent
LMS	Least mean square
LORETA	Low-resolution electromagnetic tomography algorithm
LP	Lowpass
LRT	Low-resolution tomography
LS	Least squares
LWR	Levinson–Wiggins–Robinson
MA	Moving average
MAF	Multivariate ambiguity function
MAP	Maximum *a posteriori*
MDP	Moving dipole
MEG	Magnetoencephalogram
MI	Mutual information
MIL	Matrix inversion lemma
ML	Maximum likelihood
MLE	Maximum likelihood estimation
MLE	Maximum Lyapunov exponent
MLP	Multilayered perceptron
MMN	Mismatch negativity
MP	Matching pursuits
MRI	Magnetic resonance imaging
MS	Mean square
MS	Multiple sclerosis
MSE	Mean-squared error
MTLE	Mesial temporal lobe epilepsy
MUSIC	Multichannel signal classification
MVAR	Multivariate autoregressive
Na	Sodium
NLMS	Normalized least mean square
NMF	Nonnegative matrix factorization
NN	Neural network
NREM	Nonrapid eye movement

OA	Ocular artefact
OBS	Organic brain syndrome
OP	Oddball paradigm
PARAFAC	Parallel factor
PCA	Principal component analysis
PD	Parkinson's disease
PDF	Probability density function
PET	Positron emission tomography
PLED	Periodic literalized epileptiform discharges
PMBS	Postmovement beta synchronization
PNRD	Persistent nonrhythmic delta activity
POST	Positive occipital sharp transients
PPM	Piecewise Prony method
PSDM	Phase-space dissimilarity measures
PSG	Polysomnography
PWVD	Pseudo Wigner–Ville distribution
QEEG	Quantitative EEG
QGARCH	Quadratic GARCH
QNN	Quantum neural network
QP	Quadratic programming
R&K	Rechtschtschaffen and Kales
RAP	Recursively applied and projected
RBD	REM sleep behaviour disorder
RBF	Radial basis function
REM	Rapid eye movement
RKHS	Reproducing kernel Hilbert space
RLS	Recursive least squares
RV	Residual variance
SAS	Sleep apnea syndrome
SCA	Sparse component analysis
SCP	Slow cortical potential
SCPS	Slow cortical potential shift
SDTF	Short-time DTF
SEM	Structural equation modelling
SHWA	Second half-wave amplitude
SHWD	Second half-wave duration
SHWS	Second half-wave slope
sLORETA	Standardized LORETA
SNNAP	Simulator for neural networks and action potentials
SNR	Signal-to-noise ratio
SOBI	Second-order blind identification
SPET	Single photon emission tomography
SREDA	Subclinical rhythmic EEG discharges of adults
SRNN	Sleep EEG recognition neural network
SSLOFO	Source shrinking LORETA–FOCUSS
SSPE	Subacute sclerosing panencepalities

SSVEP	Steady-state visual-evoked potential
SSVER	Steady-state visual-evoked response
STF	Space–time–frequency
STFD	Spatial time–frequency distribution
STFT	Short-time frequency transform
STL	Short-term largest Lyapunov exponent
SV	Support vector
SVD	Singular-value decomposition
SVM	Support vector machine
SWA	Slow-wave activity
SWDA	Step-wise discriminant analysis
SWS	Slow-wave sleep
TDNN	Time delay neural network
TF	Time–frequency
TGARCH	Threshold GARCH model
TLE	Temporal lobe epilepsy
TNM	Traditional nonlinear method
TTD	Thought translation device
USP	Undetected source number percentage
VEP	Visual evoked potential
WA	Wald tests on amplitudes
WL	Wald test on locations
WMN	Weighted minimum norm
WN	Wavelet network
WT	Wavelet transform
WV	Wigner–Ville

List of Symbols

$+\text{eV}$	Positive charge (electron-volt)
Hz	Hertz; cycles per second
mV	Millivolt
uF/cm^2	Microfarad per squared centimetre
Ca^{++}	Calcium with positive charge
K^+	Potassium with positive charge
Na^+	Sodium with positive charge
$a_{\text{Ca}}(.)$	Steady-state activation function
$a_K(.)$	Prediction coefficients
$\mathbf{A}^\dagger$	Moore–Penrose pseudo-inverse of $\mathbf{A}$
$\mathbf{A}^{-1}$	Inverse of matrix $\mathbf{A}$
$\hat{\mathbf{A}}$	Estimate of $\mathbf{A}$
A_k	Amplitude of the exponential function
$\mathbf{A}_\text{k}$	Matrix with known column vectors
arg	Argument
$\mathbf{A}_\text{uk}$	Matrix with unknown column vectors
$A_x(\tau, \nu)$	Ambiguity function
AE	Approximate entropy
$b^2(\omega_1, \omega_2)$	Bicoherency index
(i)	Magnetic field at electrode i
$B(\omega_1, \omega_2)$	Bispectrum
C_e	Number of interneuron cells
C_i	Number of thalamocortical neurons
$C_{ij}(\omega)$	Fourier transform of the cross-correlation coefficients between channels i and j
$C_j(k)$	Wavelet coefficient
$\mathbf{C}_\text{noise}$	Covariance of noise, i.e. $\mathbf{C}_\text{noise} = \sigma_\text{n}^2\mathbf{I}$
$\mathbf{C}_\text{sig}$	Covariance of the source signals
$\mathbf{C}_\mathbf{X}$	$\mathbf{XX}^\text{T}$
C_1	Maximum subspace correlation
$C_4^\alpha(0, 0, 0)$	Fourier coefficients of the fourth-order cyclic cumulant at zero lag
$C_{\boldsymbol{\theta}_n, \boldsymbol{x}_n}$	Cross-covariance matrix between $\boldsymbol{\theta}_n$ and $\boldsymbol{x}_n$

$\mathrm{Coh}_{ij}^2(\omega)$	Coherency of channels i and j
$\mathrm{col}\{\mathbf{F}\}_i$	$[\underbrace{0 \cdots 0}_{i-1} f_0 \cdots f_M \underbrace{0 \cdots 0}_{N-M-i}]^T$
Cum	Cumulant
d_E	Embedding dimension
$\frac{\mathrm{d}}{\mathrm{d}t}$	Differentiation with respect to time
$\mathbf{D}$	Scaling matrix
D_a	Attractor dimension
D_I	Information dimension
D_L	Lyapunov dimension
D_r	Correlation dimension
$\mathbf{D}_{ss}(.)$	STFD of the source signals
$\mathbf{D}_{xx}(.)$	STFD of the mixed signals
$\mathrm{diag}(\lambda_1, \lambda_2, \ldots, \lambda_M)$	A diagonal $M \times M$ matrix with the elements $\lambda_1, \lambda_2, \ldots, \lambda_M$
$\mathrm{D}\boldsymbol{w}(k\Delta)$	$\boldsymbol{w}(k\Delta + \Delta) - \boldsymbol{w}(k\Delta)$
$e(n)$	Residual (error)
exp(.)	Exponential
$E\ [.]$	Statistical expectation
$\breve{E}_{kb}$	Dispersion measure
µV/cm	Microvolts per centimetre
$\vert E_n/E_0 \vert$	Total error amplification factor
E_syn	Synaptic reversal potential
f_k	Discrete-time sinusoidal frequency in samples per second
f_s	Sampling frequency
$\overline{g}$	Maximal conductance
$\overline{g}_\mathrm{K}$	Potassium current
$\overline{g}_\mathrm{syn}$	Maximal synaptic conductance
g'	First derivative of g
g''	Second derivative of g
$\mathbf{G}(.)$	Gain matrix
$G_{\gamma_k}(t, \omega)$	Wigner time–frequency distribution of the kth selected function
GARCH(p,q)	GARCH model with prediction orders p and q
h_e	Excitatory postsynaptic potential
h_i	Inhibitory postsynaptic potential
$H(\omega)$	Fourier transforms of $h(t)$
$H(.)$	Entropy
$\mathbf{I}$	Identity matrix
I_i	Ionic current
$I(s)$	Entropy of s
Im(.)	Imaginary part
j	$\sqrt{-1}$
J_n	Cost function
$\mathbf{J} = [\boldsymbol{j}_1, \boldsymbol{j}_2, \ldots, \boldsymbol{j}_\mathrm{T}]$	Moments of the sources
$\Im$	Function linking the sources to the electrode potentials

$K(.,.)$	Kernel function
KL	Kulback–Laibler distance (divergence)
kurt(.)	Kurtosis
$l(i)$	Latency of the ith ERP component
l_1	Error amplification factor in the context of Lyapunov exponents
$\mathbf{L}_j$	Location of the jth dipole
$\mathbf{L}_k$	Matrix of prediction coefficients
m	Embedding dimension; number of sources
$m_k(.)$	kth moment
$M(\theta, \tau)$	Characteristic function (CF)
$\mathbf{MA}(\theta, \tau)$	Multivariate ambiguity function (MAF)
max	Maximum
min	Minimum
$\min_{\rho,\theta,\mathbf{S}}$	Minimization with respect to ρ, θ, and $\mathbf{S}$
$m \times l$	m by l
n	Discrete time index; iteration index
n_e	Number of electrodes
n_r	Number of remaining mixtures
N	Number of data samples
Neg	Negentropy
p	Prediction order
$p(\mathbf{Y})$	Joint probability distribution
p_x	Probability density function (PDF) of signal $x(n)$
$p_y(y_i(n))$	Marginal distribution of the ith output
$\mathbf{P}$	Permutation matrix
$P(.)$	Probability density function
$\mathbf{P}_\mathbf{H}^\perp$	Projection matrix to the orthogonal complement of the column; space of $\mathbf{H}$, i.e. $\mathbf{P}_\mathbf{H}^\perp = (\mathbf{I} - \mathbf{H}\mathbf{H}^\dagger)$
$P(f, n)$	Signal power at a given time–frequency
$P_\mathrm{ref}(f)$	Average power during some reference time calculated for frequency f
$q(n)$	Two-dimensional Fourier transform of the bispectrum
$r_{xy}(\tau)$	Cross-correlation between two zero mean wide-sense stationary continuous random signals $x(t)$ and $y(t)$
$\hat{r}_x(.,m)$	Estimated autocorrelation function for the mth frame
$\mathbf{R}(q)$	Covariance matrix of $\boldsymbol{x}(n)$
$\mathbf{R}_\mathrm{S}(k)$	Covariance of the sources at discrete time lag k
$\mathbf{R}_\mathrm{X}(k)$	Covariance of the mixed signals at discrete time lag k
$\mathbf{R}_\mathrm{Y}(k)$	Covariance of the outputs (estimated sources) at discrete time lag k
$\Re^M$	M-dimensional space
Re(.)	Real part
$\hat{s}(n)$	Best estimate of the actual signal $s(n)$ in the mean-square sense

$S_{xx}(\lambda_k, \boldsymbol{\theta})$	Spectral density of a Gaussian vector process $\boldsymbol{x}$ with Parameters $\boldsymbol{\theta}$
$\mathbf{S}\| \otimes \|\mathbf{D}$	Katri–Rao product
sgn(.)	Sign of (.)
subcorr	Subspace correlation
T	Signal length
T_r	Empirical threshold
tanh	Hyperbolic tangent
$\dot{u}(t)$	$du(t)/dt$
U_s	Spatial prior
U_t	Temporal prior
U_1	Likelihood probability
U_2	Prior probability
$\boldsymbol{v}(n)$	Zero-mean noise
V_{A_1}	Left earlobe reference voltage
V_{A_2}	Right earlobe reference voltage
w_k	Activation variable
$w_\infty(E)$	Steady-state activation function
$\dot{w}(t)$	$dw(t)/dt$
w_{opt}	Optimum $\boldsymbol{w}$
$W(a, b)$	Wavelet transform
x_0	Initial value of x
$\overline{x}_c$	Cluster centre
$X_i(\omega)$	Fourier transform of the ith segment of one EEG channel
$X_p(\omega)$	Power spectrum for signal x
$X_{WV}(t, \omega)$	Wigner–Ville frequency distribution of a signal $x(t)$
$\breve{X}_{WV}(t, \omega)$	Pseudo Wigner–Ville distribution (PWVD)
$X_{WV}(t, \omega)$	Spectral distribution of both auto- and cross-correlations
$\boldsymbol{x} \in R^d$	$\boldsymbol{x}$ belongs to the d-dimensional space of real values
Z	Z-transform
Z^{-1}	Inverse Z-transform
α	Alpha brain rhythm
α	Penalty term
α_j and β_j	Nonlinear model coefficients
α_k	Damping factor
$\alpha(E)$	Forward rate function
β	Penalty term
$\beta(E)$	Backward rate function
γ	Gamma brain rhythm
γ	Learning rate
δ	Delta brain rhythm
δ	*A priori* chosen tolerance
δ_1	Minimal error
δ_2	Maximal error
$\Delta w(n)$	$\boldsymbol{w}(n) - \boldsymbol{w}(n-1)$

ζ	Learning rate
$\eta\ (e(n)^2)$	Performance index
θ	Theta brain rhythm
θ_j	Initial phase in radians
κ	Kappa brain activity
λ	Lambda brain activity
λ	Largest Lyapunov exponent; regulation parameter
λ^*	Finite-time exponent
$\lambda(x_0)$	Ljapunov (Lyapunov) exponents of x starting from x_0
λ_i	Lyapunov exponent
λ_j	jth eigenvalue
λ_k	$2\pi k/N$
$\lambda_{\max}$	Maximum eigenvalue
λ_t	Temperature-dependent factor
λ_1	Maximum Lyapunov exponent (MLE)
$\hat{\mathbf{\Lambda}}_e$	Diagonal matrix of noise eigenvalues
$\hat{\mathbf{\Lambda}}_s$	Diagonal matrix containing the first m eigenvalues
μ	Mu brain rhythm
μ	Learning rate; convergence parameter; step size; mean
ξ_i	Slack parameter for feature i
ρ_{xy}	Cross-correlation coefficient
σ	Sigma brain activity
σ_i	Width of the ith ERP component
σ_x	Standard deviations of $x(n)$
σ_y	Standard deviations of $y(n)$
σ^2	Variance
σ_{ij}^2	Noise power of the ARMA model
σ_n^2	Noise variance
σ_t^2	Time-domain variance
σ_ω^2	Frequency-domain variance
τ	Tau brain activity
τ	Fractal dimension
τ_i	Latency for the ith template
$\tau_k(E)$	Time-constant function
φ	Phi brain activity
$\phi(t)$	Scaling function
$\varphi(\cdot)$	Nonlinear penalty function
$\phi(\theta, \tau)$	Kernel function acting as a mask to enhance the regions in the TF domain
$\phi(.,.)$	Cohen's class
$\mathbf{\Phi}_e$	Noise-only subspace span
$\mathbf{\Phi}_s$	Signal subspace span
$\Phi(\omega)$	Fourier transforms of $\phi(t)$
$\mathbf{\Phi} = \{\varphi_k\}$	Set of orthogonal basis functions
$\hat{\mathbf{\Phi}}_e$	Remaining (noise) eigenvectors

$\hat{\mathbf{\Phi}}_s$	First m left singular vectors
χ_{n-1}	A sequence containing all the past information up to time $n-1$
ω	Angular frequency in radians per second
$\nabla_m(.)$	Approximated gradient with respect to m
$\left.\frac{\partial J(\mathbf{W})}{\partial \mathbf{W}}\right\|_{\mathbf{W}=\mathbf{W}(n)}$	Differentiation with respect to **W** evaluated at $\mathbf{W} = \mathbf{W}(n)$
$(.)^{\mathrm{H}}$	Hermitian; complex conjugate transpose
$(.)^{\mathrm{T}}$	Transpose operation
$\|.\|$	Absolute value
$\\|.\\|_{\mathrm{F}}$	Frobenius norm
$\\|.\\|_1$	L_1-norm
$\\|.\\|^2$	Euclidean norm
$\\|.\\|_2^2$	Squared Euclidean norm
$\langle \cdot, \cdot \rangle$	Inner product
$\prod_{i=1}^{m}$	Multiplication of m components (signals)
$\sum_{i=1}^{N}$	Summation of N components (signals)
$\forall n$	For all n values
$\exists q$	There exists a 'q'
$*$	Linear convolution
$**$	Two-dimensional discrete-time convolution
$\propto$	Proportional to
∞	Infinity
$\geq$	Larger or equal to
$\leq$	Smaller or equal to
$\uparrow 2$	Up-sample by 2
$\downarrow 2$	Down-sample by 2

1

Introduction to EEG

The neural activity of the human brain starts between the 17th and 23rd week of prenatal development. It is believed that from this early stage and throughout life electrical signals generated by the brain represent not only the brain function but also the status of the whole body. This assumption provides the motivation to apply advanced digital signal processing methods to the electroencephalogram (EEG) signals measured from the brain of a human subject, and thereby underpins the later chapters of the book.

Although nowhere in this book do the authors attempt to comment on the physiological aspects of brain activities there are several issues related to the nature of the original sources, their actual patterns, and the characteristics of the medium, that have to be addressed. The medium defines the path from the neurons, as so-called signal sources, to the electrodes, which are the sensors where some form of mixtures of the sources are measured.

Understanding of neuronal functions and neurophysiological properties of the brain together with the mechanisms underlying the generation of signals and their recordings is, however, vital for those who deal with these signals for detection, diagnosis, and treatment of brain disorders and the related diseases. A brief history of EEG measurements is first provided.

1.1 History

Carlo Matteucci (1811–1868) and Emil Du Bois-Reymond (1818–1896) were the first people to register the electrical signals emitted from muscle nerves using a galvanometer and established the concept of neurophysiology [1,2]. However, the concept of *action current* introduced by Hermann Von Helmholz [3] clarified and confirmed the negative variations that occur during muscle contraction.

Richard Caton (1842–1926), a scientist from Liverpool, England, used a galvanometer and placed two electrodes over the scalp of a human subject and thereby first recorded brain activity in the form of electrical signals in 1875. Since then, the concepts of electro-(referring to registration of brain electrical activities) encephalo- (referring to emitting the signals from the head), and gram (or graphy), which means drawing or writing, were combined so that the term EEG was henceforth used to denote electrical neural activity of the brain.

EEG Signal Processing S. Sanei and J. Chambers

Fritsch (1838–1927) and Hitzig (1838–1907) discovered that the human cerebral can be electrically stimulated. Vasili Yakovlevich Danilevsky (1852–1939) followed Caton's work and finished his PhD thesis in the investigation of the physiology of the brain in 1877 [4]. In this work, he investigated the activity of the brain following electrical stimulation as well as spontaneous electrical activity in the brain of animals.

The cerebral electrical activity observed over the visual cortex of different species of animals was reported by Ernst Fleischl von Marxow (1845–1891). Napoleon Cybulski (1854–1919) provided EEG evidence of an epileptic seizure in a dog caused by electrical stimulation.

The idea of the association of epileptic attacks with abnormal electrical discharges was expressed by Kaufman [5]. Pravidch-Neminsky (1879–1952), a Russian physiologist, recorded the EEG from the brain, termed the dura, and the intact skull of a dog in 1912. He observed a 12–14 cycle/s rhythm under normal conditions, which slowed under asphyxia and later called it the *electrocerebrogram*.

The discoverer of the existence of human EEG signals was Hans Berger (1873–1941). He began his study of human EEGs in 1920 [6]. Berger is well known by almost all electroencephalographers. He started working with a string galvanometer in 1910, then migrated to a smaller Edelmann model, and after 1924, to a larger Edelmann model. In 1926, Berger started to use the more powerful Siemens double coil galvanometer (attaining a sensitivity of 130 μV/cm) [7]. His first report of human EEG recordings of one to three minutes duration on photographic paper was in 1929. In this recording he only used a one-channel bipolar method with fronto-occipital leads. Recording of the EEG became popular in 1924. The first report of 1929 by Berger included the alpha rhythm as the major component of the EEG signals, as described later in this chapter, and the alpha blocking response.

During the 1930s the first EEG recording of sleep spindles was undertaken by Berger. He then reported the effect of hypoxia on the human brain, the nature of several diffuse and localized brain disorders, and gave an inkling of epileptic discharges [8]. During this time another group established in Berlin-Buch and led by Kornmüller, provided more precise recording of the EEG [9]. Berger was also interested in cerebral localization and particularly in the localization of brain tumours. He also found some correlation between mental activities and the changes in the EEG signals.

Toennies (1902–1970) from the group in Berlin built the first biological amplifier for the recording of brain potentials. A differential amplifier for recording EEGs was later produced by the Rockefeller foundation in 1932.

The importance of multichannel recordings and using a large number of electrodes to cover a wider brain region was recognized by Kornmüller [10]. The first EEG work focusing on epileptic manifestation and the first demonstration of epileptic spikes were presented by Fischer and Löwenbach [11–13].

In England, W. Gray Walter became the pioneer of clinical electroencephalography. He discovered the foci of slow brain activity (delta waves), which initiated enormous clinical interest in the diagnosis of brain abnormalities. In Brussels, Fredric Bremer (1892–1982) discovered the influence of afferent signals on the state of vigilance [14].

Research activities related to EEGs started in North America in around 1934. In this year, Hallowell Davis illustrated a good alpha rhythm for himself. A cathode ray oscilloscope was used around this date by the group in St Louis University in Washington, in the study of peripheral nerve potentials. The work on human EEGs started at Harvard in Boston and the University of Iowa in the 1930s. The study of epileptic seizure developed by Fredric Gibbs was the major work on EEGs during these years, as the realm of epileptic seizure disorders was the domain of their greatest effectiveness. Epileptology may be divided historically into two periods [15]: before and after the advent of EEG. Gibbs and Lennox applied the idea of Fischer based on his studies about picrotoxin and its effect on the cortical EEG in animals to human epileptology. Berger [16] showed a few examples of paroxysmal EEG discharges in a case of presumed petit mal attacks and during a focal motor seizure in a patient with general paresis.

As the other great pioneers of electroencephalography in North America, Hallowel and Pauline Davis were the earliest investigators of the nature of EEG during human sleep. A. L. Loomis, E. N. Harvey, and G. A. Hobart were the first who mathematically studied the human sleep EEG patterns and the stages of sleep. At McGill University, H. Jasper studied the related behavioural disorder before he found his niche in basic and clinical epileptology [17].

The American EEG Society was founded in 1947 and the First International EEG Congress was held in London, United Kingdom, around this time. While the EEG studies in Germany were still limited to Berlin, Japan gained attention by the work of Motokawa, a researcher of EEG rhythms [18]. During these years the neurophysiologists demonstrated the thalamocortical relationship through anatomical methods. This led to the development of the concept of centrencephalic epilepsy [19].

Throughout the 1950s the work on EEGs expanded in many different places. During this time surgical operation for removing the epileptic foci became popular and the book entitled *Epilepsy and the Functional Anatomy of the Human Brain* (Penfiled and Jasper) was published. During this time microelectrodes were invented. They were made of metals such as tungsten or glass, filled with electrolytes such as potassium chloride, with diameters of less than 3 μm.

Depth electroencephalography of a human was first obtained with implanted intracerebral electrodes by Mayer and Hayne (1948). Invention of intracellular microelectrode technology revolutionized this method and was used in the spinal cord by Brock *et al.* in 1952 and in the cortex by Phillips in 1961.

Analysis of EEG signals started during the early days of EEG measurement. Berger assisted by Dietch (1932) applied Fourier analysis to EEG sequences, which was rapidly developed during the 1950s. Analysis of sleep disorders with EEGs started its development in the 1950s through the work of Kleitman at the University of Chicago.

In the 1960s analysis of the EEGs of full-term and premature newborns began its development [20]. Investigation of evoked potentials (EPs), especially visual EPs, as commonly used for monitoring mental illnesses, progressed during the 1970s.

The history of EEG, however, has been a continuous process, which started from the early 1300s and has brought daily development of clinical, experimental, and computational studies for discovery, recognition, diagnosis, and treatment of a vast number of neurological and physiological abnormalities of the brain and the rest of the central nervous system (CNS) of human beings. Nowadays, EEGs are recorded invasively and noninvasively using fully computerized systems. The EEG machines are equipped with many signal processing tools, delicate and accurate measurement electrodes, and enough memory for very long-term recordings of several hours. EEG or MEG (magnetoencephalogram) machines may be integrated with other neuroimaging systems such as functional magnetic resonance imaging (fMRI). Very delicate needle-type electrodes can also be used for recording the EEGs from over the cortex (electrocortiogram), and thereby avoid the attenuation and nonlinearity effects induced by the skull. The nature of neural activities within the human brain will be discribed next.

1.2 Neural Activities

The CNS generally consists of nerve cells and glia cells, which are located between neurons. Each nerve cell consists of axons, dendrites, and cell bodies. Nerve cells respond to stimuli and transmit information over long distances. A nerve cell body has a single nucleus and contains most of the nerve cell metabolism, especially that related to protein synthesis. The proteins created in the cell body are delivered to other parts of the nerve. An axon is a long cylinder, which transmits an electrical impulse and can be several metres long in vertebrates (giraffe axons go from the head to the tip of the spine). In humans the length can be a percentage of a millimetre to more than a metre. An axonal transport system for delivering proteins to the ends of the cell exists and the transport system has 'molecular motors', which ride upon tubulin rails.

Dendrites are connected to either the axons or dendrites of other cells and receive impulses from other nerves or relay the signals to other nerves. In the human brain each nerve is connected to approximately 10,000 other nerves, mostly through dendritic connections.

The activities in the CNS are mainly related to the synaptic currents transferred between the junctions (called synapses) of axons and dendrites, or dendrites and dendrites of cells. A potential of 60–70 mV with negative polarity may be recorded under the membrane of the cell body. This potential changes with variations in synaptic activities. If an action potential travels along the fibre, which ends in an *excitatory* synapse, an excitatory postsynaptic potential (EPSP) occurs in the following neuron. If two action potentials travel along the same fibre over a short distance, there will be a summation of EPSPs producing an action potential on the postsynaptic neuron providing a certain threshold of membrane potential is reached. If the fibre ends in an *inhibitory* synapse, then hyperpolarization will occur, indicating an inhibitory postsynaptic potential (IPSP) [21,22]. Figure 1.1 shows the above activities schematically.

Following the generation of an IPSP, there is an overflow of cations from the nerve cell or an inflow of anions into the nerve cell. This flow ultimately causes a change in potential along the nerve cell membrane. Primary transmembranous currents generate secondary inonal currents along the cell membranes in the intra- and extracellular space. The portion of these currents that flow through the extracellular space is directly responsible for the generation of field potentials. These field potentials, usually with less than 100 Hz frequency, are called EEGs when there are no changes in the signal average and DC if there are slow drifts in the average signals, which may mask the actual EEG signals. A combination of EEG and DC potentials is often observed for some abnormalities in the brain such as seizure (induced by pentylenetetrazol), hypercapnia, and asphyxia [23]. The focus will next be on the nature of active potentials.

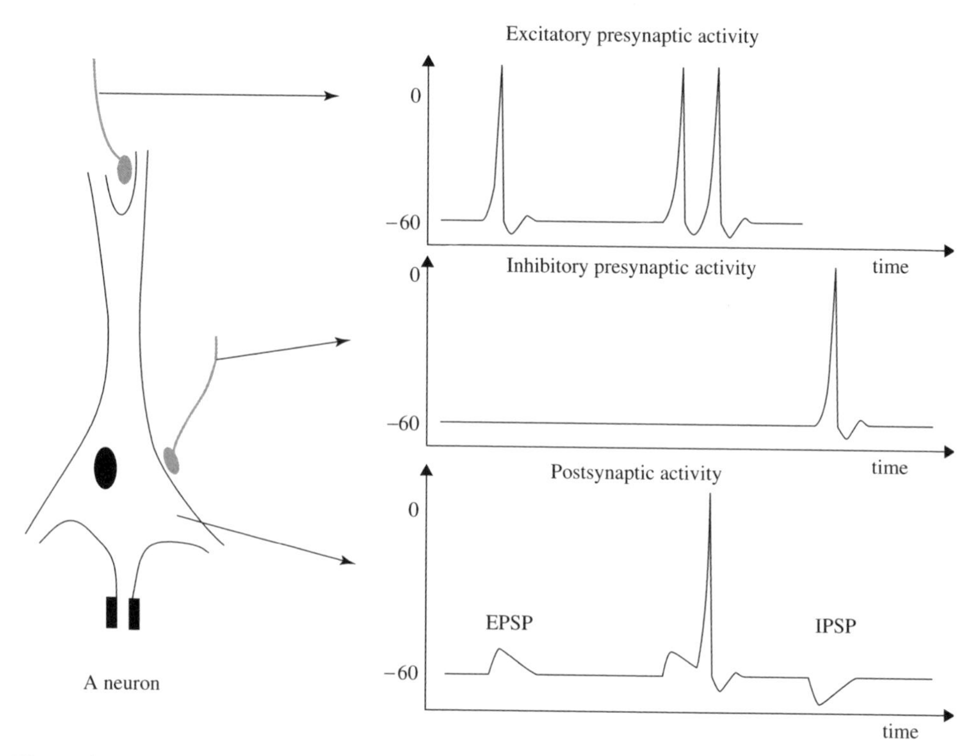

Figure 1.1 The neuron membrane potential changes and current flow during synaptic activation recorded by means of intracellular microelectrodes. Action potentials in the excitatory and inhibitory presynaptic fibre respectively lead to EPSP and IPSP in the postsynaptic neuron

1.3 Action Potentials

The information transmitted by a nerve is called an action potential (AP). APs are caused by an exchange of ions across the neuron membrane and an AP is a temporary change in the membrane potential that is transmitted along the axon. It is usually initiated in the cell body and normally travels in one direction. The membrane potential depolarizes (becomes more positive), producing a spike. After the peak of the spike the membrane repolarizes (becomes more negative). The potential becomes more negative than the resting potential and then returns to normal. The action potentials of most nerves last between 5 and 10 milliseconds. Figure 1.2 shows an example AP.

The conduction velocity of action potentials lies between 1 and 100 m/s. APs are initiated by many different types of stimuli; sensory nerves respond to many types of stimuli, such as chemical, light, electricity, pressure, touch, and stretching. On the other hand, the nerves within the CNS (brain and spinal cord) are mostly stimulated by chemical activity at synapses.

A stimulus must be above a threshold level to set off an AP. Very weak stimuli cause a small local electrical disturbance, but do not produce a transmitted AP. As soon as the stimulus strength goes above the threshold, an action potential appears and travels down the nerve.

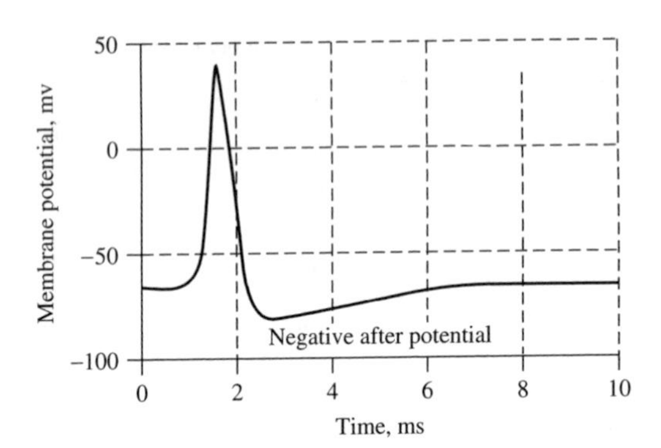

Figure 1.2 An example action potential

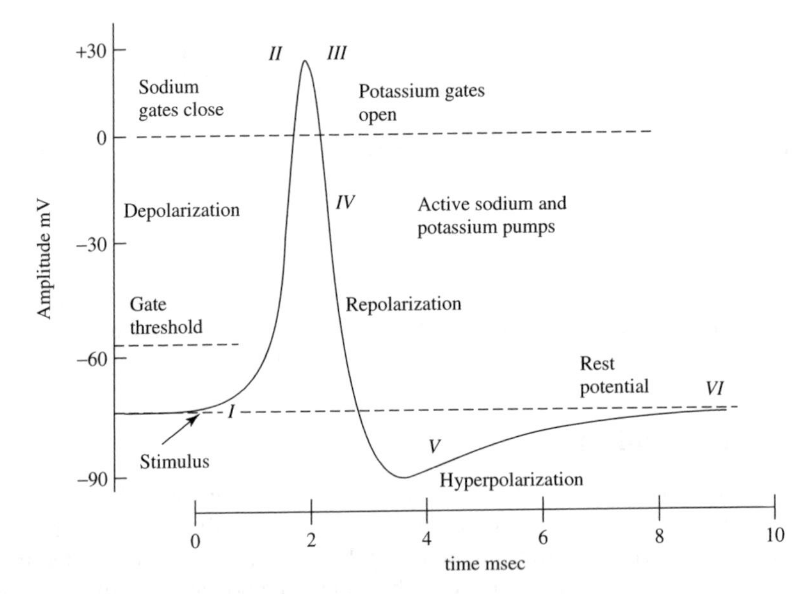

Figure 1.3 Changing the membrane potential for a giant squid by closing the Na channels and opening K channels (adopted from Ka Xiong Charand [24])

The spike of the AP is mainly caused by opening of Na (sodium) channels. The Na pump produces gradients of both Na and K (potassium) ions. Both are used to produce the action potential; Na is high outside the cell and low inside. Excitable cells have special Na and K channels with gates that open and close in response to the membrane voltage (voltage-gated channels). Opening the gates of Na channels allows Na to rush into the cell, carrying positive charge. This makes the membrane potential positive (depolarization), producing the spike. Figure 1.3 shows the stages of the process during evolution of an action potential for a giant squid. For a human being the amplitude of the AP ranges between approximately −60 mV and 10 mV. During this process [24]:

I. When the dendrites of a nerve cell receive the stimulus the Na^+ channels will open. If the opening is sufficient to drive the interior potential from −70 mV up to −55 mV, the process continues.
II. As soon as the action threshold is reached, additional Na^+ channels (sometimes called voltage-gated channels) open. The Na^+ influx drives the interior of the cell membrane up to approximately +30 mV. The process to this point is called depolarization.
III. Then Na^+ channels close and the K^+ channels open. Since the K^+ channels are much slower to open, the depolarization has time to be completed. Having both Na^+ and K^+ channels open at the same time would drive the system towards neutrality and prevent the creation of the action potential.
IV. Having the K^+ channels open, the membrane begins to repolarize back towards its rest potential.
V. The repolarization typically overshoots the rest potential to a level of approximately −90 mV. This is called hyperpolarization and would seem to be counterproductive, but it is actually important in the transmission of information. Hyperpolarization prevents the neuron from receiving another stimulus during this time, or at least raises the threshold for any new stimulus. Part of the importance of hyperpolarization is in preventing any stimulus already sent up an axon from triggering another action potential in the opposite direction. In other words, hyperpolarization ensures that the signal is proceeding in one direction.
VI. After hyperpolarization, the Na^+/K^+ pumps eventually bring the membrane back to its resting state of −70 mV.

The nerve requires approximately two milliseconds before another stimulus is presented. During this time no AP can be generated. This is called the refractory period. The generation of EEG signals is next described.

1.4 EEG Generation

An EEG signal is a measurement of currents that flow during synaptic excitations of the dendrites of many pyramidal neurons in the cerebral cortex. When brain cells (neurons) are activated, the synaptic currents are produced within the dendrites. This current generates a magnetic field measurable by electromyogram (EMG) machines and a secondary electrical field over the scalp measurable by EEG systems.

Differences of electrical potentials are caused by summed postsynaptic graded potentials from pyramidal cells that create electrical dipoles between the soma (body of a neuron) and apical dendrites, which branch from neurons (Figure 1.4). The current in the brain is generated mostly by pumping the positive ions of sodium, Na^+, potassium, K^+, calcium, Ca^{++}, and the negative ion of chlorine, Cl^-, through the neuron membranes in the direction governed by the membrane potential [25].

The human head consists of different layers including the scalp, skull, brain (Figure 1.5), and many other thin layers in between. The skull attenuates the signals approximately one hundred times more than the soft tissue. On the other hand, most of the noise is generated either within the brain (internal noise) or over the scalp (system noise or external noise). Therefore, only large populations of active neurons can generate enough potential to be recordable using the scalp electrodes. These signals are later amplified greatly for display

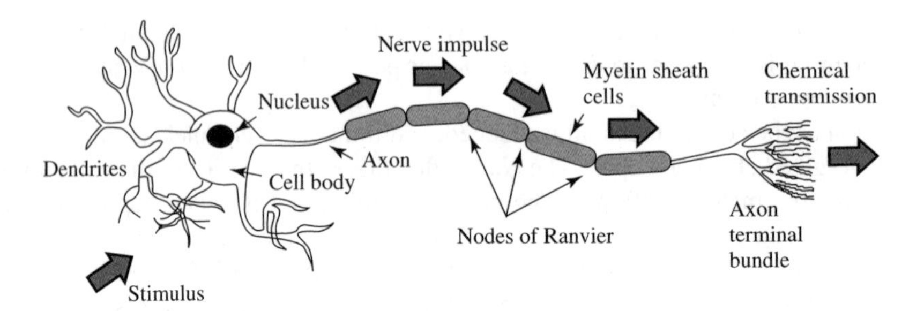

Figure 1.4 Structure of a neuron (adopted from Attwood and MacKay [25])

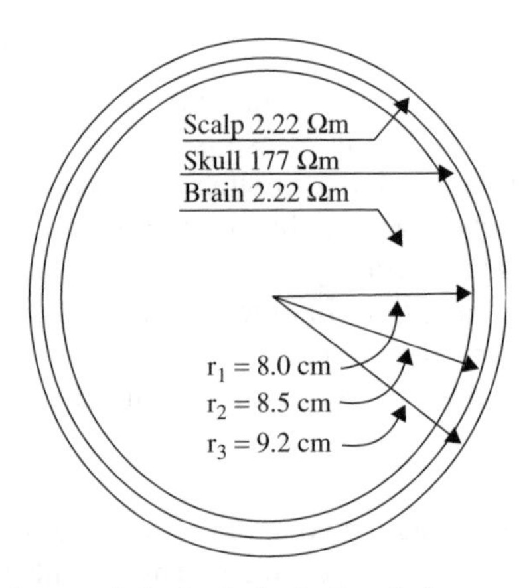

Figure 1.5 The three main layers of the brain including their approximate resistivities and thicknesses (Ω = ohm)

purposes. Approximately 10^{11} neurons are developed at birth when the central nervous system (CNS) becomes complete and functional [26]. This makes an average of 10^4 neurons per cubic mm. Neurons are interconnected into neural nets through synapses. Adults have approximately 5×10^{14} synapses. The number of synapses per neuron increases with age, whereas the number of neurons decreases with age. From an anatomical point of view the brain may be divided into three parts: the cerebrum, cerebellum, and brain stem (Figure 1.6). The cerebrum consists of both left and right lobes of the brain with highly convoluted surface layers called the cerebral cortex.

The cerebrum includes the regions for movement initiation, conscious awareness of sensation, complex analysis, and expression of emotions and behaviour. The cerebellum coordinates voluntary movements of muscles and maintains balance. The brain stem controls involuntary functions such as respiration, heart regulation, biorhythms, and neurohormone and hormone sections [27].

Based on the above section it is clear that the study of EEGs paves the way for diagnosis of many neurological disorders and other abnormalities in the human body. The acquired EEG signals from a human (and also from animals) may, for example, be used for investigation of the following clinical problems [27,28]:

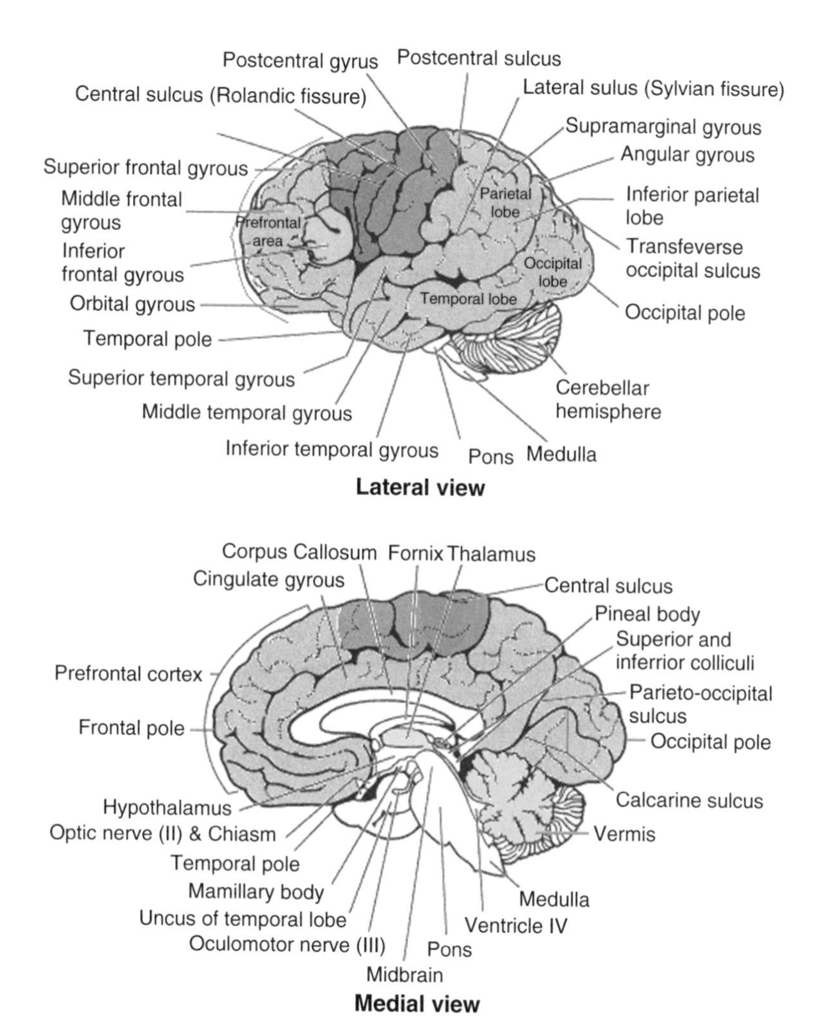

Figure 1.6 Diagrammatic representation of the major parts of the brain

(a) monitoring alertness, coma, and brain death;
(b) locating areas of damage following head injury, stroke, and tumour;
(c) testing afferent pathways (by evoked potentials);
(d) monitoring cognitive engagement (alpha rhythm);
(e) producing biofeedback situations;
(f) controlling anaesthesia depth (servo anaesthesia);
(g) investigating epilepsy and locating seizure origin;

(h) testing epilepsy drug effects;
(i) assisting in experimental cortical excision of epileptic focus;
(j) monitoring the brain development;
(k) testing drugs for convulsive effects;
(l) investigating sleep disorders and physiology;
(m) investigating mental disorders;
(n) providing a hybrid data recording system together with other imaging modalities.

This list confirms the rich potential for EEG analysis and motivates the need for advanced signal processing techniques to aid the clinician in their interpretation. The brain rhythms will next be described, which are expected to be measured within EEG signals.

1.5 Brain Rhythms

Many brain disorders are diagnosed by visual inspection of EEG signals. The clinical experts in the field are familiar with manifestation of brain rhythms in the EEG signals. In healthy adults, the amplitudes and frequencies of such signals change from one state of a human to another, such as wakefulness and sleep. The characteristics of the waves also change with age. There are five major brain waves distinguished by their different frequency ranges. These frequency bands from low to high frequencies respectively are called alpha (α), theta (θ), beta (β), delta (δ), and gamma (γ). The alpha and beta waves were introduced by Berger in 1929. Jasper and Andrews (1938) used the term 'gamma' to refer to the waves of above 30 Hz. The delta rhythm was introduced by Walter (1936) to designate all frequencies below the alpha range. He also introduced theta waves as those having frequencies within the range of 4–7.5 Hz. The notion of a theta wave was introduced by Wolter and Dovey in 1944 [29].

Delta waves lie within the range of 0.5–4 Hz. These waves are primarily associated with deep sleep and may be present in the waking state. It is very easy to confuse artefact signals caused by the large muscles of the neck and jaw with the genuine delta response. This is because the muscles are near the surface of the skin and produce large signals, whereas the signal that is of interest originates from deep within the brain and is severely attenuated in passing through the skull. Nevertheless, by applying simple signal analysis methods to the EEG, it is very easy to see when the response is caused by excessive movement.

Theta waves lie within the range of 4–7.5 Hz. The term theta might be chosen to allude to its presumed thalamic origin. Theta waves appear as consciousness slips towards drowsiness. Theta waves have been associated with access to unconscious material, creative inspiration and deep meditation. A theta wave is often accompanied by other frequencies and seems to be related to the level of arousal. It is known that healers and experienced mediators have an alpha wave that gradually lowers in frequency over long periods of time. The theta wave plays an important role in infancy and childhood. Larger contingents of theta wave activity in the waking adult are abnormal and are caused by various pathological problems. The changes in the rhythm of theta waves are examined for maturational and emotional studies [30].

Alpha waves appear in the posterior half of the head and are usually found over the occipital region of the brain. They can be detected in all parts of posterior lobes of the brain. For alpha waves the frequency lies within the range of 8–13 Hz, and commonly appears as a round or sinusoidal shaped signal. However, in rare cases it may manifest itself as sharp waves. In such cases, the negative component appears to be sharp and the positive component appears to be rounded, similar to the wave morphology of the rolandic mu (μ) rhythm. Alpha waves have been thought to indicate both a relaxed awareness without any attention or concentration. The alpha wave is the most prominent rhythm in the whole realm of brain activity and possibly covers a greater range than has been previously accepted. A peak can regularly be seen in the beta wave range in frequencies even up to 20 Hz, which has the characteristics of an alpha wave state rather than one for a beta wave. Again, very often a response is seen at 75 Hz, which appears in an alpha setting. Most subjects produce some alpha waves with their eyes closed, which is why it has been claimed that it is nothing but a waiting or scanning pattern produced by the visual regions of the brain. It is reduced or eliminated by opening the eyes, by hearing unfamiliar sounds, by anxiety, or mental concentration or attention. Albert Einstein could solve complex mathematical problems while remaining in the alpha state, although generally beta and theta waves are also present. An alpha wave has a higher amplitude over the occipital areas and has an amplitude of normally less than 50 μV. The origin and physiological significance of an alpha wave is still unknown and yet more research has to be undertaken to understand how this phenomenon originates from cortical cells [31].

A beta wave is the electrical activity of the brain varying within the range of 14–26 Hz (though in some literature no upper bound is given). A beta wave is the usual waking rhythm of the brain associated with active thinking, active attention, focus on the outside world, or solving concrete problems, and is found in normal adults. A high-level beta wave may be acquired when a human is in a panic state. Rhythmical beta activity is encountered chiefly over the frontal and central regions. Importantly, a central beta rhythm is related to the rolandic mu rhythm and can be blocked by motor activity or tactile stimulation. The amplitude of beta rhythm is normally under 30 μV. Similar to the mu rhythm, the beta wave may also be enhanced because of a bone defect [29] and also around tumoural regions.

The frequencies above 30 Hz (mainly up to 45 Hz) correspond to the gamma range (sometimes called the fast beta wave). Although the amplitudes of these rhythms are very low and their occurrence is rare, detection of these rhythms can be used for confirmation of certain brain diseases. The regions of high EEG frequencies and highest levels of cerebral blood flow (as well as oxygen and glucose uptake) are located in the frontocentral area. The gamma wave band has also been proved to be a good indication of event-related synchronization (ERS) of the brain and can be used to demonstrate the locus for right and left index finger movement, right toes, and the rather broad and bilateral area for tongue movement [32].

Waves in frequencies much higher than the normal activity range of EEG, mostly in the range of 200–300 Hz, have been found in cerebellar structures of animals, but they have not played any role in clinical neurophysiology [33,34].

Figure 1.7 shows the typical normal brain rhythms with their usual amplitude levels. In general, the EEG signals are the projection of neural activities that are attenuated by

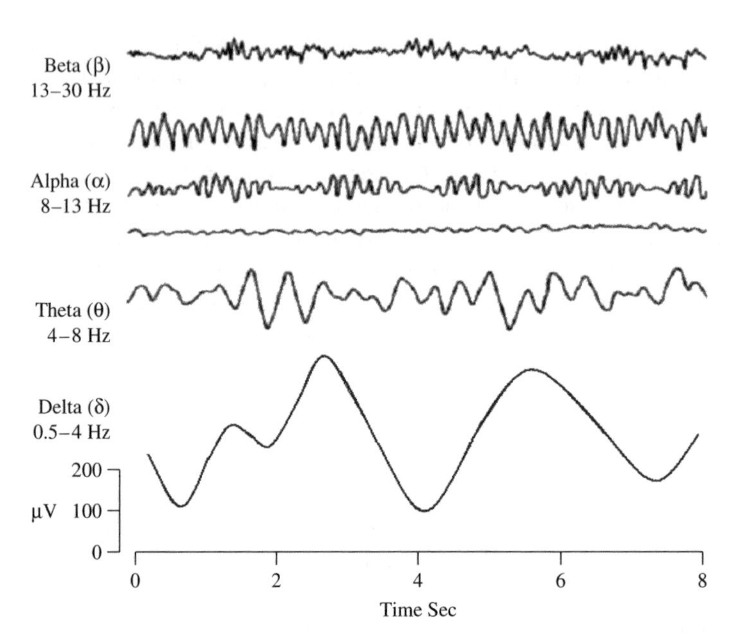

Figure 1.7 Four typical dominant brain normal rhythms, from high to low frequencies. The delta wave is observed in infants and sleeping adults, the theta wave in children and sleeping adults, the alpha wave is detected in the occipital brain region when there is no attention, and the beta wave appears frontally and parietally with low amplitude

leptomeninges, cerebrospinal fluid, dura matter, bone, galea, and the scalp. Cartographic discharges show amplitudes of 0.5–1.5 mV and up to several millivolts for spikes. However, on the scalp the amplitudes commonly lie within 10–100 μV.

The above rhythms may last if the state of the subject does not change and therefore they are approximately cyclic in nature. On the other hand, there are other brain waveforms, which may:

(a) Have a wide frequency range or appear as spiky-type signals, such as K-complexes, vertex waves (which happen during sleep), or a breach rhythm, which is an alpha-type rhythm due to a cranial bone defect [35], which does not respond to movement, and is found mainly over the midtemporal region (under electrodes T3 or T4), and some seizure signals.
(b) Be a transient such as an event-related potential (ERP) and contain positive occipital sharp transient (POST) signals (also called rho (ρ) waves).
(c) Originate from the defective regions of the brain such as tumoural brain lesions.
(d) Be spatially localized and considered as cyclic in nature, but can be easily blocked by physical movement such as mu rhythm. Mu denotes motor and is strongly related to the motor cortex. Rolandic (central) mu is related to posterior alpha in terms of amplitude and frequency. However, the topography and physiological significance are quite different. From the mu rhythm the cortical functioning and the changes in

brain (mostly bilateral) activities subject to physical and imaginary movements can be investigated. The mu rhythm has also been used in feedback training for several purposes such as treatment of epileptic seizure disorder [29].

There are also other rhythms introduced by researchers such as:

(e) Phi (φ) rhythm (less than 4 Hz) occurring within two seconds of eye closure. The phi rhythm was introduced by Daly [36].
(f) Kappa (κ) rhythm, which is an anterior temporal alpha-like rhythm. It is believed to be the result of discrete lateral oscillations of the eyeballs and is considered to be an artefact signal.
(g) The sleep spindles (also called the sigma (σ) activity) within the 11–15 Hz frequency range.
(h) Tau (τ) rhythm, which represents the alpha activity in the temporal region.
(i) Eyelid flutter with closed eyes, which gives rise to frontal artefacts in the alpha band.
(j) Chi (χ) rhythm is a mu-like activity believed to be a specific rolandic pattern of 11–17 Hz. This wave has been observed during the course of Hatha Yoga exercises [37].
(k) Lambda (λ) waves are most prominent in waking patients, but are not very common. They are sharp transients occurring over the occipital region of the head of walking subjects during visual exploration. They are positive and time-locked to saccadic eye movement with varying amplitude, generally below 90 μV [38].

It is often difficult to understand and detect the brain rhythms from the scalp EEGs, even with trained eyes. Application of advanced signal processing tools, however, should enable separation and analysis of the desired waveforms from within the EEGs. Therefore, a definition of foreground and background EEG is very subjective and entirely depends on the abnormalities and applications. Next to consider is the development in the recording and measurement of EEG signals.

1.6 EEG Recording and Measurement

Acquiring signals and images from the human body has become vital for early diagnosis of a variety of diseases. Such data can be in the form of electrobiological signals such as an electrocardiogram (ECG) from the heart, electromyogram (EMG) from muscles, electroencephalogram (EEG) from the brain, magnetoencephalogram (MEG) from the brain, electrogastrogram (EGG) from the stomach, and electrooclugram (or electrooptigram, EOG) from eye nerves. Measurements can also have the form of one type of ultrasound or radiograph such as sonograph (or ultrasound image), computerized tomography (CT), magnetic resonance imaging (MRI) or functional MRI (fMRI), positron emission tomography (PET), and single photon emission tomography (SPET).

Functional and physiological changes within the brain may be registered by either EEG, MEG, or fMRI. Application of fMRI is, however, very limited in comparison with EEG or MEG for a number of important reasons:

(a) The time resolution of fMRI image sequences is very low (for example approximately two frames/s), whereas the complete EEG bandwidth can be viewed using EEG or MEG signals.
(b) Many types of mental activities, brain disorders, and malfunctions of the brain cannot be registered using fMRI since their effect on the level of oxygenated blood is low.
(c) The accessibility to fMRI (and currently to MEG) systems is limited and costly.
(d) The spatial resolution of EEG, however, is limited to the number of recording electrodes (or number of coils for MEG).

The first electrical neural activities were registered using simple galvanometers. In order to magnify very fine variations of the pointer a mirror was used to reflect the light projected to the galvanometer on the wall. The d'Arsonval galvanometer later featured a mirror mounted on a movable coil and the light focused on the mirror was reflected when a current passed the coil. The capillary electrometer was introduced by Lippmann and Marey [39]. The string galvanometer, as a very sensitive and more accurate measuring instrument, was introduced by Einthoven in 1903. This became a standard instrument for a few decades and enabled photographic recording.

More recent EEG systems consist of a number of delicate electrodes, a set of differential amplifiers (one for each channel) followed by filters [27], and needle (pen)-type registers. The multichannel EEGs could be plotted on plane paper or paper with a grid. Soon after this system came to the market, researchers started looking for a computerized system, which could digitize and store the signals. Therefore, to analyse EEG signals it was soon understood that the signals must be in digital form. This required sampling, quantization, and encoding of the signals. As the number of electrodes grows the data volume, in terms of the number of bits, increases. The computerized systems allow variable settings, stimulations, and sampling frequency, and some are equipped with simple or advanced signal processing tools for processing the signals.

The conversion from analogue to digital EEG is performed by means of multichannel analogue-to-digital converters (ADCs). Fortunately, the effective bandwidth for EEG signals is limited to approximately 100 Hz. For many applications this bandwidth may be considered to be even half of this value. Therefore, a minimum frequency of 200 samples/s (to satisfy the Nyquist criterion) is often enough for sampling the EEG signals. In some applications where a higher resolution is required for representation of brain activities in the frequency domain, sampling frequencies of up to 2000 sample/s may be used.

In order to maintain the diagnostic information the quantization of EEG signals is normally very fine. Representation of each signal sample with up to 16 bits is very popular for the EEG recording systems. This makes the necessary memory volume for archiving the signals massive, especially for sleep EEG and epileptic seizure monitoring records. However, in general, the memory size for archiving the radiological images is often much larger than that used for archiving the EEG signals.

A simple calculation shows that for a one hour recording from 128-electrode EEG signals sampled at 500 samples/s a memory size of $128 \times 60 \times 60 \times 500 \times 16 \approx 3.68$ Gbits ≈ 0.45 Gbyte is required. Therefore, for longer recordings of a large number of patients there should be enough storage facilities such as in today's technology Zip disks, CDs, large removable hard drives, and optical disks. Although the format of reading the EEG data may be different for different EEG machines, these formats are easily convertible to spreadsheets readable by most signal processing software packages such as MATLAB.

The EEG recording electrodes and their proper function are crucial for acquiring high-quality data. Different types of electrodes are often used in the EEG recording systems, such as:

- disposable (gel-less, and pre-gelled types);
- reusable disc electrodes (gold, silver, stainless steel, or tin);
- headbands and electrode caps;
- saline-based electrodes;
- needle electrodes.

For multichannel recordings with a large number of electrodes, electrode caps are often used. Commonly used scalp electrodes consist of Ag–AgCl disks, less than 3 mm in diameter, with long flexible leads that can be plugged into an amplifier. Needle electrodes are those that have to be implanted under the skull with minimal invasive operations. High impedance between the cortex and the electrodes as well as the electrodes with high impedances can lead to distortion, which can even mask the actual EEG signals. Commercial EEG recording systems are often equipped with impedance monitors. To enable a satisfactory recording the electrode impedances should read less than 5 kΩ and be balanced to within 1 kΩ of each other. For more accurate measurement the impedances are checked after each trial.

Due to the layered and spiral structure of the brain, however, distribution of the potentials over the scalp (or cortex) is not uniform [40]. This may affect some of the results of source localization using the EEG signals.

1.6.1 Conventional Electrode Positioning

The International Federation of Societies for Electroencephalography and Clinical Neurophysiology has recommended the conventional electrode setting (also called 10–20) for 21 electrodes (excluding the earlobe electrodes), as depicted in Figure 1.8 [17]. Often the earlobe electrodes called A1 and A2, connected respectively to the left and right earlobes, are used as the reference electrodes. The 10–20 system avoids both eyeball placement and considers some constant distances by using specific anatomic landmarks from which the measurement would be made and then uses 10 or 20 % of that specified distance as the electrode interval. The odd electrodes are on the left and the even ones on the right.

For setting a larger number of electrodes using the above conventional system, the rest of the electrodes are placed in between the above electrodes with equidistance between them. For example, C_1 is placed between C_3 and C_z. Figure 1.9 represents a larger setting for 75 electrodes including the reference electrodes based on the guidelines by the American EEG Society. Extra electrodes are sometimes used for the measurement of EOG, ECG, and EMG of the eyelid and eye surrounding muscles. In some applications such as ERP analysis and brain computer interfacing a single channel may be used. In such applications, however, the position of the corresponding electrode has to be well determined.

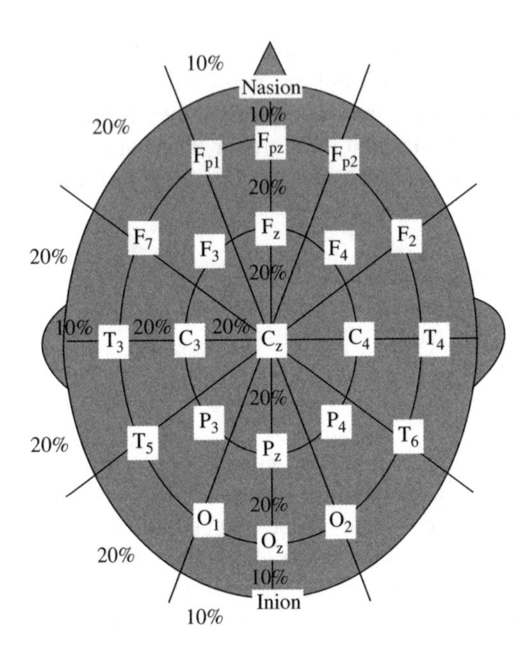

Figure 1.8 Conventional 10–20 EEG electrode positions for the placement of 21 electrodes

For example, C_3 and C_4 can be used to record the right and left finger movement related signals respectively for brain–computer interfacing (BCI) applications. Also F_3, F_4, P_3, and P_4 can be used for recording the ERP P300 signals.

Two different modes of recordings, namely differential and referential, are used. In the differential mode the two inputs to each differential amplifier are from two electrodes. In the referential mode, on the other hand, one or two reference electrodes are used. Several different reference electrode placements can be found in the literature. Physical references can be used as vertex (C_z), linked-ears, linked-mastoids, ipsilateral ear, contralateral ear, C_7, bipolar references, and tip of the nose [28]. There are also reference-free recording techniques, which actually use a common average reference. The choice of reference may produce topographic distortion if the reference is not relatively neutral. In modern instrumentation, however, the choice of a reference does not play an important role in the measurement [41]. In such systems other references such as FP_z, hand, or leg electrodes may be used [42]. The overall setting includes the active electrodes and the references.

In another similar setting, called the Maudsley electrode positioning system, the conventional 10–20 system has been modified to capture better the signals from epileptic foci in epileptic seizure recordings. The only difference between this system and the 10–20 conventional system is that the outer electrodes are slightly lowered to enable better capturing of the required signals. The advantage of this system over the conventional one is that it provides a more extensive coverage of the lower part of the cerebral convexity, increasing the sensitivity for the recording from basal subtemporal structures [43]. Other deviations from the international 10–20 system as used by researchers are found in References [44] and [45].

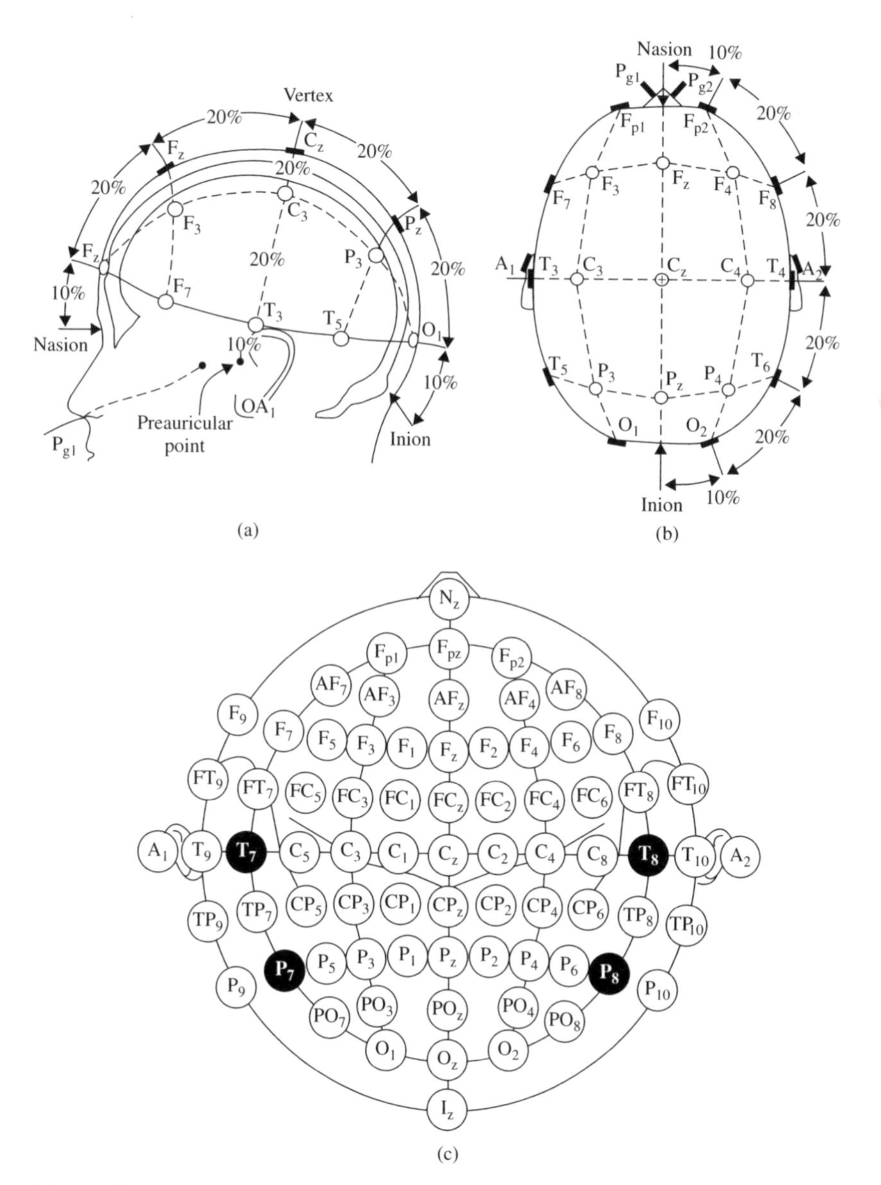

Figure 1.9 A diagrammatic representation of 10–20 electrode settings for 75 electrodes including the reference electrodes: (a) and (b) represent the three-dimensional measures, and (c) indicates a two-dimensional view of the electrode setup configuration

In many applications such as brain–computer interfacing (BCI) and study of mental activity, often a small number of electrodes around the movement-related regions are selected and used from the 10–20 setting system. Figure 1.10 illustrates a typical set of EEG signals during approximately seven seconds of normal adult brain activity.

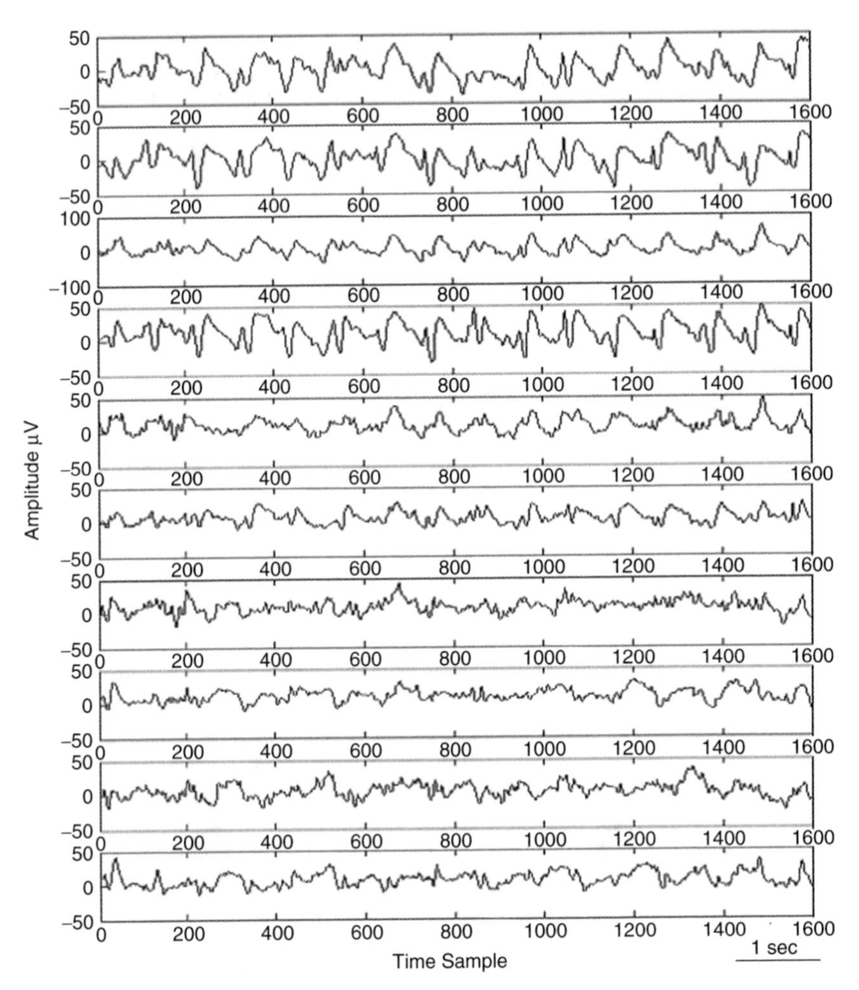

Figure 1.10 A typical set of EEG signals during approximately seven seconds of normal adult brain activity

1.6.2 Conditioning the Signals

The raw EEG signals have amplitudes of the order of μvolts and contain frequency components of up to 300 Hz. To retain the effective information the signals have to be amplified before the ADC and filtered, either before or after the ADC, to reduce the noise and make the signals suitable for processing and visualization. The filters are designed in such a way not to introduce any change or distortion to the signals. Highpass filters with a cut-off frequency of usually less than 0.5 Hz are used to remove the disturbing very low frequency components such as those of breathing. On the other hand, high-frequency noise is mitigated by using lowpass filters with a cut-off frequency of approximately 50–70 Hz. Notch filters with a null frequency of 50 Hz are often necessary to ensure perfect rejection of the strong 50 Hz power supply. In this case the sampling frequency can be as low as twice the bandwidth commonly used by most EEG systems. The commonly used sampling frequencies for EEG recordings are 100, 250, 500, 1000, and 2000 samples/s. The main artefacts can be divided into patient-related (physiological) and system artefacts. The patient-related or internal artefacts are body movement-related, EMG, ECG (and pulsation), EOG, ballistocardiogram, and sweating. The system artefacts are 50/60 Hz power supply interference, impedance fluctuation, cable defects, electrical noise from the electronic components, and unbalanced impedances of the electrodes. Often in the preprocessing stage these artefacts are highly mitigated and the informative information is restored. Some methods for removing the EEG artefacts will be discussed in the related chapters of this book. Figure 1.11 shows a set of normal EEG signals affected by the eye-blinking artefact. Similarly, Figure 1.12 represents a multichannel EEG set with the clear appearance of ECG signals over the electrodes in the occipital region.

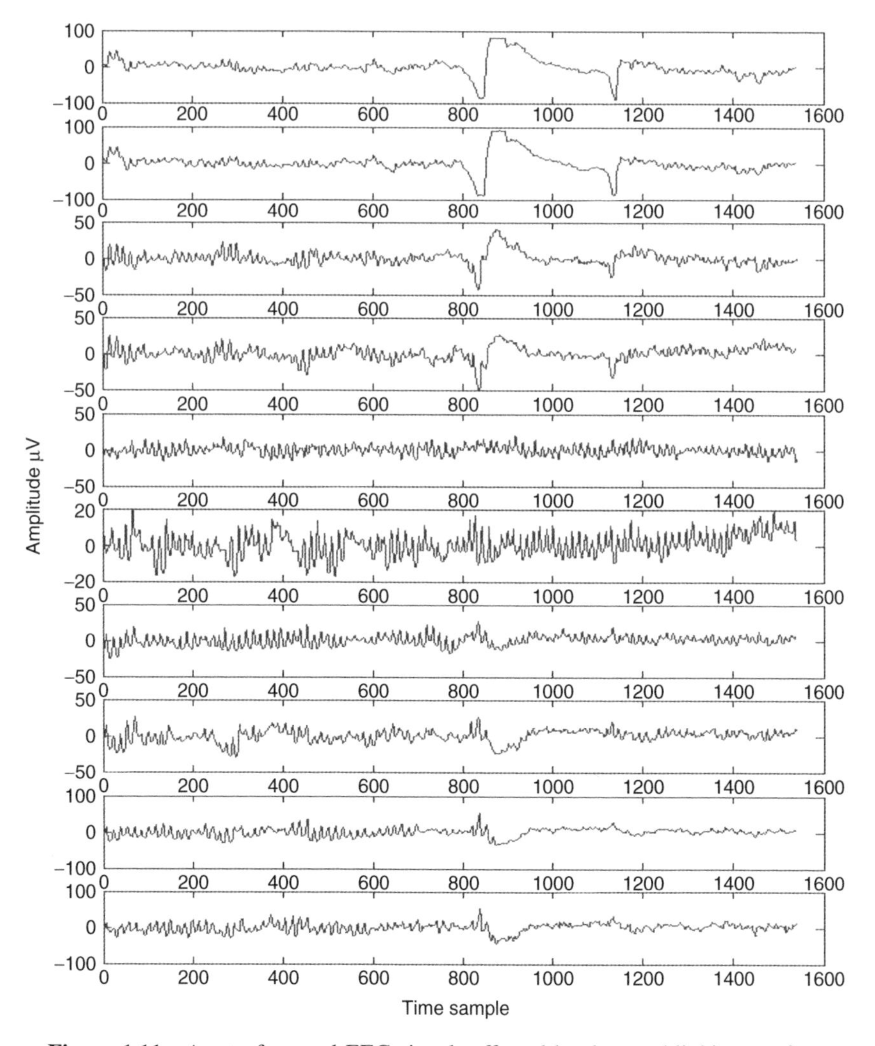

Figure 1.11 A set of normal EEG signals affected by the eye-blinking artefact

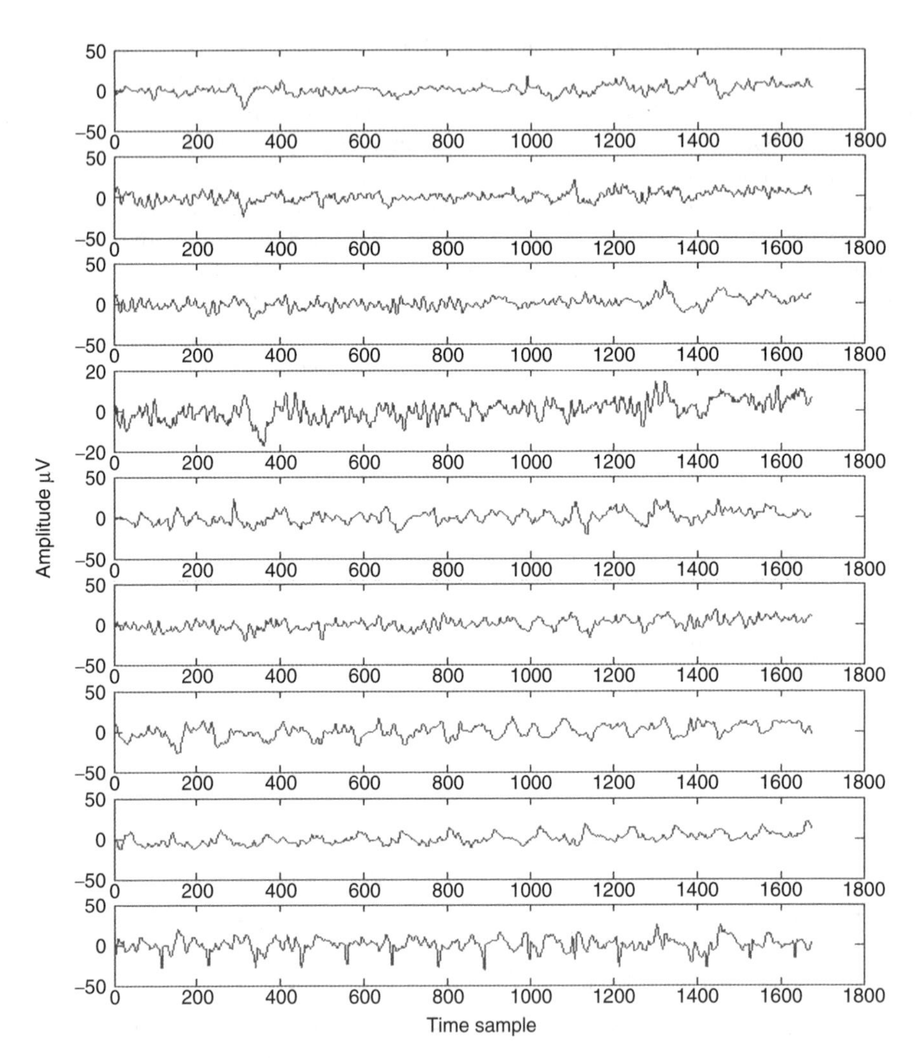

Figure 1.12 A multichannel EEG set with the clear appearance of ECG signals over the electrodes in the occipital region

The next section highlights the changes in EEG measurements that correlate with physiological and mental abnormalities in the brain.

1.7 Abnormal EEG Patterns

Variations in the EEG patterns for certain states of the subject indicate abnormality. This may be due to distortion and the disappearance of abnormal patterns, appearance and increase of abnormal patterns, or disappearance of all patterns. Sharbrough [46] divided the nonspecific abnormalities in the EEGs into three categories: (a) widespread intermittent slow wave abnormalities, often in the delta wave range and associated with brain dysfunction; (b) bilateral persistent EEG, usually associated with impaired conscious cerebral reactions; and (c) focal persistent EEG usually associated with focal cerebral disturbance.

The first category is a burst-type signal, which is attenuated by alerting the individual and eye opening, and accentuated with eye closure, hyperventilation, or drowsiness. The peak amplitude in adults is usually localized in the frontal region and influenced by age. In children, however, it appears over the occipital or posterior head region. Early findings showed that this abnormal pattern frequently appears with an increased intracranial pressure with tumour or aqueductal stenosis. Also, it correlates with grey matter disease, both in cortical and subcortical locations. However, it can be seen in association with a wide variety of pathological processes varying from systemic toxic or metabolic disturbances to focal intracranial lesions.

Regarding the second category, i.e. bilateral persistent EEG, the phenomenon in different stages of impaired, conscious, purposeful responsiveness are etiologically nonspecific and the mechanisms responsible for their generation are only partially understood. However, the findings in connection with other information concerning etiology and chronicity may be helpful in arriving more quickly at an accurate prognosis concerning the patient's chance of recovering previous conscious life.

As for the third category, i.e. focal persistent EEG, these abnormalities may be in the form of distortion and disappearance of normal patterns, appearance and increase of abnormal patterns, or disappearance of all patterns, but such changes are seldom seen at the cerebral cortex. The focal distortion of normal rhythms may produce an asymmetry of amplitude, frequency, or reactivity of the rhythm. The unilateral loss of reactivity of a physiological rhythm, such as the loss of reactivity of the alpha rhythm to eye opening [47] or to mental alerting [48], may reliably identify the focal side of abnormality. A focal lesion may also distort or eliminate the normal activity of sleep-inducing spindles and vertex waves.

Focal persistent nonrhythmic delta activity (PNRD) may be produced by focal abnormalities. This is one of the most reliable findings of a focal cerebral disturbance. The more persistent, the less reactive, and the more nonrhythmic and polymorphic is such focal slowing, the more reliable an indicator it becomes for the appearance of a focal cerebral disturbance [49–51]. There are other cases such as focal inflammation, trauma, vascular disease, brain tumour, or almost any other cause of focal cortical disturbance, including an asymmetrical onset of CNS degenerative diseases that may result in similar abnormalities in the brain signal patterns.

The scalp EEG amplitude from cerebral cortical generators underlying a skull defect is also likely to increase unless acute or chronic injury has resulted in significant depression of underlying generator activity. The distortions in cerebral activities are because focal abnormalities may alter the interconnections, number, frequency, synchronicity, voltage output, and access orientation of individual neuron generators, as well as the location and amplitude of the source signal itself.

With regards to the three categories of abnormal EEGs, their identification and classification requires a dynamic tool for various neurological conditions and any other available information. A precise characterization of the abnormal patterns leads to a clearer insight into some specific pathophysiologic reactions, such as epilepsy, or specific disease processes, such as subacute sclerosing panencephalitis (SSPE) or Creutzfeldt–Jakob disease (CJD) [46].

Over and above the reasons mentioned above there are many other causes for abnormal EEG patterns. The most common abnormalities are briefly described in the following sections.

1.8 Ageing

The ageing process affects the normal cerebral activity in waking and sleep, and changes the response of the brain to stimuli. The changes stem from reducing the number of neurons and due to a general change in the brain pathology. This pathology indicates that the frontal and temporal lobes of the brain are more affected than the parietal lobes, resulting in shrinkage of large neurons and increasing the number of small neurons and glia [52]. A diminished cortical volume indicates that there is age-related neuronal loss. A general cause for ageing of the brain may be the decrease in cerebral blood flow [52].

A reduction of the alpha frequency is probably the most frequent abnormality in EEG. This often introduces a greater anterior spread to frontal regions in the elderly and reduces the alpha wave blocking response and reactivity. The diminished mental function is somehow related to the degree of bilateral slowing in the theta and delta waves [52].

Although the changes in high-frequency brain rhythms have not been well established, some researchers have reported an increase in beta wave activity. This change in beta wave activity may be considered as an early indication of intellectual loss [52].

As for the sleep EEG pattern, older adults enter into drowsiness with a more gradual decrease in EEG amplitude. Over the age of sixty, the frontocentral waves become slower, the frequency of the temporal rhythms also decreases, frequency lowering with slow eye movements become more prominent, and spindles appear in the wave pattern after the dropout of the alpha rhythm. The amplitudes of both phasic and tonic nonrapid eye movement (NREM) sleep EEG [52] reduce with age. There is also a significant change in rapid eye movement (REM) sleep organization with age; the REM duration decreases during the night and there is a significant increase in sleep disruption [52].

Dementia is the most frequent mental disorder that occurs predominantly in the elderly. Therefore, the prevalence of dementia increases dramatically with ageing of the society. Generally, EEGs are a valuable diagnostic tool in differentiation between organic brain syndromes (OBSs) and functional psychiatric disorders [52], and together with evoked potentials (EPs) play an important role in the assessment of normal and pathological ageing. Ageing is expected to change most neurophysiological parameters. However, the variability of these parameters must exceed the normal degree of spontaneous variability to become a diagnostic factor in acute and chronic disease conditions. Automatic analysis of the EEG during sleep and wakefulness may provide a better contrast in the data and enable a robust diagnostic tool. Next particular and very common mental disorders are described, whose early onset may be diagnosed with EEG measurements.

1.9 Mental Disorders

1.9.1 Dementia

Dementia is a syndrome that consists of a decline in intellectual and cognitive abilities. This consequently affects the normal social activities, mode, and the relationship and interaction with other people [53]. EEG is often used to study the effect of dementia. In most cases, such as in primary degenerative dementia, e.g. Alzheimer's, and psychiatric disorder, e.g. depression with cognitive impairment, the EEG can be used to detect the abnormality [54].

In Reference [54] dementia is classified into cortical and subcortical forms. The most important cortical dementia is Alzheimer's disease (AD), which accounts for approximately 50 % of the cases. Other known cortical abnormalities are Pick's disease and Creutzfeldt–Jakob diseases (CJD). They are characterized clinically by findings such as aphasia, apraxia, and agnosia. CJD can often be diagnosed using the EEG signals. Figure 1.13 shows a set of EEG signals from a CJD patient. On the other hand, the most common subcortical diseases are Parkinson's disease, Huntington's disease, lacunar state, normal pressure hydrocephalus, and progressive supranuclear palsy. These diseases are characterized by forgetfulness, slowing of thought processes, apathy, and depression. Generally, subcortical dementias introduce less abnormality to the EEG patterns than the cortical ones.

In AD the EEG posterior rhythm (alpha rhythm) slows down and the delta and theta wave activities increase. On the other hand, beta wave activity may decrease. In severe cases epileptiform discharges and triphasic waves can appear. In such cases, cognitive impairment often results. The spectral power also changes; the power increases in delta and theta bands and decreases in beta and alpha bands and also in mean frequency.

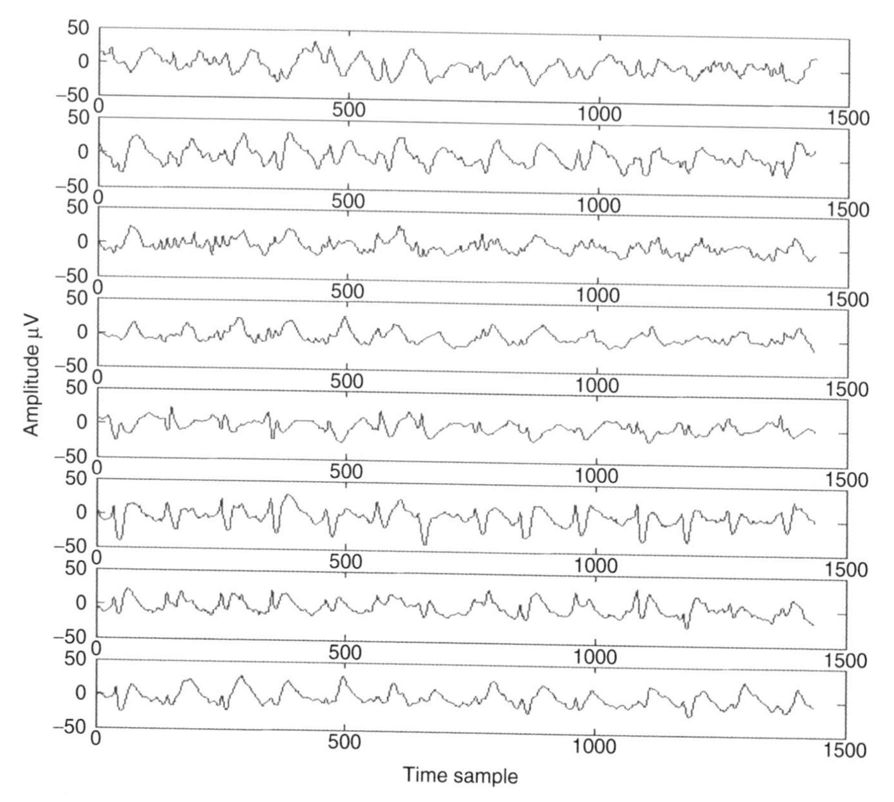

Figure 1.13 A set of multichannel EEG signals from a patient suffering from CJD

The EEG wave morphology is almost the same for AD and Pick's disease. Pick's disease involves the frontal and temporal lobes. An accurate analysis followed by an efficient classification of the cases may discriminate these two diseases. CJD is a mixed cortical and subcortical dementia. This causes slowing of the delta and theta wave activities and, after approximately three months of the onset of the disease, periodic sharp wave complexes are generated that occur almost every second, together with a decrease in the background activity [54]. Parkinson's disease is a subcortical dementia, which causes slowing down of the background activity and an increase of the theta and delta wave activities. Some works have been undertaken using spectral analysis to confirm the above changes [55]. Some other disorders such as depression have a lesser effect on the EEGs and more accurate analysis of the EEGs has to be performed to detect the signal abnormalities for these brain disorders.

Generally, EEG is usually used in the diagnosis and evaluation of many cortical and subcortical dementias. Often it can help to differentiate between a degenerative disorder such as AD and pseudodementia due to psychiatric illness [54]. The EEG may also show whether the process is focal or diffuse (i.e. involves the background delta and theta wave activities). The EEG may also reveal the early CJD-related abnormalities. However, more advanced signal processing and quantitative techniques may be implemented to achieve robust diagnostic and monitoring performance.

1.9.2 Epileptic Seizure and Nonepileptic Attacks

Often the onset of a clinical seizure is characterized by a sudden change of frequency in the EEG measurement. It is normally within the alpha wave frequency band with a slow decrease in frequency (but increase in amplitude) during the seizure period. It may or may not be spiky in shape. Sudden desynchronization of electrical activity is found in electrodecremental seizures. The transition from the preictal to the ictal state, for a focal epileptic seizure, consists of a gradual change from chaotic to ordered waveforms. The amplitude of the spikes does not necessarily represent the severity of the seizure. Rolandic spikes in a child of 4–10 years, for example, are very prominent; however, the seizure disorder is usually quite benign or there may not be clinical seizure [56].

In terms of spatial distribution, in childhood the occipital spikes are very common. Rolandic central–midtemporal–parietal spikes are normally benign, whereas frontal spikes or multifocal spikes are more epileptogenic. The morphology of the spikes varies significantly with age. However, the spikes may occur in any level of awareness including wakefulness and deep sleep.

The distinction of seizure from common artefacts is not difficult. Seizure artefacts within an EEG measurement have a prominent spiky but repetitive (rhythmical) nature, whereas the majority of other artefacts are transients or noise-like in shape. For the case of the ECG, the frequency of occurrence of the QRS waveforms (an element of the ECG) is approximately 1 Hz. These waveforms have a certain shape which is very different from that of seizure signals.

The morphology of an epileptic seizure signal slightly changes from one type to another. The seizure may appear in different frequency ranges. For example, a petit mal discharge often has a slow spike at around 3 Hz, lasting for approximately 70 ms, and normally has its maximum amplitude around the frontal midline. On the other hand, higher frequency spike wave complexes occur for patients over 15 years old. Complexes at 4 Hz and 6 Hz may appear in the frontal region of the brain of epileptic patients. As for the 6 Hz complex (also called benign EEG variants and patterns), patients with anterior 6 Hz spike waves are more likely to have epileptic seizures and those with posterior discharges tend to have neuroautonomic disturbances [57]. The experiments do not always result in the same conclusion [56]. It was also found that the occipital 6 Hz spikes can be seen and are often drug related (due to hypoanalgetics or barbiturates) and due to withdrawal [58].

Among nonepileptics, the discharges may occur in patients with cerebrovascular disorder, syncopal attacks, and psychiatric problems [56]. Fast and needle-like spike discharges may be seen over the occipital region in most congenitally blind children. These spikes are unrelated to epilepsy and normally disappear in older age patients.

Bursts of 13–16 Hz or 5–7 Hz, as shown in Figure 1.14 (also called 14 and 6 Hz waves), with amplitudes less than 75 μV and arch shapes may be seen over the posterior temporal and the nearby regions of the head during sleep. These waves are positive with respect to the background waves. The 6 and 14 Hz waves may appear independently and

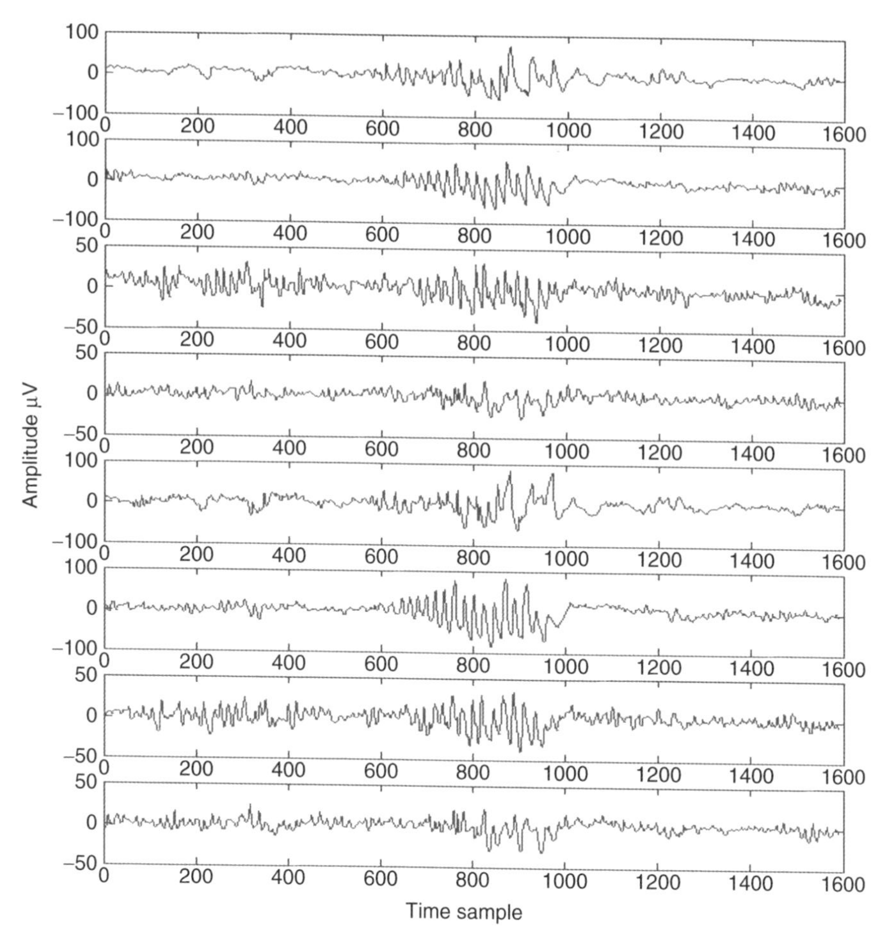

Figure 1.14 Bursts of 3–7 Hz seizure activity in a set of adult EEG signals

be found respectively in younger and older children. These waves may be confined to the regions lying beneath a skull defect. Despite the 6 Hz wave, there are rhythmical theta bursts of wave activities relating to drowsiness around the midtemporal region, with a morphology very similar to ictal patterns. In old age patients other similar patterns may occur, such as *subclinical rhythmic EEG discharges of adults* (SREDA), over the 4–7 Hz frequency band around the centroparietal region, and a wide frequency range (2–120 Hz) *temporal minor sharp transient* and *wicket spikes* over the anterior temporal and midtemporal lobes of the brain. These waves are also nonepileptic but with a seizure-type waveform [56].

The epileptic seizure patterns, called ictal wave patterns, appear during the onset of epilepsy. Although Chapter 4 of this book focuses on an analysis of these waveforms from a signal processing point of view, here a brief explanation of morphology of these waveforms is given. Researchers in signal processing may exploit these concepts in the development of their algorithms. Although these waveform patterns are often highly obscured by muscle movements, they normally maintain certain key characteristics.

Tonic–clonic seizure (also called grand mal) is the most common type of epileptic seizure. It appears in all electrodes but more towards the frontal electrodes (Figure 1.15). It has a rhythmic but spiky pattern in the EEG and occurs within the frequency range of 6–12 Hz. Petit mal is another interictal paroxysmal seizure pattern which occurs at approximately 3 Hz with a generalized synchronous spike wave complex of prolonged bursts. A temporal lobe seizure (also called a psychomotor seizure or complex partial seizure) is presented by bursts of serrated slow waves with a relatively high amplitude of above 60 μV and frequencies of 4–6 Hz. Cortical (focal) seizures have contralateral distribution with rising amplitude and diminishing frequency during the ictal period. The attack is usually initiated by local desynchronization, i.e. very fast and very low voltage spiky activity, which gradually rises in amplitude with diminishing frequency. Myoclonic seizures have concomitant polyspikes, seen clearly in the EEG signals. They can have generalized or bilateral spatial distribution that is more dominant in the frontal region [59]. Tonic seizures occur in patients with the Lennox–Gastaut syndrome [60] and have spikes that repeat with a frequency of approximately 10 Hz. Atonic seizures may appear in the form of a few seconds drop attack or be inhibitory, lasting for a few minutes. They show a few polyspike waves or spike waves with generalized spatial distribution of approximately 10 Hz followed by large slow waves of 1.5–2 Hz [61]. Akinetic seizures are rare and characterized by arrest of all motion, which, however, is not caused by sudden loss of tone as in atonic seizure and the patient is in an absent-like state. They are rhythmic with a frequency of 1–2 Hz. Jackknife seizures, also called salaam attacks, are common in children with hypsarrhythmia (infantile spasms, West syndrome) and are either in the form of sudden generalized flattening desynchronization or have rapid spike discharges [60].

There are generally several varieties of recurring or quasirecurring discharges, which may or may not be related to epileptic seizure. These abnormalities may be due to psychogenic changes, variation in body metabolism, or circulatory insufficiency (which often appears as acute cerebral ischemia). Of these, the most important ones are: periodic or quasiperiodic discharges related to severe CNS diseases; periodic complexes in subacute sclerosing panencephalitis (SSPE); periodic complexes in herpes simplex encephalitis; syncopal attacks; breath holding attacks; hypoglycemia and hyperventilation syndrome due

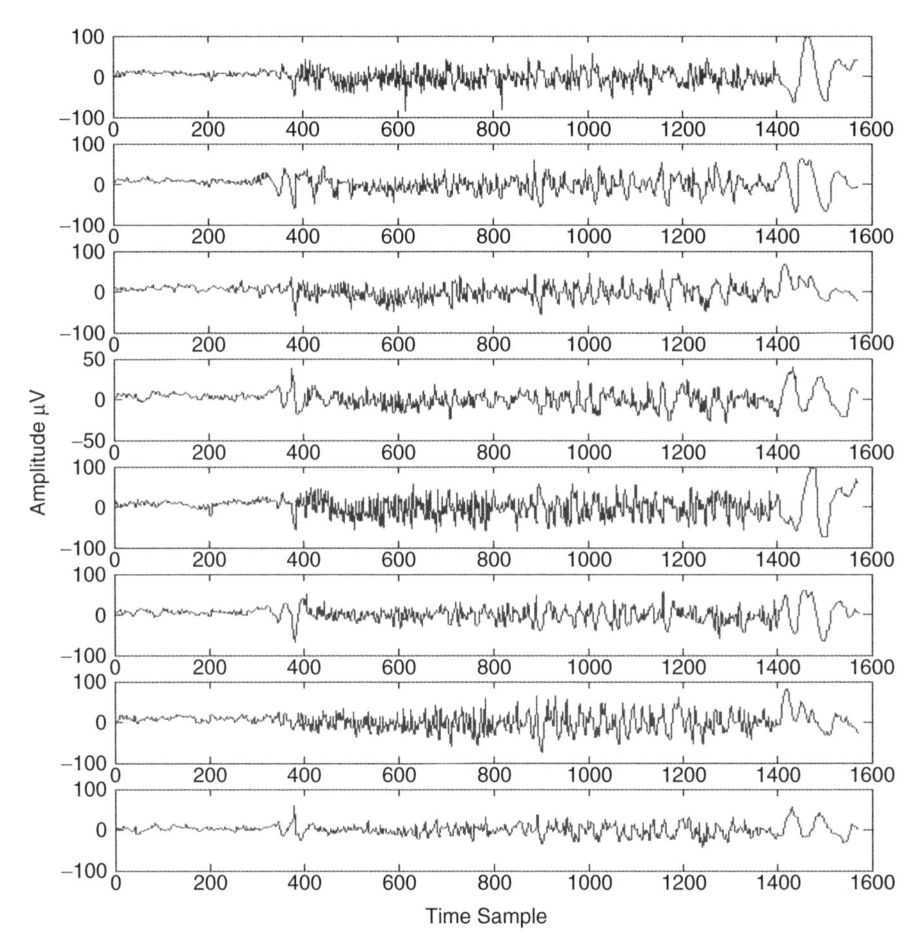

Figure 1.15 Generalized tonic–clonic (grand mal) seizure. The seizure appears in almost all of the electrodes

to sudden changes in blood chemistry [62]; and periodic discharges in Creutzfeldt–Jakob (mad cow) disease [63,64]. The waveforms for this latter abnormality consist of a sharp wave or a sharp triphasic transient signal of 100–300 ms duration, with a frequency of 0.5–2 Hz. The periodic activity usually shows a maximum over the anterior region except for the Heidenhain form, which has a posterior maximum [56]. Other epileptic waveforms include periodic literalized epileptiform discharges (PLED), periodic discharges in acute cerebral anoxia, and periodic discharges of other etiologies.

Despite the above epileptiform signals there are spikes and other paroxysmal discharges in healthy nonepileptic persons. These discharges may be found in healthy individuals without any other symptoms of diseases. However, they are often signs of certain cerebral dysfunctions that may or may not develop into an abnormality. They may appear during periods of particular mental challenge on individuals, such as soldiers in the war front line, pilots, and prisoners.

A comprehensive overview of epileptic seizure disorders and nonepileptic attacks can be found in many books and publications such as References [62] and [65]. In Chapter 5 some recent attempts in application of advanced signal processing techniques to the automatic detection and prediction of epileptic seizures are explained.

1.9.3 Psychiatric Disorders

Not only can functional and certain anatomical brain abnormalities be investigated using EEG signals, pathophysiological brain disorders can also be studied by analysing such signals. According to the *Diagnostic and Statistical Manual (DSM) of Mental Disorders* of the American Psychiatric Association, changes in psychiatric education have evolved considerably since the 1970s. These changes have mainly resulted from physical and neurological laboratory studies based upon EEG signals [66].

There have been evidences from EEG coherence measures suggesting differential patterns of maturation between normal and learning-disabled children [67]. This finding can lead to the establishment of some methodology in monitoring learning disorders. Several psychiatric disorders are diagnosed by analysis of evoked potentials (EPs) achieved by simply averaging a number of consecutive trails having the same stimuli.

A number of pervasive mental disorders cause significant losses in multiple functioning areas [66]. Examples of these are dyslexia, which is a developmental reading disorder; autistic disorder, which is related to abnormal social interaction, communication, and restricted interests and activities, and starts appearing from the age of three; Rett's disorder, characterized by the development of multiple deficits following a period of normal postnatal functioning; and Asperger's disorder, which leads to severe and sustained impairments in social interaction and restricted repetitive patterns of behaviour, interests, and activities.

Attention-deficit hyperactivity disorder (ADHD) and attention-deficit disorder (ADD), conduct disorder, oppositional defiant disorder, and disruptive behaviour disorder have also been under investigation and considered within the DSM. Most of these abnormalities appear during childhood and often prevent children from learning and socializing well. The associated EEG features have been rarely analytically investigated, but the EEG observations are often reported in the literature [68–72]. However, most of such abnormalities tend to disappear with advancing age.

EEG has also been analysed recently for the study of delirium [73,74], dementia [75,76], and many other cognitive disorders [77]. In EEGs, characteristics of delirium include slowing or dropout of the posterior dominant rhythm, generalized theta or delta slow-wave activity, poor organization of the background rhythm, and loss of reactivity of the EEG to eye opening and closing. In parallel with that, the quantitative EEG (QEEG) shows increased absolute and relative slow-wave (theta and delta) power, reduced ratio of fast-to-slow band power, reduced mean frequency, and reduced occipital peak frequency [74].

Dementia includes a group of neurodegenerative diseases that cause acquired cognitive and behavioural impairment of sufficient severity to interfere significantly with social and occupational functioning. Alzheimer disease is the most common of the diseases that cause dementia. At present, the disorder afflicts approximately 5 million people in the United States and more than 30 million people worldwide. A larger number of individuals have lesser levels of cognitive impairment, which frequently evolves into full-blown dementia.

The prevalence of dementia is expected to nearly triple by 2050, since the disorder preferentially affects the elderly, who constitute the fastest-growing age bracket in many countries, especially in industrialized nations [76].

Among other psychiatric and mental disorders, amnestic disorder (or amnesia), mental disorder due to a general medical condition, substance-related disorder, schizophrenia, mood disorder, anxiety disorder, somatoform disorder, dissociative disorder, sexual and gender identity disorder, eating disorders, sleep disorders, impulse-controlled disorder, and personality disorders have often been addressed in the literature [66]. However, the corresponding EEGs have seldom been analysed by means of advanced signal processing tools.

1.9.4 External Effects

EEG signal patterns may significantly change when using drugs for the treatment and suppression of various mental and CNS abnormalities. Variations in EEG patterns may also arise by just looking at the TV screen or listening to music without any attention. However, among the external effects the most significant ones are the pharmacological and drug effects. Therefore, it is important to know the effects of these drugs on the changes of EEG waveforms due to chronic overdosage, and the patterns of overt intoxication [78].

The effect of administration of drugs for anesthesia on EEGs is of interest to clinicians. The related studies attempt to find the correlation between the EEG changes and the stages of anesthesia. It has been shown that in the initial stage of anesthesia a fast frontal activity appears. In deep anesthesia this activity becomes slower with higher amplitudes. In the last stage, a burst-suppression pattern indicates the involvement of brainstem functions, including respiration, and finally the EEG activity ceases [78]. In cases of acute intoxication, the EEG patterns are similar to those of anesthesia [78].

Barbiturate is commonly used as an anticonvulsant and antiepileptic drug. With small dosages of barbiturate the activities within the 25–35 Hz frequency band around the frontal cortex increases. This changes to 15–25 Hz and spreads to the parietal and occipital regions. Dependence and addiction to barbiturates are common. Therefore, after a long-term ingestion of barbiturates, its abrupt withdrawal leads to paroxysmal abnormalities. The major complications are myoclonic jerks, generalized tonic–clonic seizures, and delirium [78].

Many other drugs are used in addition to barbiturates as sleeping pills, such as melatonin and bromides. Very pronounced EEG slowing is found in chronic bromide encephalopathies [78]. Antipsychotic drugs also influence the EEG patterns. For example, neuroleptics increase the alpha wave activity but reduce the duration of beta wave bursts and their average frequency. As another example, clozapine increases the delta, theta, and above 21 Hz beta wave activities. As another antipsychotic drug, tricyclic antidepressants such as imipramine, amitriptyline, doxepin, desipramine, notryptiline, and protriptyline increase the amount of slow and fast activity along with instability of frequency and voltage, and also slow down the alpha wave rhythm. After administration of tricyclic antidepressants the seizure frequency in chronic epileptic patients may increase. With high dosages, this may further lead to single or multiple seizures occurring in nonepileptic patients [78].

During acute intoxication, a widespread, poorly reactive, irregular 8–10 Hz activity and paroxysmal abnormalities including spikes, as well as unspecific coma patterns, are observed in the EEGs [78]. Lithium is often used in the prophylactic treatment of bipolar mood disorder. The related changes in the EEG pattern consist of slowing of the beta rhythm and of paroxysmal generalized slowing, occasionally accompanied by spikes. Focal slowing also occurs, which is not necessarily a sign of a focal brain lesion. Therefore, the changes in the EEG are markedly abnormal with lithium administration [78]. The beta wave activity is highly activated by using benzodiazepines as an anxiolytic drug. These activities persist in the EEG as long as two weeks after ingestion. Benzodiazepine leads to a decrease in an alpha wave activity and its amplitude, and slightly increases the 4–7 Hz frequency band activity. In acute intoxication the EEG shows prominent fast activity with no response to stimuli [78]. The psychotogentic drugs such as lysergic acid diethylamide and mescaline decrease the amplitude and possibly depress the slow waves [78]. The CNS stimulants increase the alpha and beta wave activities and reduce the amplitude and the amount of slow waves and background EEGs [78].

The effect of many other drugs, especially antiepileptic drugs, is investigated and new achievements are published frequently. One of the significant changes of the EEG of epileptic patients with valproic acid consists of reduction or even disappearance of generalized spikes along with seizure reduction. Lamotrigine is another antiepileptic agent that blocks voltage-gated sodium channels, thereby preventing excitatory transmitter glutamate release. With the intake of lamotrigine a widespread EEG attenuation occurs [78]. Penicillin if administered in high dosage may produce jerks, generalized seizures, or even status epilepticus [78].

1.10 Summary and Conclusions

In this chapter the fundamental concepts in the generation of action potentials and consequently the EEG signals have been briefly explained. The conventional measurement setups for EEG recording and the brain rhythms present in normal or abnormal EEGs have also been described. In addition, the effects of popular brain abnormalities such as mental diseases, ageing, and epileptic and nonepileptic attacks have been pointed out. Despite the known neurological, physiological, pathological, and mental abnormalities of the brain mentioned in this chapter, there are many other brain disorders and dysfunctions that may or may not manifest some kinds of abnormalities in the related EEG signals. Degenerative disorders of the CNS [79], such as a variety of lysosomal disorders, several peroxisomal disorders, a number of mitochondrial disorders, inborn disturbances of the urea cycle, many aminoacidurias, and other metabolic and degenerative diseases, as well as chromosomal aberrations, have to be evaluated and their symptoms correlated with the changes in the EEG patterns. The similarities and differences within the EEGs of these diseases have to be well understood. On the other hand, the developed mathematical algorithms need to take the clinical observations and findings into account in order to enhance the outcome of such processing further. Although a number of technical methods have been well established for the processing of the EEGs with relation to the above abnormalities, there is still a long way to go and many questions to be answered.

The following chapters of this book introduce new digital signal processing techniques employed mainly for analysis of EEG signals followed by a number of examples in the applications of such methods.

References

[1] Caton, R., 'The electric currents of the brain', *Br. Med. J.*, **2**, 1875, 278.

[2] Walter, W. G., 'Slow potential waves in the human brain associated with expectancy, attention and decision', *Arch. Psychiat. Nervenkr.*, **206**, 1964, 309–322.

[3] Cobb, M., 'Exorcizing the animal spirits: Jan Swammerdam on nerve function', *Neuroscience*, **3**, 2002, 395–400.

[4] Danilevsky, V. Y., 'Investigation into the physiology of the brain' [in Russian], Doctoral Thesis, University of Kharkov, 1877, Zit. Nach: Brazier MAB; A history of Neurophysiology in the 19th Century, New York: Raven; 1988, 208.

[5] Brazier, M. A. B., *A History of the Electrical Activity of the Brain; The First Half-Century*, Macmillan, New York, 1961.

[6] Massimo, A., 'In Memoriam Pierre Gloor (1923–2003): an appreciation', *Epilepsia*, **45**(7), July 2004, 882.

[7] Grass, A. M., and Gibbs, F. A., 'A Fourier transform of the electroencephalogram', *J. Neurophysiol.*, **1**, 1938, 521–526.

[8] Haas, L. F., 'Hans Berger (1873–1941), Richard Caton (1842–1926), and electroencephalography', *J. Neurol. Neurosurg. Psychiat.*, **74**, 2003, 9.

[9] Spear, J. H., 'Cumulative change in scientific production: research technologies and the structuring of new knowledge', *Perspectives on Sci.*, **12**(1), 2004, 55–85.

[10] Shipton, H. W., 'EEG analysis: a history and prospectus', *Annual Rev., Univ. of Iowa, USA*, 1975, 1–15.

[11] Fischer, M. H., 'Elektrobiologische Auswirkungen von Krampfgiften am Zentralnervensystem', *Med. Klin.*, **29**, 1933, 15–19.

[12] Fischer, M. H., and Lowenbach, H., 'Aktionsstrome des Zentralnervensystems unter der Einwirkung von Krampfgiften, 1. Mitteilung Strychnin und Pikrotoxin', *Arch. F. Exp. Pathol. und Pharmakol.*, **174**, 1934, 357–382.

[13] Kornmuller, A. E., 'Der Mechanismus des Epileptischen Anfalles auf Grund Bioelektrischer Untersuchungen am Zentralnervensystem', *Fortschr. Neurol. Psychiatry*, **7**, 1935, 391–400; 414–432.

[14] Bremer, F., 'Cerveau isole' et physiologie du sommeil', *C.R. Soc. Biol. (Paris)*, **118**, 1935, 1235–1241.

[15] Niedermeyer, E., 'Historical aspects', Chapter 1, *Electroencephalography, Basic Principles, Clinical Applications, and Related Fields*, Eds E. Niedermeyer and F. Lopes da Silva, 4th edn., Lippincott, Williams and Wilkins, Philadelphia, Pennsylvania, 1999, 1–14.

[16] Berger, H., 'Uber das Elektrenkephalogramm des Menschen', *Arch. Psychiatr. Nervenkr.*, **87**, 1929, 527–580.

[17] Jasper, H., 'Report of committee on methods of clinical exam in EEG', *Electroencephalogr. Clin. Neurophysiol.*, **10**, 1958, 370–375.

[18] Motokawa, K., 'Electroencephalogram of man in the generalization and differentiation of condition reflexes', *Tohoku J. Expl. Medicine*, **50**, 1949, 225.

[19] Niedermeyer, E., 'Common generalized epilepsy. The so-called idiopathic or centrencephalic epilepsy', *Eur. Neurol.*, **9**(3), 1973, 133–156.

[20] Aserinsky, E., and Kleitman, N., 'Regularly occurring periods of eye motility, and concomitant phenomena, during sleep', *Science*, **118**, 1953, 273–274.

[21] Speckmann, E.-J., and Elger, C. E., 'Introduction to the neurophysiological basis of the EEG and DC potentials', in *Electroencephalography Basic Principles, Clinical Applications, and Related Fields*, Eds E. Niedermeyer and F. Lopes da Silva, 4th edn, Lippincott, Williams and Wilkins, Philadelphia, Pennsylvania, 1999.

[22] Shepherd, G. M., *The Synaptic Organization of the Brain*, Oxford University Press, London, 1974.

[23] Caspers, H., Speckmann E.-J., and Lehmenkühler, A., 'DC potentials of the cerebral cortex, seizure activity and changes in gas pressures', *Rev. Physiol., Biochem. Pharmacol.*, **106**, 1986, 127–176.

[24] Ka Xiong Charand, http://hyperphysics.phy-astr.gsu.edu/hbase/biology/actpot.html.

[25] Attwood, H. L., and MacKay, W. A., *Essentials of Neurophysiology*, B. C. Decker, Hamilton, Canada, 1989.

[26] Nunez, P. L., *Neocortical Dynamics and Human EEG Rhythms*, Oxford University Press, New York, 1995.

[27] Teplan, M., 'Fundamentals of EEG measurements', *Measmt Sci. Rev.*, **2**(2), 2002.

[28] Bickford, R. D., 'Electroencephalography', in *Encyclopedia of Neuroscience*, Ed. G. Adelman, Birkhauser, Cambridge (USA), 1987, 371–373.

[29] Sterman, M. B., MacDonald, L. R., and Stone, R. K., 'Biofeedback training of sensorimotor EEG in man and its effect on epilepsy', *Epilepsia*, **15**, 1974, 395–416.

[30] Ashwal, S., and Rust, R., 'Child neurology in the 20th century', *Pedia. Res.*, **53**, 2003, 345–361.

[31] Niedermeyer, E., 'The normal EEG of the waking adult', Chapter 10, in *Electroencephalography, Basic Principles, Clinical Applications, and Related Fields*, Eds E. Niedermeyer and F. Lopes da Silva, 4th edn, Lippincott, Williams and Wilkins, Philadelphia, Pennsylvania, 1999, 174–188.

[32] Pfurtscheller, G., Flotzinger, D., and Neuper, C., 'Differentiation between finger, toe and tongue movement in man based on 40 Hz EEG', *Electroencephalogr. Clin. Neurophysiol.*, **90**, 1994, 456–460.

[33] Adrian, E. D., and Mattews, B. H. C., 'The Berger rhythm, potential changes from the occipital lob in man', *Brain*, **57**, 1934, 345–359.

[34] Trabka, J., 'High frequency components in brain waves', *Electroencephalogr. Clin. Neurophysiol.*, **14**, 1963, 453–464.

[35] Cobb, W. A., Guiloff, R. J., and Cast, J., 'Breach rhythm: the EEG related to skull defects', *Electroencephalogr. Clin. Neurophysiol.*, **47**, 1979, 251–271.

[36] Silbert, P. L., Radhakrishnan, K., Johnson, J., and Class, D. W., 'The significance of the phi rhythm', *Electroencephalogr. Clin. Neurophysiol.*, **95**, 1995, 71–76.

[37] Roldan, E., Lepicovska, V., Dostalek, C., and Hrudova, L., 'Mu-like EEG rhythm generation in the course of Hatha-yogi exercises', *Electroencephalogr. Clin. Neurophysiol.*, **52**, 1981, 13.

[38] IFSECN, 'A glossary of terms commonly used by clinical electroencephalographers', *Electroencephalogr. Clin. Neurophysiol.*, **37**, 1974, 538–548.

[39] O'Leary, J. L., and Goldring, S., *Science and Epilepsy*, Raven Press, New York, 1976, pp. 19–152.

[40] Gotman, J., Ives, J. R., and Gloor, R., 'Automatic recognition of interictal epileptic activity in prolonged EEG recordings', *Electroencephalogr. Clin. Neurophysiol.*, **46**, 1979, 510–520.

[41] 'Effects of electrode placement', http://www.focused-technology.com/electrod.htm, California.

[42] Collura, T., *A Guide to Electrode Selection, Location, and Application for EEG Biofeedback*, Ohio, BrainMaster Technologies, Inc. 1998.

[43] Nayak, D., Valentin, A., Alarcon, G., Seoane, J. J. G., Brunnhuber, F., Juler, J., Polkey, C. E., and Binnie, C. D., 'Characteristics of scalp electrical fields associated with deep medial temporal epileptiform discharges', *Clin. Neurophysiol.*, **115**, 2004, 1423–1435.

[44] Barrett, G., Blumhardt, L., Halliday, L., Halliday, A. M., and Kriss, A., 'A paradox in the lateralization of the visual evoked responses', *Nature*, **261**, 1976, 253–255.

[45] Halliday, A. M., Evoked potentials in neurological disorders', in *Event-Related Brain Potentials in Man*, Eds E. Calloway, P. Tueting, and S. H. Coslow, Academic Press, New York, 1978, 197–210.

[46] Sharbrough, F. W., 'Nonspecific abnormal EEG patterns', Chapter. 12, in *Electroencephalography, Basic Principles, Clinical Applications, and Related Fields*, Eds E. Niedermeyer and F. Lopes Da Silva, 4th edn., Lippincott, Williams and Wilkins, Philadelphia, Pennsylvania, 1999.

[47] Bancaud, J., Hecaen, H., and Lairy, G. C., 'Modification de la reactivite EEG, troubles des functions symboliques et troubles con fusionels dans les lesions hemispherigues localisees', *Electroencephalogr. Clin. Neurophysiol.*, **7**, 1955, 179.

[48] Westmoreland, B., and Klass, D., 'Asymmetrical attention of alpha activity with arithmetical attention', *Electroencephalogr. Clin. Neurophysiol.*, **31**, 1971, 634–635.

[49] Cobb, W., 'EEG interpretation in clinical medicine', Part B, in *Handbook of Electroencephalography and Clinical Neurophysiology*, Ed. A. Remond, Amsterdam, Vol. 11, Elsevier, 1976.

[50] Hess, R., 'Brain tumors and other space occupying processing', Part C, in *Handbook of Electroencephalography and Clinical Neurophysiology*, Ed. A. Remond, Amsterdam, Vol. 14, Elsevier, 1975.

[51] Klass, D., and Daly, D. (Eds), *Current Practice of Clinical Electroencephalography*, 1st edn. Raven Press, 1979.

[52] Van Sweden, B., Wauquier, A., and Niedermeyer, E., 'Normal aging and transient cognitive disorders in the elderly', Chapter 18, in *Electroencephalography, Basic Principles, Clinical Applications, and Related Fields*, Eds E. Niedermeyer and F. Lopes da Silva, 4th edn, Lippincott, Williams and Wilkins, Philadelphia, Pennsylvania, 1999, 340–348.

[53] America Psychiatric Association, Committee on Nomenclature and Statistics, *Diagnostic and Statistical Manual of Mental Disorder: DSM-IV*, 4th edn., American Psychiatric Association, Washington, DC, 1994.

[54] Brenner, R. P., 'EEG and dementia', Chapter 19, in *Electroencephalography, Basic Principles, Clinical Applications, and Related Fields*, Eds E. Niedermeyer and F. Lopes da Silva, 4th edn., Lippincott, Williams and Wilkins, Philadelphia, Pennsylvania, 1999, 349–359.

[55] Neufeld, M. Y., Bluman, S., Aitkin, I., Parmet, Y., and Korczyn, A. D., 'EEG frequency analysis in demented and nondemented Parkinsonian patients', *Dementia*, **5**, 1994, 23–28.

[56] Niedermeyer, E., 'Abnormal EEG patterns: epileptic and paroxysmal', Chapter 13, in *Electroencephalography, Basic Principles, Clinical Applications, and Related Fields*, Eds E. Niedermeyer and F. Lopes da Silva, 4th edn, Lippincott, Williams and Wilkins, Philadelphia, Pennsylvania, 1999, 235–260.

[57] Hughes, J. R., and Gruener, G. T., 'Small sharp spikes revisited: further data on this controversial pattern', *Electroencephalogr. Clin. Neurophysiol.*, **15**, 1984, 208–213.

[58] Hecker, A., Kocher, R., Ladewig, D., and Scollo-Lavizzari, G., 'Das Minature-Spike-Wave', *Das EEG Labor*, **1**, 1999 51–56.

[59] Geiger, L. R., and Harner, R. N., 'EEG patterns at the time of focal seizure onset', *Arch. Neurol.*, **35**, 1978, 276–286.

[60] Gastaut, H., and Broughton, R., *Epileptic Seizure*, Charles C. Thomas, Springfield, Illinois, 1972.

[61] Oller-Daurella, L., and Oller-Ferrer-Vidal, L., *Atlas de Crisis Epilepticas*, Geigy Division Farmaceut, Spain, Barcelona 1977.

[62] Niedermeyer, E., 'Nonepileptic Attacks', Chapter 28, in *Electroencephalography, Basic Principles, Clinical Applications, and Related Fields*, Eds E. Niedermeyer and F. Lopes da Silva, 4th edn, Lippincott, Williams and Wilkins, Philadelphia, Pennsylvania, 1999, 586–594.

[63] Creutzfeldt, H. G., 'Uber eine Eigenartige Herdformige Erkrankung des Zentralnervensystems', *Z. Ges. Neurol. Psychiatr.*, **57**, 1968, 1, Quoted after W. R. Kirschbaum, 1920.

[64] Jakob, A., 'Uber Eigenartige Erkrankung des Zentralnervensystems mit Bemerkenswerten Anatomischen Befunden (Spastistische Pseudosklerose, Encephalomyelopathie mit Disseminerten Degenerationsbeschwerden)', *Deutsch. Z. Nervenheilk*, **70**, 1968, 132, Quoted after W. R. Kirschbaum, 1921.

[65] Niedermeyer, E., 'Epileptic seizure disorders', Chapter 27, in *Electroencephalography, Basic Principles, Clinical Applications, and Related Fields*, Eds E. Niedermeyer and F. Lopes da Silva, 4th edn, Lippincott, Williams and Wilkins, Philadelphia, Pennsylvania, 1999, 476–585.

[66] Small, J. G., 'Psychiatric disorders and EEG', Chapter 30, in *Electroencephalography, Basic Principles, Clinical Applications, and Related Fields*, Eds E. Niedermeyer and F. Lopes da Silva, 4th ed., Lippincott, Williams and Wilkins, Philadelphia, Pennsylvania, 1999, 235–260.

[67] Marosi, E., Harmony, T., Sanchez, L., Becker, J., Bernal, J., Reyes, A., Diaz de Leon, A. E., Rodriguez, M., and Fernandez, T., 'Maturation of the coherence of EEG activity in normal and learning disabled children', *Electroencephalogr. Clin. Neurophysiol.* **83**, 1992, 350–357.

[68] Linden, M., Habib, T., and Radojevic, V., 'A controlled study of the effects of EEG biofeedback on cognition and behavior of children with attention deficit disorder and learning disabilities', *Biofeedback Self Regul.*, **21**(1), 1996, pp. 35–49.

[69] Hermens, D. F., Soei, E. X., Clarke, S. D., Kohn, M. R., Gordon, E., and Williams, L. M., 'Resting EEG theta activity predicts cognitive performance in attention-deficit hyperactivity disorder', *Pediatr. Neurol.*, **32**(4), 2005, 248–256.

[70] Swartwood, J. N., Swartwood, M. O., Lubar, J. F., and Timmermann, D. L., 'EEG differences in ADHD-combined type during baseline and cognitive tasks', *Pediatr. Neurol.*, **28**(3), 2003, 199–204.

[71] Clarke, A. R., Barry, R. J., McCarthy, R., and Selikowitz, M., 'EEG analysis of children with attention-deficit/hyperactivity disorder and comorbid reading disabilities', *J. Learn. Disabil.*, **35**(3), 2002, 276–285.

[72] Yordanova, J., Heinrich, H., Kolev, V., and Rothenberger, A., 'Increased event-related theta activity as a psychophysiological marker of comorbidity in children with tics and attention-deficit/hyperactivity disorders', *Neuroimage*, **32**(2), 2006, 940–955.

[73] Jacobson, S., and Jerrier, H., 'EEG in delirium', *Semin. Clin. Neuropsychiat.*, **5**(2), 2000, 86–92.

[74] Onoe, S., and Nishigaki, T., 'EEG spectral analysis in children with febrile delirium', *Brain Devel.*, **26**(8), 2004, 513–518.

[75] Brunovsky, M., Matousek, M., Edman, A., Cervena, K., and Krajca, V., 'Objective assessment of the degree of dementia by means of EEG', *Neuropsychobiology*, **48**(1), 2003, 19–26.

[76] Koenig, T., Prichep, L., Dierks, T., Hubl, D., Wahlund, L. O., John, E. R., and Jelic, V., 'Decreased EEG synchronization in Alzheimer's disease and mild cognitive impairment', *Neurobiol. Aging*, **26**(2), 2005, 165–171.

[77] Babiloni, C., Binetti, G., Cassetta, E., Dal Forno, G., Del Percio, C., Ferreri, F., Ferri, R., Frisoni, G., Hirata, K., Lanuzza, B., Miniussi, C., Moretti, D. V., Nobili, F., Rodriguez, G., Romani, G. L., Salinari, S., and Rossini, P. M., 'Sources of cortical rhythms change as a function of cognitive impairment in pathological aging: a multicenter study', *Clin. Neurophysiol.*, **117**(2), 2006, 252–268.

[78] Bauer, G., and Bauer, R., 'EEG, drug effects, and central nervous system poisoning', Chapter 35, in *Electroencephalography, Basic Principles, Clinical Applications, and Related Fields*, Eds E. Niedermeyer and F. Lopes da Silva, 4th edn, Lippincott, Williams and Wilkins, Philadelphia, Pennsylvania, 1999, 671–691.

[79] Naidu, S. and Niedermeyer, E., 'Degenerative disorders of the central nervous system', Chapter 20, in *Electroencephalography, Basic Principles, Clinical Applications, and Related Fields*, Eds E. Niedermeyer and F. Lopes da Silva, 4th edn., Lippincott, Williams and Wilkins, Philadelphia, Pennsylvania, 1999, 360–382.

2

Fundamentals of EEG Signal Processing

EEG signals are the signatures of neural activities. They are captured by multiple-electrode EEG machines either from inside the brain, over the cortex under the skull, or certain locations over the scalp, and can be recorded in different formats. The signals are normally presented in the time domain, but many new EEG machines are capable of applying simple signal processing tools such as the Fourier transform to perform frequency analysis and equipped with some imaging tools to visualize EEG topographies (maps of the brain activities in the spatial domain).

There have been many algorithms developed so far for processing EEG signals. The operations include, but are not limited to, time-domain analysis, frequency-domain analysis, spatial-domain analysis, and multiway processing. Also, several algorithms have been developed to visualize the brain activity from images reconstructed from only the EEGs. Separation of the desired sources from the multisensor EEGs has been another research area. This can later lead to the detection of brain abnormalities such as epilepsy and the sources related to various physical and mental activities. In Chapter 7 of this book it can be seen that the recent works in brain–computer interfacing (BCI) [1] have been focused upon the development of advanced signal processing tools and algorithms for this purpose.

Modelling of neural activities is probably more difficult than modelling the function of any other organ. However, some simple models for generating EEG signals have been proposed. Some of these models have also been extended to include generation of abnormal EEG signals.

Localization of brain signal sources is another very important field of research [2]. In order to provide a reliable algorithm for localization of the sources within the brain sufficient knowledge about both propagation of electromagnetic waves and how the information from the measured signals can be exploited in separation and localization of the sources within the brain is required. The sources might be considered as magnetic dipoles for which the well-known inverse problem has to be solved, or they can be considered as distributed current sources.

Patient monitoring and sleep monitoring require real-time processing of (up to a few days) long EEG sequences. The EEG provides important and unique information about the sleeping brain. Major brain activities during sleep can be captured using the developed algorithms [3], such as the method of matching pursuits (MPs) discussed [4] later in this chapter.

Epilepsy monitoring, detection, and prediction have also attracted many researchers. Dynamical analysis of a time series together with the application of blind separation of the signal sources has enabled prediction of focal epilepsies from the scalp EEGs. On the other hand, application of time–frequency-domain analysis for detection of the seizure in neonates has paved the way for further research in this area.

In the following sections most of the tools and algorithms for the above objectives are explained and the mathematical foundations discussed. The application of these algorithms to analysis of the normal and abnormal EEGs, however, will follow in later chapters of this book. The reader should also be aware of the required concepts and definitions borrowed from linear algebra, further details of which can be found in Reference [5]. Throughout this chapter and the reminder of this book continuous time is denoted by t and discrete time, with normalized sampling period $T = 1$, by n.

2.1 EEG Signal Modelling

Most probably the earliest physical model is based on the Hodgkin and Huxley's Nobel Prize winning model for the squid axon published in 1952 [6–8]. A nerve axon may be stimulated and the activated sodium (Na^+) and potassium (K^+) channels produced in the vicinity of the cell membrane may lead to the electrical excitation of the nerve axon. The excitation arises from the effect of the membrane potential on the movement of ions, and from interactions of the membrane potential with the opening and closing of voltage-activated membrane channels. The membrane potential increases when the membrane is polarized with a net negative charge lining the inner surface and an equal but opposite net positive charge on the outer surface. This potential may be simply related to the amount of electrical charge Q, using

$$E = Q/C_{\mathrm{m}} \tag{2.1}$$

where Q is in terms of coulombs/cm^2, C_{m} is the measure of the capacity of the membrane in units of farads/cm^2, and E is in units of volts. In practice, in order to model the action potentials (APs) the amount of charge Q^+ on the inner surface (and Q^- on the outer surface) of the cell membrane has to be mathematically related to the stimulating current I_{stim} flowing into the cell through the stimulating electrodes. The electrical potential (often called the electrical force) E is then calculated using Equation (2.1). The Hodgkin and Huxley model is illustrated in Figure 2.1. In this figure I_{memb} is the result of positive charges flowing out of the cell. This current consists of three currents, namely Na, K, and leak currents. The leak current is due to the fact that the inner and outer Na and K ions are not exactly equal.

Hodgkin and Huxley estimated the activation and inactivation functions for the Na and K currents and derived a mathematical model to describe an AP similar to that of a

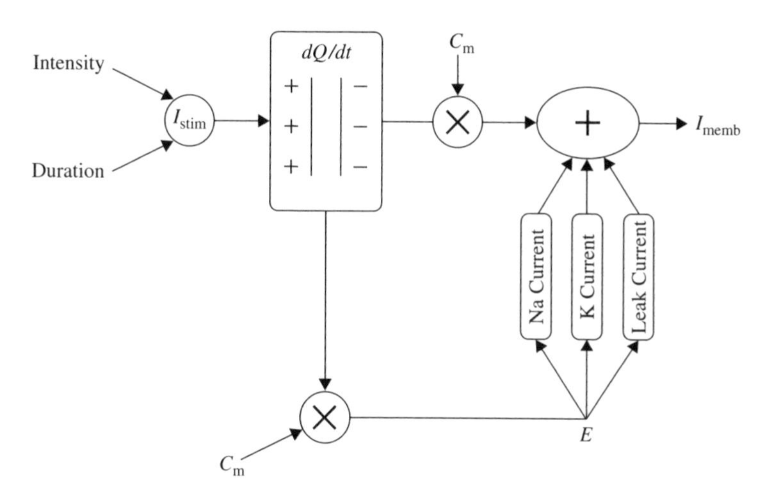

Figure 2.1 The Hodgkin–Huxley excitation model

giant squid. The model is a neuron model that uses voltage-gated channels. The space-clamped version of the Hodgkin–Huxley model may be well described using four ordinary differential equations [9]. This model describes the change in the membrane potential (E) with respect to time and is described in Reference [10]. The overall membrane current is the sum of capacity current and ionic current, i.e.

$$I_{\mathrm{memb}} = C_{\mathrm{m}} \frac{\mathrm{d}E}{\mathrm{d}t} + I_{\mathrm{i}} \tag{2.2}$$

where I_{i} is the ionic current and, as indicated in Figure 2.1, can be considered as the sum of three individual components: Na, K, and leak currents:

$$I_{\mathrm{i}} = I_{\mathrm{Na}} + I_{\mathrm{K}} + I_{\mathrm{leak}} \tag{2.3}$$

I_{Na} can be related to the maximal conductance $\overline{g}_{\mathrm{Na}}$, activation variable a_{Na}, inactivation variable h_{Na}, and a driving force $(E - E_{\mathrm{Na}})$ through

$$I_{\mathrm{Na}} = \overline{g}_{\mathrm{Na}} a_{\mathrm{Na}}^3 h_{\mathrm{Na}} (E - E_{\mathrm{Na}}) \tag{2.4}$$

Similarly, I_{K} can be related to the maximal conductance $\overline{g}_{\mathrm{K}}$, activation variable a_{Na}, inactivation variable a_{K}, and a driving force $(E - E_{\mathrm{K}})$ as

$$I_{\mathrm{K}} = \overline{g}_{\mathrm{K}} a_{\mathrm{K}} (E - E_{\mathrm{K}}) \tag{2.5}$$

and I_{leak} is related to the maximal conductance $\overline{g}_{\mathrm{l}}$ and a driving force $(E - E_{\mathrm{l}})$ as

$$I_{\mathrm{l}} = \overline{g}_{\mathrm{l}} (E - E_{\mathrm{l}}) \tag{2.6}$$

The changes in the variables a_{Na}, a_{K}, and h_{Na} vary from 0 to 1 according to the following equations:

$$\frac{\mathrm{d}a_{\mathrm{Na}}}{\mathrm{d}t} = \lambda_t[\alpha_{\mathrm{Na}}(E)(1 - a_{\mathrm{Na}}) - \beta_{\mathrm{Na}}(E)a_{\mathrm{Na}}] \tag{2.7}$$

$$\frac{\mathrm{d}h_{\mathrm{Na}}}{\mathrm{d}t} = \lambda_t[\alpha_{\mathrm{h}}(E)(1 - h_{\mathrm{Na}}) - \beta_{\mathrm{h}}(E)h_{\mathrm{Na}}] \tag{2.8}$$

$$\frac{\mathrm{d}a_{\mathrm{K}}}{\mathrm{d}t} = \lambda_t[\alpha_{\mathrm{K}}(E)(1 - a_{\mathrm{K}}) - \beta_{\mathrm{K}}(E)a_{\mathrm{K}}] \tag{2.9}$$

where $\alpha(E)$ and $\beta(E)$ are respectively forward and backward rate functions and λ_t is a temperature-dependent factor. The forward and backward parameters depend on voltage and were empirically estimated by Hodgkin and Huxley as

$$\alpha_{\mathrm{Na}}(E) = \frac{3.5 + 0.1E}{1 - \mathrm{e}^{-(3.5+0.1E)}} \tag{2.10}$$

$$\beta_{\mathrm{Na}}(E) = 4\mathrm{e}^{-(E+60)/18} \tag{2.11}$$

$$\alpha_{\mathrm{h}}(E) = 0.07\mathrm{e}^{-(E+60)/20} \tag{2.12}$$

$$\beta_{\mathrm{h}}(E) = \frac{1}{1 + \mathrm{e}^{-(3+0.1E)}} \tag{2.13}$$

$$\alpha_{\mathrm{K}}(E) = \frac{0.5 + 0.01E}{1 - \mathrm{e}^{-(5+0.1E)}} \tag{2.14}$$

$$\beta_{\mathrm{K}}(E) = 0.125\mathrm{e}^{-(E+60)/80} \tag{2.15}$$

As stated in the Simulator for Neural Networks and Action Potentials (SNNAP) literature [9], the $\alpha(E)$ and $\beta(E)$ parameters have been converted from the original Hodgkin–Huxley version to agree with the present physiological practice, where depolarization of the membrane is taken to be positive. In addition, the resting potential has been shifted to −60 mV (from the original 0 mV). These equations are used in the model described in the SNNAP. In Figure 2.2 an AP has been simulated. For this model the parameters are set to $C_{\mathrm{m}} = 1.1$ uF/cm^2, $\overline{g}_{\mathrm{Na}} = 100$ ms/cm^2, $\overline{g}_{\mathrm{K}} = 35$ ms/cm^2, $\overline{g}_{\mathrm{l}} = 0.35$ ms/cm^2, and $E_{\mathrm{Na}} = 60$ mV.

The simulation can run to generate a series of action potentials, as happens in practice in the case of ERP signals. If the maximal ionic conductance of the potassium current, $\overline{g}_{\mathrm{K}}$, is reduced the model will show a higher resting potential. Also, for $\overline{g}_{\mathrm{K}} = 16$ ms/cm^2, the model will begin to exhibit oscillatory behaviour. Figure 2.3 shows the result of a Hodgkin–Huxley oscillatory model with reduced maximal potassium conductance.

The SNNAP can also model bursting neurons and central pattern generators. This stems from the fact that many neurons show cyclic spiky activities followed by a period of inactivity. Several invertebrate as well as mammalian neurons are bursting cells and exhibit alternating periods of high-frequency spiking behaviour followed by a period of no spiking activity.

A simpler model than that due to Hodgkin–Huxley for simulating spiking neurons is the Morris–Lecar model [11]. This model is a minimal biophysical model, which generally exhibits single action potential. This model considers that the oscillation of a slow

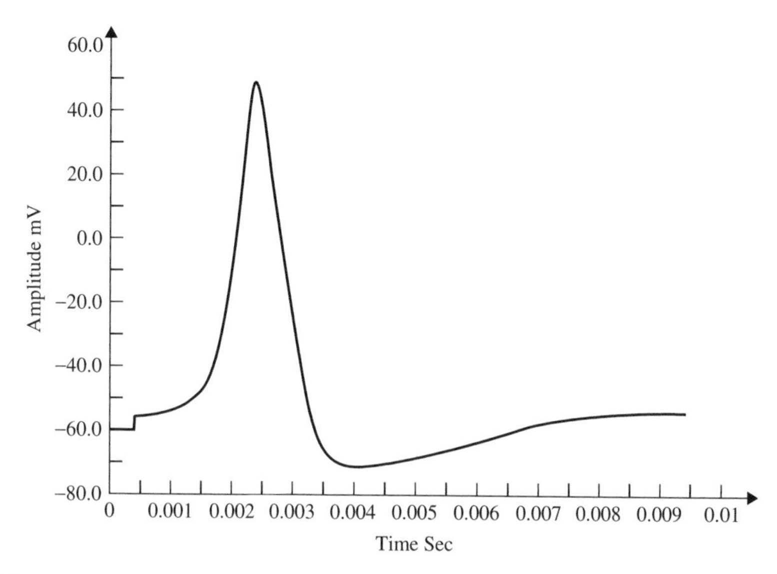

Figure 2.2 A single AP in response to a transient stimulation based on the Hodgkin–Huxley model. The initiated time is at $t = 0.4$ ms and the injected current is 80 μA/cm^2 for a duration of 0.1 ms. The selected parameters are $C_m = 1.2$ uF/cm^2, $\overline{g}_{Na} = 100$ mS/cm^2, $\overline{g}_K = 35$ ms/cm^2, $\overline{g}_l = 0.35$ ms/cm^2, and $E_{Na} = 60$ mV

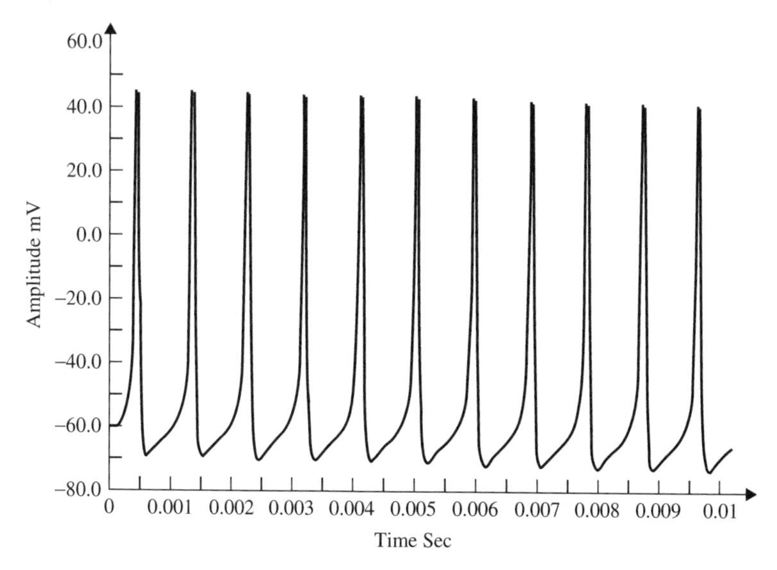

Figure 2.3 The AP from a Hodgkin–Huxley oscillatory model with reduced maximal potassium conductance

calcium wave that depolarizes the membrane leads to a bursting state. The Morris–Lecar model was initially developed to describe the behaviour of barnacle muscle cells. The governing equations relating the membrane potential (E) and potassium activation w_K to the activation parameters are given as

$$C\frac{dE}{dt} = I_i - \overline{g}_{Ca}a_{Ca}(E)(E - E_{Ca}) - \overline{g}_K w_K(E - E_K) - \overline{g}_l(E - E_l) \tag{2.16}$$

$$\frac{dw_K}{dt} = \lambda_t\left(\frac{w_\infty(E) - w_K}{\tau_K(E)}\right) \tag{2.17}$$

where I_i is the combination of three ionic currents, calcium (Ca), potassium (K), and leak (l), and, similar to the Hodgkin–Huxley model, are products of a maximal conductance $\overline{g}$, activation components (in such as a_{Ca}, w_K), and the driving force E. The changes in the potassium activation variable w_K is proportional to a steady-state activation function $w_K(E)$ (a sigmoid curve) and a time-constant function $\tau_K(E)$ (a bell-shaped curve). These functions are respectively defined as

$$w_\infty(E) = \frac{1}{1 + e^{-(E-h_w)/S_w}} \tag{2.18}$$

$$\tau_K(E) = \frac{1}{e^{(E-h_w)/(2S_w)} + e^{-(E-h_w)/(2S_w)}} \tag{2.19}$$

The steady-state activation function $a_{Ca}(E)$, involved in calculation of the calcium current, is defined as

$$a_{Ca}(E) = \frac{1}{1 + e^{-(E-h_{Ca})/s_m}} \tag{2.20}$$

Similar to the sodium current in the Hodgkin–Huxley model, the calcium current is an inward current. Since the calcium activation current is a fast process in comparison with the potassium current, it is modelled as an instantaneous function. This means that for each voltage E, the steady-state function $a_{Ca}(E)$ is calculated. The calcium current does not incorporate any inactivation process. The activation variable w_K here is similar to a_K in the Hodgkin–Huxley model, and finally the leak currents for both models are the same [9]. A simulation of the Morris–Lecar model is presented in Figure 2.4.

Calcium-dependent potassium channels are activated by intracellular calcium; the higher the calcium concentration the higher the channel activation [9]. For the Morris–Lecar model to exhibit bursting behaviour, the two parameters of maximal time constant and the input current have to be changed [9]. Figure 2.5 shows the bursting behaviour of the Morris–Lecar model. The basic characteristics of a bursting neuron are the duration of the spiky activity, the frequency of the action potentials during a burst, and the duration of the quiescence period. The period of an entire bursting event is the sum of both active and quiescence duration [9].

Neurons communicate with each other across synapses through axon–dendrites or dendrites–dendrites connections, which can be excitatory, inhibitory, or electric [9]. By combining a number of the above models a neuronal network can be constructed. The network exhibits oscillatory behaviour due to the synaptic connection between the neurons.

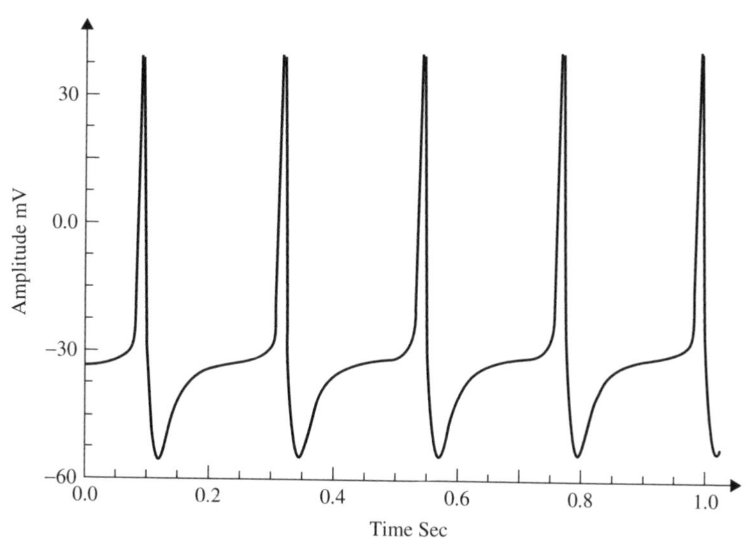

Figure 2.4 Simulation of an AP within the Morris–Lecar model. The model parameters are: $C_m = 22$ uF/cm^2, $\overline{g}_{Ca} = 3.8$ ms/cm^2, $\overline{g}_K = 8.0$ ms/cm^2, $\overline{g}_l = 1.6$ ms/cm^2, $E_{Ca} = 125$ mV, $E_K = -80$ mV, $E_l = -60$ mV, $\lambda_t = 0.06$, $h_{Ca} = -1.2$, and $S_m = 8.8$

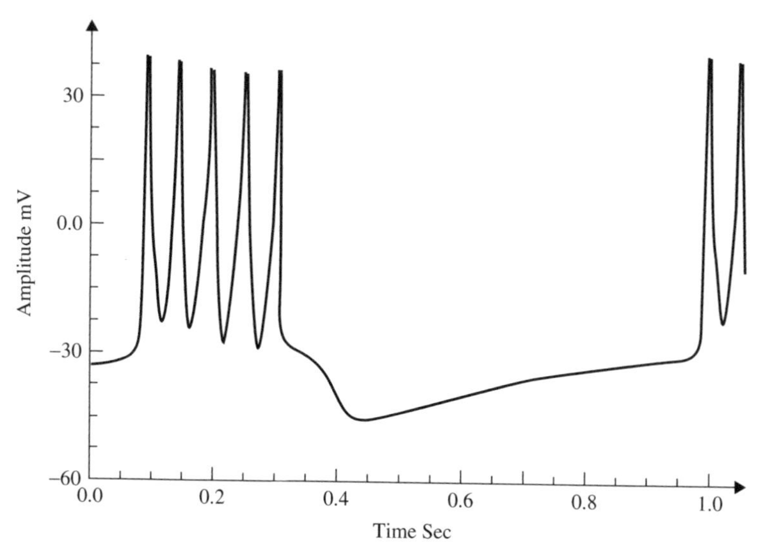

Figure 2.5 An illustration of the bursting behaviour that can be generated by the Morris–Lecar model

A synaptic current is produced as soon as a neuron fires an AP. This current stimulates the connected neuron and may be modelled by an alpha function multiplied by a maximal conductance and a driving force as

$$I_{\mathrm{syn}} = \overline{g}_{\mathrm{syn}}\, g_{\mathrm{syn}}(t)[E(t) - E_{\mathrm{syn}}] \tag{2.21}$$

where

$$g_{\mathrm{syn}}(t) = t\ \mathrm{e}^{(-t/u)} \tag{2.22}$$

and t is the latency or time since the trigger of the synaptic current, u is the time to reach to the peak amplitude, E_{syn} is the synaptic reversal potential, and $\overline{g}_{\mathrm{syn}}$ is the maximal synaptic conductance. The parameter u alters the duration of the current while $\overline{g}_{\mathrm{syn}}$ changes the strength of the current. This concludes the treatment of the modelling of APs.

As the nature of the EEG sources cannot be determined from the electrode signals directly, many researchers have tried to model these processes on the basis of information extracted using signal processing techniques. The method of linear prediction described in the later sections of this chapter is frequently used to extract a parametric description.

2.1.1 Linear Models

2.1.1.1 Prediction Method

The main objective of using prediction methods is to find a set of model parameters that best describe the signal generation system. Such models generally require a noise-type input. In autoregressive (AR) modelling of signals each sample of a single-channel EEG measurement is defined to be linearly related with respect to a number of its previous samples, i.e.

$$y(n) = -\sum_{k=1}^{p} a_k y(n-k) + x(n) \tag{2.23}$$

where $a_k, k = 1, 2, \ldots, p$, are the linear parameters, n denotes the discrete sample time normalized to unity, and $x(n)$ is the noise input. In an autoregressive moving average (ARMA) linear predictive model each sample is obtained based on a number of its previous input and output sample values, i.e.

$$y(n) = -\sum_{k=1}^{p} a_k y(n-k) + \sum_{k=0}^{q} b_k x(n-k) \tag{2.24}$$

where $b_k, k = 1, 2, \ldots, q$, are the additional linear parameters. The parameters p and q are the model orders. The Akaike criterion can be used to determine the order of the appropriate model of a measurement signal by minimizing the following equation [12] with respect to the model order:

$$\mathrm{AIC}(i, j) = N \ln(\sigma_{ij}^2) + 2(i + j) \tag{2.25}$$

where i and j represent the assumed AR and MA (moving average) model prediction orders respectively, N is the number of signal samples, and σ_{ij}^2 is the noise power of the ARMA model at the ith and jth stage. Later in this chapter it will be shown how the model parameters are estimated either directly or by employing some iterative optimization techniques.

In a multivariate AR (MVAR) approach a multichannel scheme is considered. Therefore, each signal sample is defined versus both its previous samples and the previous samples of the other channels, i.e. for channel i,

$$y_i(n) = -\sum_{k=1}^{p} a_{ik} y_i(n-k) - \sum_{\substack{j=1 \\ j\neq i}}^{m} \sum_{k=1}^{p} a_{jk} y_j(n-k) + x_i(n) \tag{2.26}$$

where m represents the number of channels and $x_i(n)$ represents the noise input to channel i. Similarly, the model parameters can be calculated iteratively in order to minimize the error between the actual and predicted values [13].

These linear models will be described further later in this chapter and some of their applications are discussed in other chapters. Different algorithms have been developed to find the model coefficients efficiently. In the maximum likelihood estimation (MLE) method [14–16] the likelihood function is maximized over the system parameters formulated from the assumed real, Gaussian distributed, and sufficiently long input signals of approximately 10–20 seconds (consider a sampling frequency of $f_s = 250$ samples/s as often used for EEG recordings). Using Akaike's method, the gradient of the squared error is minimized using the Newton–Raphson approach applied to the resultant nonlinear equations [16,17]. This is considered as an approximation to the MLE approach. In the Durbin method [18] the Yule–Walker equations, which relate the model coefficients to the autocorrelation of the signals, are iteratively solved. The approach and the results are equivalent to those using a least-squares-based scheme [19]. The MVAR coefficients are often calculated using the Levinson–Wiggins–Robinson (LWR) algorithm [20]. The MVAR model and its application in representation of what is called a direct transfer function (DTF), and its use in the quantification of signal propagation within the brain, will come in the following section. After the parameters are estimated the synthesis filter can be excited with wide-sense stationary noise to generate the EEG signal samples. Figure 2.6 illustrates the simplified system.

2.1.1.2 Prony's Method

Prony's method has been previously used to model evoked potentials (EPs) [21,22]. Based on this model an EP, which is obtained by applying a short audio or visual stimulation to the brain, can be considered as the impulse response (IR) of a linear infinite impulse

Figure 2.6 A linear model for the generation of EEG signals

response (IIR) system. The original attempt in this area was to fit an exponentially damped sinusoidal model to the data [23]. This method was later modified to model sinusoidal signals [24]. Prony's method is used to calculate the linear prediction (LP) parameters. The angles of the poles in the z plane of the constructed LP filter are then referred to the frequencies of the damped sinusoids of the exponential terms used for modelling the data. Consequently, both the amplitude of the exponentials and the initial phase can be obtained following the methods used for an AR model, as follows.

Based on the original method the output of an AR system with zero excitation can be considered to be related to its IR as

$$y(n) = \sum_{k=1}^{p} a_k y(n-k) = \sum_{j=1}^{p} w_j \sum_{k=1}^{p} a_k r_j^{n-k-1} \tag{2.27}$$

where $y(n)$ represents the exponential data samples, p is the prediction order,$w_j = A_j \mathrm{e}^{\mathrm{j}\theta_j}$, $r_k = \exp[(\alpha_k + \mathrm{j}2\pi f_k)T_s]$, T_s is the sampling period normalized to 1, A_k is the amplitude of the exponential, α_k is the damping factor, f_k is the discrete-time sinusoidal frequency in samples/s, and θ_j is the initial phase in radians. Therefore, the model coefficients are first calculated using one of the methods previously mentioned in this section, i.e. $\boldsymbol{a} = -\mathbf{Y}^{-1}\breve{\mathbf{y}}$, where

$$\boldsymbol{a} = \begin{bmatrix} a_0 \\ a_1 \\ \vdots \\ a_p \end{bmatrix}, \quad \mathbf{Y} = \begin{bmatrix} y(p)\ldots y(1) \\ y(p-1)\ldots y(2) \\ \vdots \\ y(2p-1)\cdots y(p) \end{bmatrix}, \quad \text{and} \quad \breve{\mathbf{y}} = \begin{bmatrix} y(p+1) \\ y(p+2) \\ \vdots \\ y(2p) \end{bmatrix} \tag{2.28}$$

where $a_0 = 1$. The prediction filter output, i.e. on the basis of Equation (2.27), $y(n)$ is calculated as the weighted sum of p past values of $y(n)$, and the parameters f_k and r_k are estimated. Hence, the damping factors are obtained as

$$\alpha_k = \ln |r_k| \tag{2.29}$$

and the resonance frequencies as

$$f_k = \frac{1}{2\pi} \tan^{-1}\left[\frac{\mathrm{Im}(r_k)}{\mathrm{Re}(r_k)}\right] \tag{2.30}$$

where Re(.) and Im(.) denote the real and imaginary parts of a complex quantity respectively. The w_k parameters are calculated using the fact that $y(n) = \sum_{k=1}^{p} w_k r_k^{n-1}$ or

$$\begin{bmatrix} r_1^0 & r_2^0 & \cdots & r_p^0 \\ r_1^1 & r_2^1 & \cdots & r_p^1 \\ \vdots & \vdots & & \vdots \\ r_1^{p-1} & r_2^{p-1} & \cdots & r_p^{p-1} \end{bmatrix} \begin{bmatrix} w_1 \\ w_2 \\ \vdots \\ w_p \end{bmatrix} = \begin{bmatrix} y(1) \\ y(2) \\ \vdots \\ y(p) \end{bmatrix} \tag{2.31}$$

In vector form this can be illustrated as $\mathbf{R}\boldsymbol{w} = \boldsymbol{y}$, where $[\mathbf{R}]_{k,l} = r_l^k$, $k = 0, 1, \ldots, p-1$, $l = 1, \ldots, p$, denoting the elements of the matrix in the above equation. Therefore, $\boldsymbol{w} = \mathbf{R}^{-1}\boldsymbol{y}$, assuming $\mathbf{R}$ is a full-rank matrix, i.e. there are no repeated poles. Often, this is simply carried out by implementing the Cholesky decomposition algorithm [25]. Finally, using w_k, the amplitude and initial phases of the exponential terms are calculated as follows:

$$A_k = |w_k| \tag{2.32}$$

and

$$\theta_k = \tan^{-1}\left[\frac{\mathrm{Im}(w_k)}{\mathrm{Re}(w_k)}\right] \tag{2.33}$$

In the above solution it was considered that the number of data samples N is equal to $N = 2p$, where p is the prediction order. For cases where $N > 2p$, a least-squares (LS) solution for $\boldsymbol{w}$ can be obtained as

$$\boldsymbol{w} = (\mathbf{R}^{\mathrm{H}}\mathbf{R})^{-1}\mathbf{R}^{\mathrm{H}}\boldsymbol{y} \tag{2.34}$$

where $(.)^{\mathrm{H}}$ denotes the conjugate transpose. This equation can also be solved using the Cholesky decomposition method. For real data such as EEG signals this equation changes to $\boldsymbol{w} = (\mathbf{R}^{\mathrm{T}}\mathbf{R})^{-1}\mathbf{R}^{\mathrm{T}}\boldsymbol{y}$, where $(.)^{\mathrm{T}}$ represents the transpose operation. A similar result can be achieved using principal component analysis (PCA) [15].

In the cases where the data are contaminated with white noise the performance of Prony's method is reasonable. However, for nonwhite noise the noise information is not easily separable from the data and therefore the method may not be sufficiently successful.

In a later chapter of this book it will be seen that, Prony's algorithm has been used in modelling and analysis of audio and visual evoked potentials (AEP and VEP) [26,27].

2.1.2 *Nonlinear Modelling*

An approach similar to AR or MVAR modelling in which the output samples are nonlinearly related to the previous samples may be followed based on the methods developed for forecasting financial growth in economical studies. In the generalized autoregressive conditional heteroskedasticity (GARCH) method [28] each sample relates to its previous samples through a nonlinear (or sum of nonlinear) function(s). This model was originally introduced for time-varying volatility (honoured with the Nobel Prize in Economic Sciences in 2003). Nonlinearities in the time series are declared with the aid of the McLeod–Li [29] and BDS (Brock, Dechert, and Scheinkman) tests [30]. However, both tests lack the ability to reveal the actual kind of nonlinear dependency.

Generally, it is not possible to discern whether the nonlinearity is deterministic or stochastic in nature, and nor can a distinction be made between multiplicative and additive dependencies. The type of stochastic nonlinearity can be determined on the basis of the Hseih test [31]. Both additive and multiplicative dependencies can be discriminated by using this test. However, the test itself is not used to obtain the model parameters.

Considering the input to a nonlinear system to be $u(n)$ and the generated signal as the output of such a system to be $x(n)$, a restricted class of nonlinear models suitable for the analysis of such a process is given by

$$x(n) = g(u(n-1), u(n-2), \ldots) + u_n\, h(u(n-1), u(n-2), \ldots) \tag{2.35}$$

Multiplicative dependence means nonlinearity in the variance, which requires the function $h(.)$ to be nonlinear; additive dependence, on the other hand, means nonlinearity in the mean, which holds if the function $g(.)$ is nonlinear. The conditional statistical mean and variance are respectively defined as

$$E[x(n)|\chi_{n-1}] = g(u(n-1), u(n-2), \ldots) \tag{2.36}$$

and

$$\mathrm{Var}[x(n)|\chi_{n-1}] = h^2(u(n-1), u(n-2), \ldots) \tag{2.37}$$

where χ_{n-1} contains all the past information up to time $n-1$. The original GARCH(p, q) model, where p and q are the prediction orders, considers a zero mean case, i.e. $g(.) = 0$. If $e(n)$ represents the residual (error) signal using the above nonlinear prediction system, then

$$\mathrm{Var}[e(n)|\chi_{n-1}] = \sigma^2(n) = \alpha_0 + \sum_{j=1}^{q} \alpha_j \mathrm{e}^2(n-j) + \sum_{j=1}^{p} \beta_j \sigma^2(n-1) \tag{2.38}$$

where α_j and β_j are the nonlinear model coefficients. The second term (first sum) in the right-hand side corresponds to a qth-order moving average (MA) dynamical noise term and the third term (second sum) corresponds to an autoregressive (AR) model of order p. It is seen that the current conditional variance of the residual at time sample n depends on both its previous sample values and previous variances.

Although in many practical applications such as forecasting of stock prices the orders p and q are set to small fixed values such as $(p, q) = (1, 1)$, for a more accurate modelling of natural signals such as EEGs the orders have to be determined mathematically. The prediction coefficients for various GARCH models or even the nonlinear functions g and h are estimated recursively as for the linear ARMA models [28,29].

Clearly, such simple GARCH models are only suitable for multiplicative nonlinear dependence. In addition, additive dependencies can be captured by extending the modelling approach to the class of GARCH-M (GARCH-in-mean) models [32].

Another limitation of the above simple GARCH model is failing to accommodate sign asymmetries. This is because the squared residual is used in the update equations. Moreover, the model cannot cope with rapid transitions such as spikes. Considering these shortcomings, numerous extensions to the GARCH model have been proposed. For example, the model has been extended and refined to include the asymmetric effects of positive and negative jumps such as the exponential GARCH (EGARCH) model [33], the Glosten, Jagannathan, and Runkle GARCH (GJR-GARCH) model [34], the threshold GARCH (TGARCH) model [35], the asymmetric power GARCH (APGARCH) model

[36], and the quadratic GARCH (QGARCH) model [37]. In the EGARCH model, for example, the above equation changes to

$$\log[\sigma^2(n)] = \log(\alpha_0) + \sum_{j=1}^{q} \alpha_j \log[e^2(n-j)] + \sum_{j=1}^{p} \beta_j \log[\sigma^2(n-j)] \tag{2.39}$$

where log[·] denotes natural logarithm. This logarithmic expression has the advantage of preventing the variance from becoming negative.

In these models different functions for $g(.)$ and $h(.)$ are defined. For example, in the EGARCH model proposed by Glosten *et al.* [34] $h(n)$ is iteratively computed as

$$h_{(n)} = b + \alpha_1 u^2(n-1)(1-\eta_{(n-1)}) + \alpha_2 u^2(n-1)\eta_{(n-1)} + \kappa h_{(n-1)} \tag{2.40}$$

where b, α_1, α_2, and κ are constants and $\eta_{(n)}$ is an indicator function that is zero when $u_{(n)}$ is negative and one otherwise.

Despite modelling the signals, the GARCH approach has many other applications. In some recent works [38] the concept of GARCH modelling of covariance is combined with Kalman filtering to provide a more flexible model with respect to space and time for solving the inverse problem. There are several alternatives for solution to the inverse problem. Many approaches fall into the category of constrained least-squares methods employing Tikhonov regularization [39]. Localization of the sources within the brain using the EEG information is as an example. This approach has become known as low-resolution electromagnetic tomography (LORETA) [40]. Among numerous possible choices for the GARCH dynamics, the EGARCH [33] has been used to estimate the variance parameter of the Kalman filter sequentially.

The above methods are used to model the existing data, but to generate the EEG signals accurately a very complex model that exploits the physiological dynamics and various mental activities of the brain has to be constructed. Such a model should also incorporate the changes in the brain signals due to abnormalities and the onset of diseases. The next section considers the interaction among various brain components to establish a more realistic model for generation of the EEG signals.

2.1.3 *Generating EEG Signals Based on Modelling the Neuronal Activities*

The objective in this section is to introduce some established models for generating normal and some abnormal EEGs. These models are generally nonlinear, some have been proposed [41] for modelling a normal EEG signal and some others for the abnormal EEGs.

A simple distributed model consisting of a set of simulated neurons, thalamocortical relay cells, and interneurons was proposed [42,43] that incorporates the limited physiological and histological data available at that time. The basic assumptions were sufficient to explain the generation of the alpha rhythm, i.e. the EEGs within the frequency range of 8–13 Hz.

A general nonlinear lumped model may take the form shown in Figure 2.7. Although the model is analogue in nature, all the blocks are implemented in a discrete form. This model can take into account the major characteristics of a distributed model and it is easy

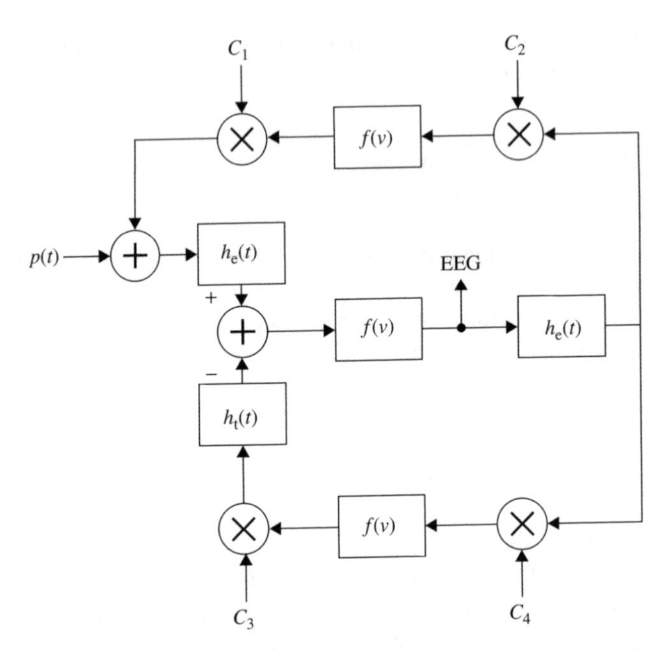

Figure 2.7 A nonlinear lumped model for generating the rhythmic activity of the EEG signals; $h_e(t)$ and $h_i(t)$ are the excitatory and inhibitory postsynaptic potentials, $f(v)$ is normally a simplified nonlinear function, and the C_is are respectively the interaction parameters representing the interneurons and thalamocortical neurons

to investigate the result of changing the range of excitatory and inhibitory influences of thalamocortical relay cells and interneurons.

In this model [42] there is a feedback loop including the inhibitory postsynaptic potentials, the nonlinear function, and the interaction parameters C_3 and C_4. The other feedback includes mainly the excitatory potentials, nonlinear function, and the interaction parameters C_1 and C_2. The role of the excitatory neurons is to excite one or two inhibitory neurons. The latter, in turn, serve to inhibit a collection of excitatory neurons. Thus, the neural circuit forms a feedback system. The input $p(t)$ is considered as a white noise signal. This is a general model; more assumptions are often needed to enable generation of the EEGs for the abnormal cases. Therefore, the function $f(v)$ may change to generate the EEG signals for different brain abnormalities. Accordingly, the C_i coefficients can be varied. In addition, the output is subject to environment and measurement noise. In some models, such as the local EEG model (LEM) [42] the noise has been considered as an additive component in the output.

Figure 2.8 shows the LEM model. This model uses the formulation by Wilson and Cowan [44] who provided a set of equations to describe the overall activity (not specifically the EGG) in a cartel of excitatory and inhibitory neurons having a large number of interconnections [45]. Similarly, in the LEM the EEG rhythms are assumed to be generated by distinct neuronal populations, which possess frequency selective properties. These populations are formed by the interconnection of the individual neurons and are assumed to be driven by a random input. The model characteristics, such as the neural

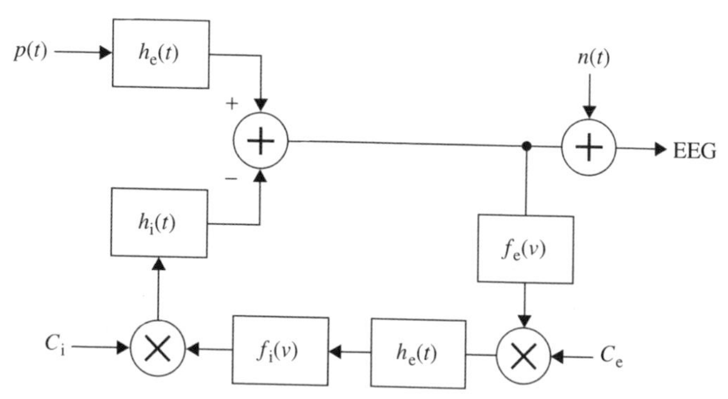

Figure 2.8 The local EEG model (LEM). The thalamocortical relay neurons are represented by two linear systems having impulse responses $h_e(t)$, on the upper branch, and the inhibitory postsynaptic potential represented by $h_i(t)$. The nonlinearity of this system is denoted by $f_e(v)$, representing the spike-generating process. The interneuron activity is represented by another linear filter $h_e(t)$ in the lower branch, which generally can be different from the first linear system, and a nonlinearity function $f_i(v)$. C_e and C_i represent respectively the number of interneuron cells and the thalamocortical neurons

interconnectivity, synapse pulse response, and threshold of excitation, are presented by the LEM parameters. The changes in these parameters produce the relevant EEG rhythms.

In Figure 2.8, as in Figure 2.7, the notation 'e' and 'i' refer to excitatory and inhibitory respectively. The input $p(t)$ is assumed to result from the summation of a randomly distributed series of random potentials which drive the excitatory cells of the circuit, producing the ongoing background EEG signal. Such signals originate from other deeper brain sources within the thalamus and brain stem and constitute part of the ongoing or spontaneous firing of the central nervous system (CNS). In the model, the average number of inputs to an inhibitory neuron from the excitatory neurons is designated by C_e and the corresponding average number from inhibitory neurons to each individual excitatory neuron is C_i. The difference between two decaying exponentials is used for modelling each postsynaptic potential h_e or h_i:

$$h_e(t) = A[\exp(-a_1 t) - \exp(-a_2 t)] \tag{2.41}$$

$$h_i(t) = B[\exp(-b_1 t) - \exp(-b_2 t)] \tag{2.42}$$

where A, B, a_k, and b_k are constant parameters, which control the shape of the pulse waveforms. The membrane potentials are related to the axonal pulse densities via the static threshold functions f_e and f_i. These functions are generally nonlinear, but to ease the manipulations they are considered linear for each short time interval. Using this model, the normal brain rhythms, such as the alpha wave, are considered as filtered noise.

The main problem with such a model is due to the fact that only a single-channel EEG is generated and there is no modelling of interchannel relationships. Therefore, a more accurate model has to be defined to enable simulation of a multichannel EEG generation system. This is still an open question and remains an area of research.

2.2 Nonlinearity of the Medium

The head as a mixing medium combines EEG signals which are locally generated within the brain at the sensor positions. As a system, the head may be more or less susceptible to such sources in different situations. Generally, an EEG signal can be considered as the output of a nonlinear system, which may be characterized deterministically.

The changes in brain metabolism as a result of biological and physiological phenomena in the human body can change the mixing process. Some of these changes are influenced by the activity of the brain itself. These effects make the system nonlinear. Analysis of such a system is very complicated and up to now nobody has fully modelled the system to aid in the analysis of brain signals.

On the other hand, some measures borrowed from chaos theory and analysis of the dynamics of time series such as dissimilarity, attractor dimension, and largest Lyapunov exponents (LLE) can characterize the nonlinear behaviour of EEG signals. These concepts are discussed in Section 2.7 and some of their applications are given in Chapter 5.

2.3 Nonstationarity

Nonstationarity of the signals can be quantified by measuring some statistics of the signals at different time lags. The signals can be deemed stationary if there is no considerable variation in these statistics.

Although generally the multichannel EEG distribution is considered as multivariate Gaussian, the mean and covariance properties generally change from segment to segment. Therefore EEGs are considered stationary only within short intervals, i.e. quasistationarity. This Gaussian assumption holds during a normal brain condition, but during mental and physical activities this assumption is not valid. Some examples of nonstationarity of the EEG signals can be observed during the change in alertness and wakefulness (where there are stronger alpha oscillations), during eye blinking, during the transitions between various ictal states, and in the event-related potential (ERP) and evoked potential (EP) signals.

The change in the distribution of the signal segments can be measured in terms of both the parameters of a Gaussian process and the deviation of the distribution from Gaussian. The non-Gaussianity of the signals can be checked by measuring or estimating some higher-order moments such as skewness, kurtosis, negentropy, and Kulback–Laibler (KL) distance.

Skewness is a measure of symmetry or, more precisely, the lack of symmetry of the distribution. A distribution, or data set, is symmetric if it looks the same to the left and right of the centre point. The skewness is defined for a real signal as

$$\text{Skewness} = \frac{E[(x(n) - \mu)^3]}{\sigma^3} \tag{2.43}$$

where μ and σ are the mean and standard deviation respectively, and E denotes statistical expectation. If the distribution is more to the right of the mean point the skewness is negative, and vice versa. For a symmetric distribution such as Gaussian, the skewness is zero.

Kurtosis is a measure of whether the data are peaked or flat relative to a normal distribution; i.e. data sets with high kurtosis tend to have a distinct peak near the mean, decline rather rapidly, and have heavy tails. Data sets with low kurtosis tend to have a flat top near the mean rather than a sharp peak. A uniform distribution would be the extreme case. The kurtosis for a real signal $x(n)$ is defined as

$$\text{kurt} = \frac{m_4[x(n)]}{m_2^2[x(n)]} \tag{2.44}$$

where $m_i[x(n)]$ is the ith central moment of the signal $x(n)$, i.e. $m_i[x(n)] = E[(x(n) - \mu)^i]$. The kurtosis for signals with normal distributions is three. Therefore, an excess or normalized kurtosis is often used and defined as

$$\text{Ex kurt} = \frac{m_4[x(n)]}{m_2^2[x(n)]} - 3 \tag{2.45}$$

which is zero for Gaussian distributed signals. Often the signals are considered ergodic; hence the statistical averages can be assumed identical to time averages and so can be estimated with time averages.

The negentropy of a signal $x(n)$ [46] is defined as

$$J_{\text{neg}}[x(n)] = H[x_{\text{Gauss}}(n)] - H[x(n)] \tag{2.46}$$

where, $x_{\text{Gauss}}(n)$ is a Gaussian random signal with the same covariance as $x(n)$ and $H(.)$ is the differential entropy [47], defined as

$$H[x(n)] = \int_{-\infty}^{\infty} p[x(n)] \log \frac{1}{p[x(n)]} \, \mathrm{d}x(n) \tag{2.47}$$

and $p[x(n)]$ is the signal distribution. Negentropy is always nonnegative.

The KL distance between two distributions p_1 and p_2 is defined as

$$\text{KL} = \int_{-\infty}^{\infty} p_1(x_1(n)) \log \frac{p_1(x_1(n))}{p_2(x_2(n))} \, \mathrm{d}z \tag{2.48}$$

It is clear that the KL distance is generally asymmetric, therefore by changing the position of p_1 and p_2 in this equation the KL distance changes. The minimum of the KL distance occurs when $p_1(x_1(n)) = p_2(x_2(n))$.

2.4 Signal Segmentation

Often it is necessary to label the EEG signals by segments of similar characteristics that are particularly meaningful to clinicians and for assessment by neurophysiologists. Within each segment, the signals are considered statistically stationary, usually with similar time and frequency statistics. As an example, an EEG recorded from an epileptic patient may be divided into three segments of preictal, ictal, and postictal segments. Each

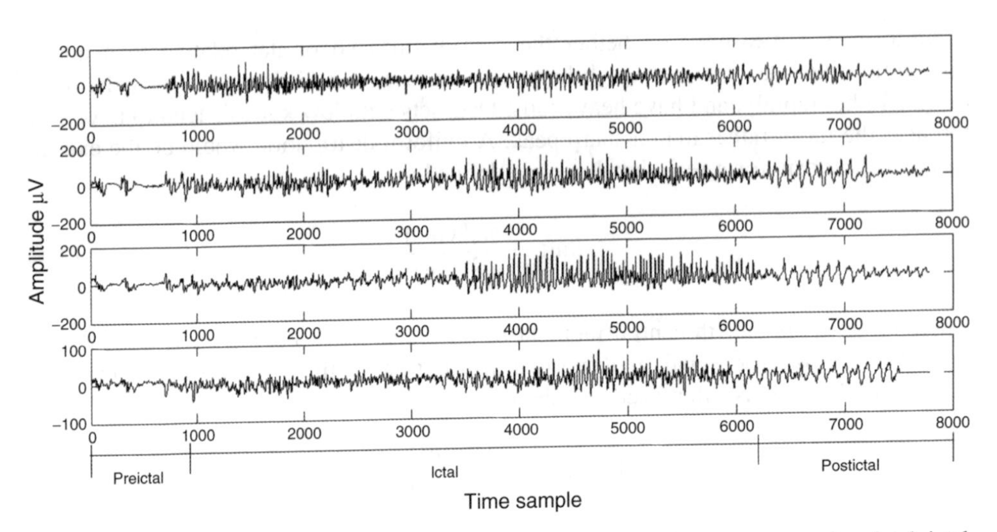

Figure 2.9 An EEG set of tonic–clonic seizure signals including three segments of preictal, ictal, and postictal behaviour

may have a different duration. Figure 2.9 represents an EEG sequence including all the above segments.

In segmentation of an EEG the time or frequency properties of the signals may be exploited. This eventually leads to a dissimilarity measurement denoted as $d(m)$ between the adjacent EEG frames, where m is an integer value indexing the frame and the difference is calculated between the m and $(m-1)$th (consecutive) signal frames. The boundary of the two different segments is then defined as the boundary between the m and $(m-1)$th frames provided $d(m) > \eta_{\mathrm{T}}$, and η_{T} is an empirical threshold level. An efficient segmentation is possible by highlighting and effectively exploiting the diagnostic information within the signals with the help of expert clinicians. However, access to such experts is not always possible and therefore algorithmic methods are required.

A number of different dissimilarity measures may be defined based on the fundamentals of digital signal processing. One criterion is based on the autocorrelations for segment m, defined as

$$r_x(k,m) = E[x(n,m)x(n+k,m)] \tag{2.49}$$

The autocorrelation function of the mth length N frame for an assumed time interval $n, n+1, \ldots, n+(N-1)$ can be approximated as

$$\hat{r}_x(k,m) = \begin{cases} \dfrac{1}{N}\displaystyle\sum_{l=0}^{N-1-k} x(l+m+k)x(l+m), & k = 0,\ldots,N-1 \\ 0, & k = N, N+1, \ldots \end{cases} \tag{2.50}$$

Then the criterion is set to

$$d_1(m) = \frac{\sum_{k=-\infty}^{\infty} [\hat{r}_x(k, m) - \hat{r}_x(k, m-1)]^2}{\hat{r}_x(0, m)\hat{r}_x(0, m-1)} \tag{2.51}$$

A second criterion can be based on higher-order statistics. The signals with more uniform distributions such as normal brain rhythms have a low kurtosis, whereas seizure signals or event related potentials (ERP signals) often have high kurtosis values. Kurtosis is defined as the fourth-order cumulant at zero time lags and is related to the second- and fourth-order moments as given in Equations (2.43) to (2.45). A second level discriminant $d_2(m)$ is then defined as

$$d_2(m) = \text{kurt}_x(m) - \text{kurt}_x(m-1) \tag{2.52}$$

where m refers to the mth frame of the EEG signal $x(n)$. A third criterion is defined from the spectral error measure of the periodogram. A periodogram of the mth frame is obtained by discrete time Fourier transforming of the correlation function of the EEG signal

$$S_x(\omega, m) = \sum_{k=-\infty}^{\infty} \hat{r}_x(k, m)\mathrm{e}^{-\mathrm{j}\omega k}, \quad \omega \in [-\pi, \pi] \tag{2.53}$$

where $\hat{r}_x(., m)$ is the autocorrelation function for the mth frame as defined above. The criterion is then defined based on the normalized periodogram as

$$d_3(m) = \frac{\int_{-\pi}^{\pi} [S_x(\omega, m) - S_x(\omega, m-1)]^2 \, \mathrm{d}\omega}{\int_{-\pi}^{\pi} S_x(\omega, m)\mathrm{d}\omega \int_{-\pi}^{\pi} S_x(\omega, m-1) \, \mathrm{d}\omega} \tag{2.54}$$

The test window sample autocorrelation for the measurement of both $d_1(m)$ and $d_3(m)$ can be updated through the following recursive equation over the test windows of size N:

$$\hat{r}_x(k, m) = \hat{r}_x(k, m-1) + \frac{1}{N} [x(m-1+N)x(m-1+N-k) - x(m-1+k)x(m-1)] \tag{2.55}$$

and thereby computational complexity can be reduced in practice. A fourth criterion corresponds to the error energy in autoregressive (AR)-based modelling of the signals. The prediction error in the AR model of the mth frame is simply defined as

$$e(n, m) = x(n, m) - \sum_{k=1}^{p} a_k(m)x(n-k, m) \tag{2.56}$$

where p is the prediction order and $a_k(m)$, $k = 1, 2, \ldots, p$, are the prediction coefficients. For certain p the coefficients can be found directly (e.g. Durbin's method) in such a way as to minimize the error (residual) signal energy. In this approach it is assumed that the frames of length N are overlapped by one sample. The prediction coefficients estimated for the $(m-1)$th frame are then used to predict the first sample in the mth frame, which is denoted as $\hat{e}(1, m)$. If this error is small, it is likely that the statistics of the mth frame are similar to those of the $(m-1)$th frame. On the other hand, a large value is likely to indicate a change. An indicator for the fourth criterion can then be the differencing of this prediction signal, which gives a peak at the segment boundary, i.e.

$$d_4(m) = \max[\nabla_m \hat{e}(1, m)] \tag{2.57}$$

where $\nabla_m(.)$ denotes the gradient with respect to m, approximated by a first-order difference operation. Figure 2.10 shows the residual and the gradient defined in Equation (2.57)

Finally, a fifth criterion $d_5(m)$ may be defined by using the AR-based spectrum of the signals in the same way as the short-term frequency transform (STFT) for $d_3(m)$. The above AR model is a univariate model, i.e. it models a single-channel EEG. A similar criterion may be defined when multichannel EEGs are considered [20]. In such cases a multivariate AR (MVAR) model is analysed. The MVAR can also be used for characterization and quantification of the signal propagation within the brain and is discussed in the next section.

Although the above criteria can be effectively used for segmentation of EEG signals, better systems may be defined for the detection of certain abnormalities. In order to do that, the features that best describe the behaviour of the signals have to be identified and used. Therefore the segmentation problem becomes a classification problem for which different classifiers can be used.

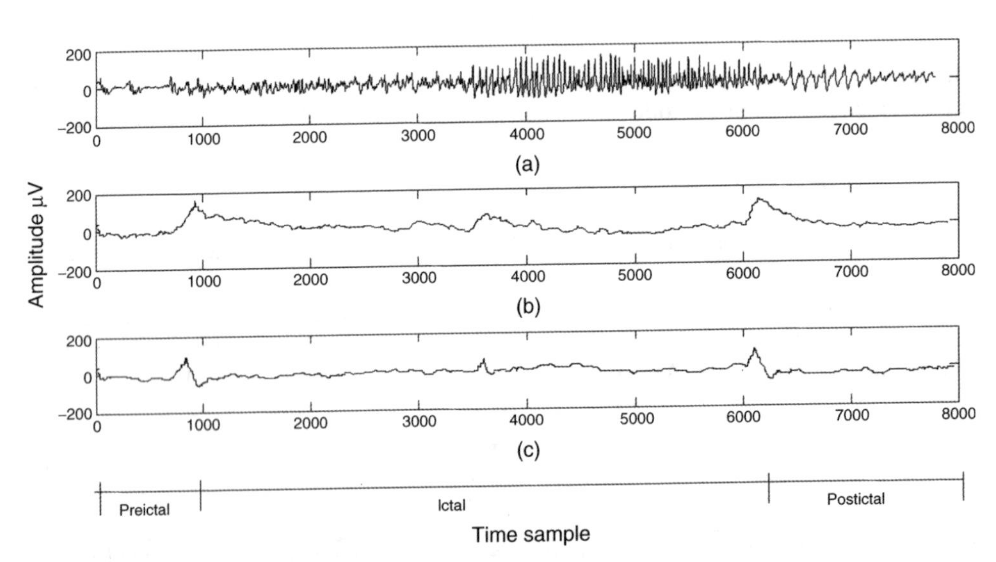

Figure 2.10 (a) An EEG seizure signal including preictal ictal and postictal segments, (b) the error signal, and (c) the approximate gradient of the signal, which exhibits a peak at the boundary between the segments. The number of prediction coefficients $p = 12$

2.5 Signal Transforms and Joint Time–Frequency Analysis

If the signals are statistically stationary it is straightforward to characterize them in either the time or frequency domains. The frequency-domain representation of a finite-length signal can be found by using linear transforms such as the (discrete) Fourier transform (DFT), (discrete) cosine transform (DCT), or other semi-optimal transform, which have kernels independent of the signal. However, the results of these transforms can be degraded by spectral smearing due to the short-term time-domain windowing of the signals and fixed transform kernels. An optimal transform such as the Karhunen–Loéve transform (KLT) requires complete statistical information, which may not be available in practice.

Parametric spectrum estimation methods such as those based on AR or ARMA modelling can outperform the DFT in accurately representing the frequency-domain characteristics of a signal, but they may suffer from poor estimation of the model parameters mainly due to the limited length of the measured signals. For example, in order to model the EEGs using an AR model, accurate values for the prediction order and coefficients are necessary. A high prediction order may result in splitting the true peaks in the frequency spectrum and a low prediction order results in combining peaks in close proximity in the frequency domain.

For an AR model of the signal $x(n)$ the error or driving signal is considered to be zero mean white noise. Therefore, by applying a z-transform to Equation (2.56), dropping the block index m, and replacing z by $\mathrm{e}^{\mathrm{j}\omega}$ gives

$$\frac{X_p(\omega)}{E(\omega)} = \frac{1}{1 - \sum_{k=1}^{p} a_k \mathrm{e}^{-\mathrm{j}k\omega}} \tag{2.58}$$

where, $E(\omega) = K_\omega$ (constant) is the power spectrum of the white noise and $X_p(\omega)$ is used to denote the signal power spectrum. Hence,

$$X_p(\omega) = \frac{K_\omega}{1 - \sum_{k=1}^{p} a_k \mathrm{e}^{-\mathrm{j}k\omega}} \tag{2.59}$$

and the parameters K_ω, a_k, $k = 1, \ldots, p$, are the exact values. In practical AR modelling these would be estimated from the finite length measurement, thereby degrading the estimate of the spectrum. Figure 2.11 provides a comparison of the spectrum of an EEG segment of approximately 1550 samples of a single-channel EEG using both DFT analysis and AR modelling.

The fluctuations in the DFT result as shown in Figure 2.11(b) are a consequence of the statistical inconsistency of periodogram-like power spectral estimation techniques. The result from the AR technique (Figure 2.11(c)) overcomes this problem provided the model fits the actual data. EEG signals are often statistically nonstationary, particularly where there is an abnormal event captured within the signals. In these cases the frequency-domain components are integrated over the observation interval and do not show the characteristics of the signals accurately. A time–frequency (TF) approach is the solution to the problem.

In the case of multichannel EEGs, where the geometrical positions of the electrodes reflect the spatial dimension, a space–time–frequency (STF) analysis through multiway processing methods has also become popular [48]. The main concepts in this area, together

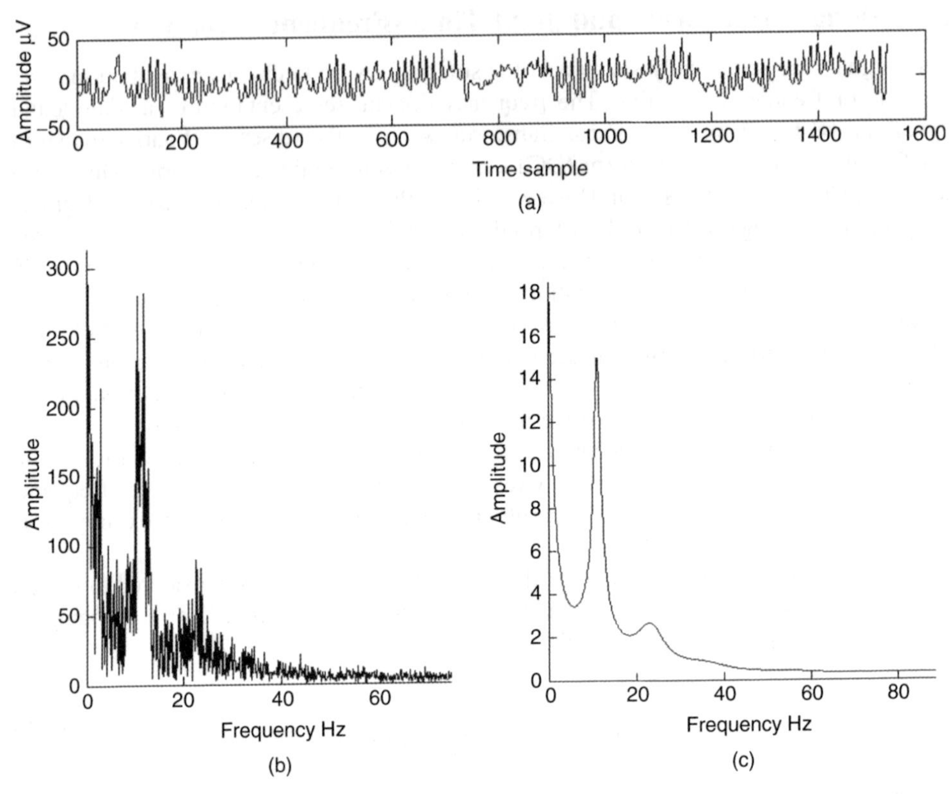

Figure 2.11 Single-channel EEG spectrum: (a) a segment of the EEG signal with a dominant alpha rhythm, (b) the spectrum of the signal in (a) using the DFT, and (c) the spectrum of the signal in (a) using a 12-order AR model

with the parallel factor analysis (PARAFAC) algorithm, will be reviewed in Chapter 7 where its major applications will be discussed.

The short-time Fourier transform (STFT) is defined as the discrete-time Fourier transform evaluated over a sliding window. The STFT can be performed as

$$X(n, \omega) = \sum_{\tau=-\infty}^{\infty} x(\tau)w(n-\tau)\mathrm{e}^{-\mathrm{j}\omega\tau} \tag{2.60}$$

where the discrete-time index n refers to the position of the window $w(n)$. Analogous with the periodogram, a spectrogram is defined as

$$S_x(n, \omega) = |X(n, \omega)|^2 \tag{2.61}$$

Based on the uncertainity principle, i.e. $\sigma_t^2\sigma_\omega^2 \geq \frac{1}{4}$, where σ_t^2 and σ_ω^2 are respectively the time- and frequency-domain variances, perfect resolution cannot be achieved in both time and frequency domains. Windows are typically chosen to eliminate discontinuities at block edges and to retain positivity in the power spectrum estimate. The choice also impacts upon the spectral resolution of the resulting technique, which, put simply, corresponds to the minimum frequency separation required to resolve two equal amplitude frequency components [49].

Figure 2.12 shows the TF representation of an EEG segment during the evolution from preictal to ictal and to postictal stages. In this figure the effect of time resolution has been illustrated using a Hanning window of different durations of 1 and 2 seconds. Importantly, in this figure the drift in frequency during the ictal period is observed clearly.

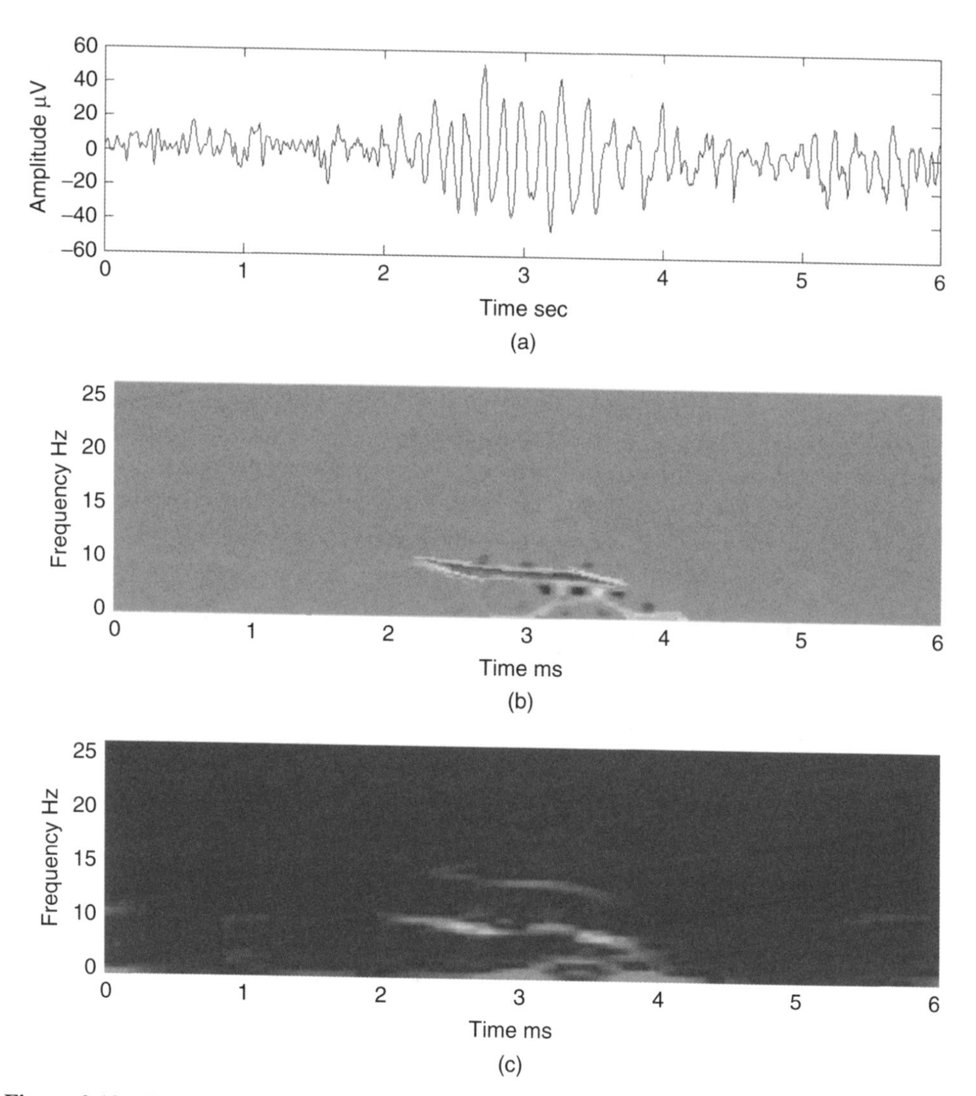

Figure 2.12 TF representation of an epileptic waveform (a) for different time resolutions using a Hanning window of (b) 1 ms and (c) 2 ms duration

2.5.1 *Wavelet Transform*

The wavelet transform (WT) is another alternative for a time–frequency analysis. There is already a well-established literature detailing the WT, such as References [50] and [51]. Unlike the STFT, the time–frequency kernel for the WT-based method can better localize the signal components in time–frequency space. This efficiently exploits the dependency between time and frequency components. Therefore, the main objective of introducing the WT by Morlet [50] was likely to have a coherence time proportional to the sampling period. To proceed, consider the context of a continuous time signal.

2.5.1.1 Continuous Wavelet Transform

The Morlet–Grossmann definition of the continuous wavelet transform for a one-dimensional signal $f(t)$ is

$$W(a,b) = \frac{1}{\sqrt{a}}\int_{-\infty}^{\infty} f(t)\psi^*\left(\frac{t-b}{a}\right)\mathrm{d}t \tag{2.62}$$

where $(.)^*$ denotes the complex conjugate, $\psi(t)$ is the analysing wavelet, $a(>0)$ is the scale parameter (inversely proportional to frequency), and b is the position parameter. The transform is linear and is invariant under translations and dilations, i.e.

$$\text{If} \quad f(t) \to W(a,b) \text{ then } f(t-\tau) \to W(a,b-\tau) \tag{2.63}$$

and

$$f(\sigma t) \to \frac{1}{\sqrt{\sigma}} W(\sigma a, \sigma b) \tag{2.64}$$

The last property makes the wavelet transform very suitable for analysing hierarchical structures. It is similar to a mathematical microscope with properties that do not depend on the magnification. Consider a function $W(a,b)$ which is the wavelet transform of a given function $f(t)$. It has been shown [52,53] that $f(t)$ can be recovered according to

$$f(t) = \frac{1}{C_\varphi}\int_0^{\infty}\int_{-\infty}^{\infty}\frac{1}{\sqrt{a}} W(a,b)\varphi\left(\frac{t-b}{a}\right)\frac{\mathrm{d}a\mathrm{d}b}{a^2} \tag{2.65}$$

where

$$C_\varphi = \int_0^{\infty}\frac{\hat{\psi}^*(v)\hat{\varphi}(v)}{v}\mathrm{d}v = \int_{-\infty}^{0}\frac{\hat{\psi}^*(v)\hat{\varphi}(v)}{v}\mathrm{d}v \tag{2.66}$$

Although often it is considered that $\psi(t) = \varphi(t)$, other alternatives for $\varphi(t)$ may enhance certain features for some specific applications [54]. The reconstruction of $f(t)$ is subject to having C_φ defined (admissibility condition). The case $\psi(t) = \varphi(t)$ implies $\hat{\psi}(0) = 0$; i.e. the mean of the wavelet function is zero.

2.5.1.2 Examples of Continuous Wavelets

Different waveforms/wavelets/kernels have been defined for the continuous wavelet transforms. The most popular ones are given below.

Morlet's wavelet is a complex waveform defined as

$$\psi(t) = \frac{1}{\sqrt{2\pi}} e^{-t^2/2 + j2\pi b_0 t} \tag{2.67}$$

This wavelet may be decomposed into its constituent real and imaginary parts as

$$\psi_r(t) = \frac{1}{\sqrt{2\pi}} e^{-t^2/2} \cos(2\pi b_0 t) \tag{2.68}$$

$$\psi_i(t) = \frac{1}{\sqrt{2\pi}} e^{-t^2/2} \sin(2\pi b_0 t) \tag{2.69}$$

where b_0 is a constant, and it is considered that $b_0 > 0$ to satisfy the admissibility condition. Figure 2.13 shows respectively the real and imaginary parts.

The Mexican hat defined by Murenzi *et al.* [51] is

$$\psi(t) = (1 - t^2)e^{-0.5t^2} \tag{2.70}$$

which is the second derivative of a Gaussian waveform (see Figure 2.14).

2.5.1.3 Discrete-Time Wavelet Transform

In order to process digital signals a discrete approximation of the wavelet coefficients is required. The discrete wavelet transform (DWT) can be derived in accordance with the sampling theorem if a frequency band-limited signal is processed.

The continuous form of the WT may be discretized with some simple considerations on the modification of the wavelet pattern by dilation. Since generally the wavelet function $\psi(t)$ is not band-limited, it is necessary to suppress the values of the frequency components above half the sampling frequency to avoid aliasing (overlapping in frequency) effects.

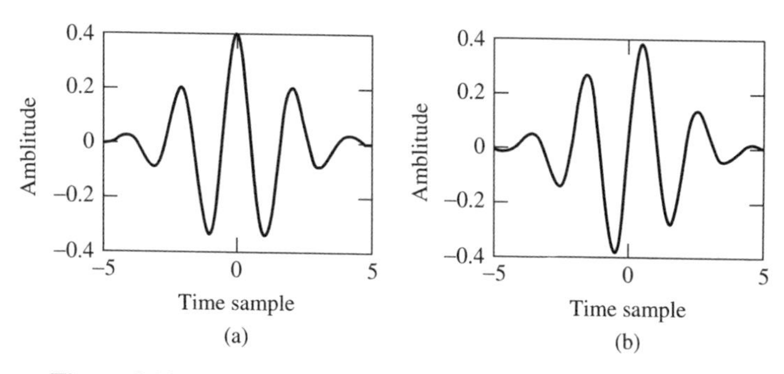

Figure 2.13 Morlet's wavelet: (a) real and (b) imaginary parts

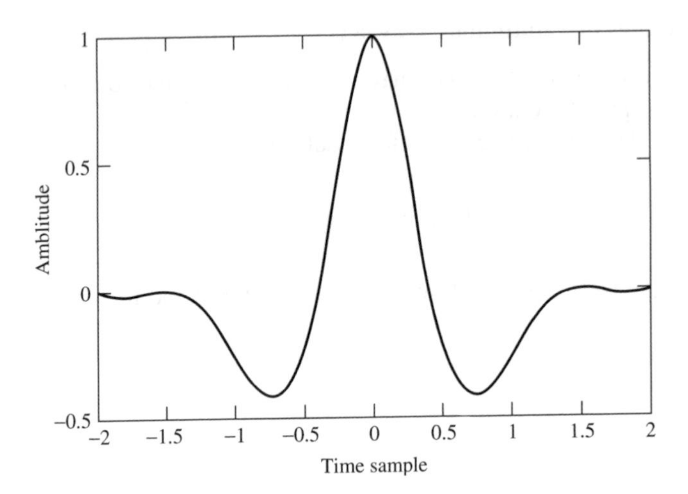

Figure 2.14 Mexican hat wavelet

A Fourier space may be used to compute the transform scale-by-scale. The number of elements for a scale can be reduced if the frequency bandwidth is also reduced. This requires a band-limited wavelet. The decomposition proposed by Littlewood and Paley [55] provides a very informative illustration of the reduction of elements scale-by-scale. This decomposition is based on an stagewise dichotomy of the frequency band. The associated wavelet is well localized in Fourier space, where it allows a reasonable analysis to be made, although not in the original space. The search for a discrete transform that is well localized in both spaces leads to a multiresolution analysis.

2.5.1.4 Multiresolution Analysis

Multiresolution analysis results from the embedded subsets generated by the interpolations (or down-sampling and filtering) of the signal at different scales. A function $f(t)$ is projected at each step j on to the subset V_j. This projection is defined by the scalar product $c_j(k)$ of $f(t)$ with the scaling function $\phi(t)$, which is dilated and translated as

$$C_j(k) = \langle f(t), 2^{-j}\phi(2^{-j}t - k)\rangle \tag{2.71}$$

where $\langle\cdot,\cdot\rangle$ denotes an inner product and $\phi(t)$ has the property

$$\frac{1}{2}\phi\left(\frac{t}{2}\right) = \sum_{n=-\infty}^{\infty} h(n)\phi(t-n) \tag{2.72}$$

where the right-hand side is convolution of h and ϕ. By taking the Fourier transform of both sides,

$$\Phi(2\omega) = H(\omega)\Phi(\omega) \tag{2.73}$$

where $H(\omega)$ and $\Phi(\omega)$ are the Fourier transforms of $h(t)$ and $\phi(t)$ respectively. For a discrete frequency space (i.e. using the DFT) the above equation permits the computation of the wavelet coefficient $C_{j+1}(k)$ from $C_j(k)$ directly. If a start is made from $C_0(k)$ and all $C_j(k)$, with $j > 0$, are computed without directly computing any other scalar product, then

$$C_{j+1}(k) = \sum_{n=0}^{N-1} C_j(n)h(n-2k) \tag{2.74}$$

where k is the discrete frequency index and N is the signal length.

At each step, the number of scalar products is divided by two and consequently the signal is smoothed. Using this procedure the first part of a filter bank is built up. In order to restore the original data, Mallat uses the properties of orthogonal wavelets, but the theory has been generalized to a large class of filters by introducing two other filters $\tilde{h}$ and $\tilde{g}$, also called conjugate filters. The restoration is performed with

$$C_j(k) = 2\sum_{l=0}^{\frac{N}{2}-1} [C_{j+1}(l)\tilde{h}(k+2l) + w_{j+1}(l)\tilde{g}(k+2l)] \tag{2.75}$$

where $w_{j+1}(.)$ are the wavelet coefficients at the scale $j+1$ defined later in this section. For an exact restoration, two conditions have to be satisfied for the conjugate filters:

Anti-aliasing condition:

$$H\left(\omega+\tfrac{1}{2}\right)\tilde{H}(\omega) + G\left(\omega+\tfrac{1}{2}\right)\tilde{G}(\omega) = 0 \quad \forall\omega \tag{2.76}$$

Exact restoration:

$$H(\omega)\tilde{H}(\omega) + G(\omega)\tilde{G}(\omega) = 1 \quad \forall\omega \tag{2.77}$$

In the decomposition, the input is successively convolved with the time domain forms of the two filters H (low frequencies) and G (high frequencies). Each resulting function is decimated by suppression of one sample out of two. The high-frequency signal is left untouched, and the decomposition continues with the low-frequency signal (left-hand side of Figure 2.15). In the reconstruction, the sampling is restored by inserting a zero between each sample; then the conjugate filters $\tilde{H}$ and $\tilde{G}$ are applied, the resulting outputs are added and the result is multiplied by 2. Reconstruction continues to the smallest scale (right-hand side of Figure 2.15). Orthogonal wavelets correspond to the restricted case where

$$G(\omega) = \mathrm{e}^{-2\pi\omega}H^*\left(\omega+\tfrac{1}{2}\right) \tag{2.78}$$

$$\tilde{H}(\omega) = H^*(\omega) \tag{2.79}$$

$$\tilde{G}(\omega) = G^*(\omega) \tag{2.80}$$

and

$$|H(\omega)|^2 + \left|H\left(\omega+\tfrac{1}{2}\right)\right|^2 = 1 \quad \forall\omega \tag{2.81}$$

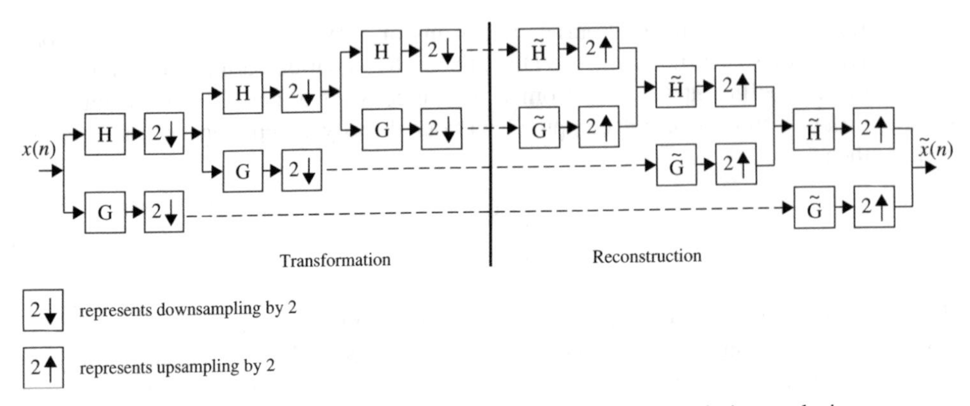

Figure 2.15 The filter bank associated with the multiresolution analysis

It can easily be seen that this set satisfies the two basic relations (2.72) and (2.73). Among various wavelets, Daubechie's wavelets are the only compact solutions to satisfy the above conditions. For biorthogonal wavelets, then

$$G(\omega) = \mathrm{e}^{-2\pi\omega}\tilde{H}^*\left(\omega + \tfrac{1}{2}\right) \tag{2.82}$$

$$\tilde{G}(\omega) = \mathrm{e}^{2\pi\omega}H^*\left(\omega + \tfrac{1}{2}\right) \tag{2.83}$$

and

$$H(\omega)\tilde{H}(\omega) + H^*\left(\omega + \tfrac{1}{2}\right)\tilde{H}^*\left(\omega + \tfrac{1}{2}\right) = 1 \quad \forall\omega \tag{2.84}$$

The relations (2.76) and (2.77) have also to be satisfied. A large class of compact wavelet functions can be used. Many sets of filters have been proposed, especially for coding [56]. It has been shown that the choice of these filters must be guided by the regularity of the scaling and the wavelet functions. The complexity is proportional to N. The algorithm provides a pyramid of N elements.

2.5.1.5 The Wavelet Transform using the Fourier Transform

Consider the scalar products $c_0(k) = \langle f(t)\ \phi(t-k)\rangle$ for continuous wavelets. If $\phi(t)$ is band-limited to half of the sampling frequency, the data can be correctly sampled. The data at the resolution $j = 1$ are

$$c_1(k) = \left\langle f(t)\ \tfrac{1}{2}\phi\left(\tfrac{t}{2} - k\right)\right\rangle \tag{2.85}$$

and the set $c_1(k)$ can be computed from $c_0(k)$ with a discrete-time filter with the frequency response $H(\omega)$:

$$H(\omega) = \begin{cases} \dfrac{\Phi(2\omega)}{\Phi(\omega)} & \text{if } |\omega| < \omega_c \\ 0 & \text{if } \omega_c \le |\omega| < \tfrac{1}{2} \end{cases} \tag{2.86}$$

and for $\forall\omega$ and $\forall$ integer m

$$H(\omega + m) = H(\omega) \tag{2.87}$$

Therefore, the coefficients at the next scale can be found from

$$C_{j+1}(\omega) = C_j(\omega)H(2^j\omega) \tag{2.88}$$

The cut-off frequency is reduced by a factor 2 at each step, allowing a reduction of the number of samples by this factor. The wavelet coefficients at the scale $j+1$ are

$$w_{j+1} = \langle f(t), 2^{-(j+1)}\psi(2^{-(j+1)}t - k)\rangle \tag{2.89}$$

and can be computed directly from C_j by

$$W_{j+1}(\omega) = C_j(\omega)G(2^j\omega) \tag{2.90}$$

where G is the following discrete-time filter:

$$G(\omega) = \begin{cases} \dfrac{\Psi(2\omega)}{\Phi(\omega)} & \text{if } |\omega| < \omega_c \\ 0 & \text{if } \omega_c \le |\omega| < \frac{1}{2} \end{cases} \tag{2.91}$$

and for $\forall\omega$ and $\forall$ integer m

$$G(\omega + m) = G(\omega) \tag{2.92}$$

The frequency band is also reduced by a factor of two at each step. These relationships are also valid for DWT, following Section 2.5.1.4.

2.5.1.6 Reconstruction

The reconstruction of the data from its wavelet coefficients can be performed step-by-step, starting from the lowest resolution. At each scale,

$$C_{j+1} = H(2^j\omega)C_j(\omega) \tag{2.93}$$

$$W_{j+1} = G(2^j\omega)C_j(\omega) \tag{2.94}$$

when a search is made for C_j knowing C_{j+1}, W_{j+1}, h, and g. Then $C_j(\omega)$ is restored by minimizing

$$P_h(2^j\omega)|C_{j+1}(\omega) - H(2^j\omega)C_j(\omega)|^2 + P_g(2^j\omega)|W_{j+1}(\omega) - G(2^j\omega)C_j(\omega)|^2 \tag{2.95}$$

using a least squares estimator. $P_h(\omega)$ and $P_g(\omega)$ are weight functions that permit a general solution to the restoration of $C_j(\omega)$. The relationship of $C_j(\omega)$ is in the form of

$$C_j(\omega) = C_{j+1}(\omega)\tilde{H}(2^j\omega) + W_{j+1}(\omega)\tilde{G}(2^j\omega) \tag{2.96}$$

where the conjugate filters have the expressions

$$\tilde{H}(\omega) = \frac{P_h(\omega)H^*(\omega)}{P_h(\omega)|H(\omega)|^2 + P_g(\omega)|G(\omega)|^2} \tag{2.97}$$

$$\tilde{H}(\omega) = \frac{P_g(\omega)G^*(\omega)}{P_h(\omega)|H(\omega)|^2 + P_g(\omega)|G(\omega)|^2} \tag{2.98}$$

It is straightforward to see that these filters satisfy the exact reconstruction condition given in Equation (2.77). In fact, Equations (2.97) and (2.98) give the general solutions to this equation. In this analysis, the Shannon sampling condition is always respected. No aliasing exists, so that the antialiasing condition (2.76) is not necessary. The denominator is simplified if

$$G(\omega) = \sqrt{1 - |H(\omega)|^2} \tag{2.99}$$

This corresponds to the case where the wavelet is the difference between the squares of two resolutions:

$$|\Psi(2\omega)|^2 = |\Phi(\omega)|^2 - |\Phi(2\omega)|^2 \tag{2.100}$$

The reconstruction algorithm then carries out the following steps:

1. Compute the fast Fourier transform (FFT) of the signal at the low resolution.
2. Set j to n_p and perform the following iteration steps.
3. Compute the FFT of the wavelet coefficients at the scale j.
4. Multiply the wavelet coefficients W_j by $\tilde{G}$.
5. Multiply the signal coefficients at the lower resolution C_j by $\tilde{H}$.
6. The inverse Fourier transform of $W_j\tilde{G} + C_j\tilde{H}$ gives the coefficients C_{j-1}.
7. Then $j = j - 1$ and return to step 3.

The use of a band-limited scaling function allows a reduction of sampling at each scale and limits the computation complexity.

The wavelet transform has been widely used in EEG signal analysis. Its application to seizure detection, especially for neonates, modelling of the neuron potentials, and the detection of evoked potentials (EP) and event-related potentials (ERP) will be discussed in the corresponding chapters of this book.

2.5.2 *Ambiguity Function and the Wigner–Ville Distribution*

The ambiguity function for a continuous time signal is defined as

$$A_x(\tau, \nu) = \int_{-\infty}^{\infty} x^*\left(t - \frac{\tau}{2}\right) x\left(t + \frac{\tau}{2}\right) e^{j\nu t}\, dt \tag{2.101}$$

This function has its maximum value at the origin as

$$A_x(0,0) = \int_{-\infty}^{\infty} |x(t)|^2 \, dt \tag{2.102}$$

As an example, if a continuous time signal is considered to consist of two modulated signals with different carrier frequencies such as

$$\begin{aligned} x(t) &= x_1(t) + x_2(t) \\ &= s_1(t)e^{j\omega_1 t} + s_2(t)e^{j\omega_2 t} \end{aligned} \tag{2.103}$$

The ambiguity function $A_x(\tau, \nu)$ will be in the form of

$$A_x(\tau,\nu) = A_{x_1}(\tau,\nu) + A_{x_2}(\tau,\nu) + \text{cross terms} \tag{2.104}$$

This concept is very important in the separation of signals using the TF domain. This will be addressed in the context of blind source separation (BSS) later in this chapter. Figure 2.16 demonstrates this concept.

The Wigner–Ville frequency distribution of a signal $x(t)$ is then defined as the two-dimensional Fourier transform of the ambiguity function

$$\begin{aligned} X_{WV}(t,\omega) &= \frac{1}{2\pi}\int_{-\infty}^{\infty}\int_{-\infty}^{\infty} A_x(\tau,\nu)e^{-j\nu t}e^{-j\omega t}\, d\nu\, d\tau \\ &= \frac{1}{2\pi}\int_{-\infty}^{\infty}\int_{-\infty}^{\infty}\int_{-\infty}^{\infty} x^*\left(\beta - \frac{\tau}{2}\right) x\left(\beta + \frac{\tau}{2}\right) e^{-j\nu(t-\beta)} \\ &\quad e^{-j\omega\tau}\, d\beta\, d\nu\, d\tau \end{aligned} \tag{2.105}$$

which changes to the dual form of the ambiguity function as

$$X_{WV}(t,\omega) = \int_{-\infty}^{\infty} x^*\left(t - \frac{\tau}{2}\right) x\left(t + \frac{\tau}{2}\right) e^{-j\omega\tau}\, d\tau \tag{2.106}$$

A quadratic form for the TF representation with the Wigner–Ville distribution can also be obtained using the signal in the frequency domain as

$$X_{WV}(t,\omega) = \int_{-\infty}^{\infty} X^*\left(\omega - \frac{\nu}{2}\right) X\left(\omega + \frac{\nu}{2}\right) e^{-j\nu t}\, d\nu \tag{2.107}$$

The Wigner–Ville distribution is real and has very good resolution in both the time and frequency domains. Also it has time and frequency support properties; i.e. if $x(t) = 0$ for $|t| > t_0$, then $X_{WV}(t,\omega) = 0$ for $|t| > t_0$, and if $X(\omega) = 0$ for $|\omega| > \omega_0$, then $X_{WV}(t,\omega) = 0$ for $|\omega| > \omega_0$. It has also both time-marginal and frequency-marginal conditions of the form

$$\frac{1}{2\pi}\int_{-\infty}^{\infty} X_{WV}(t,\omega)\, d\omega = |X(t)|^2 \tag{2.108}$$

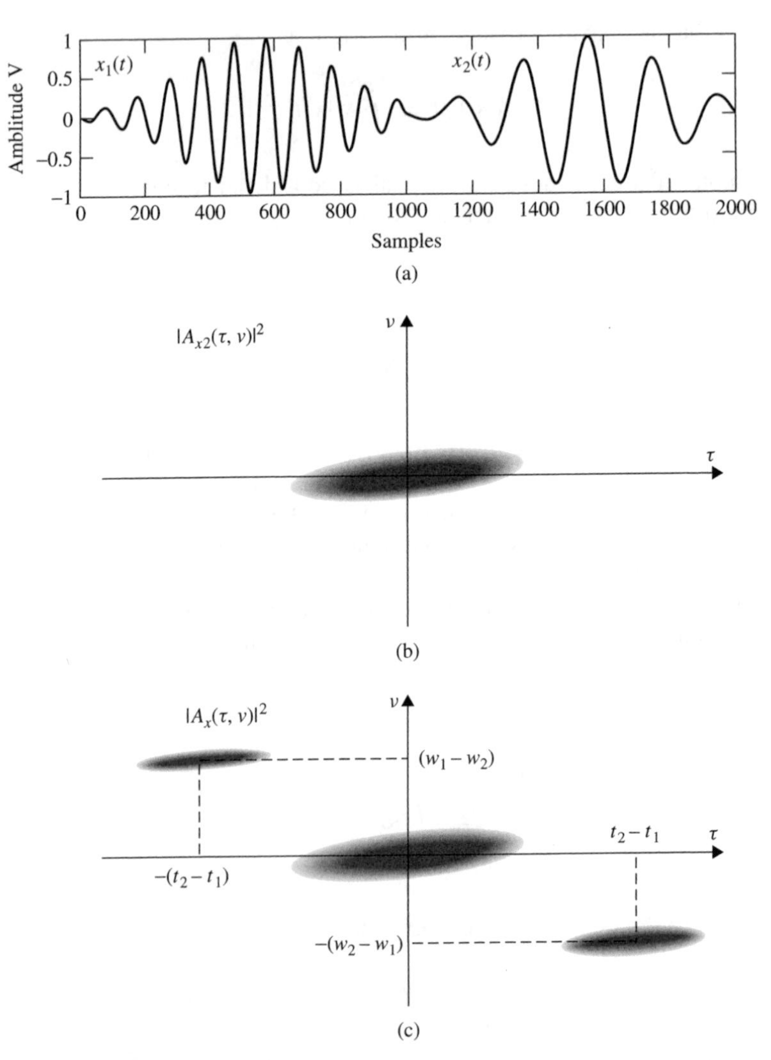

Figure 2.16 (a) A segment of a signal consisting of two modulated components, (b) an ambiguity function for $x_1(t)$ only, and (c) the ambiguity function for $x(t) = x_1(t) + x_2(t)$

and

$$\int_{-\infty}^{\infty} X_{WV}(t, \omega)\, dt = |X(\omega)|^2 \tag{2.109}$$

If $x(t)$ is the sum of two signals $x_1(t)$ and $x_2(t)$, i.e. $x(t) = x_1(t) + x_2(t)$, the Wigner–Ville distribution of $x(t)$ with respect to the distributions of $x_1(t)$ and $x_2(t)$ will be

$$X_{WV}(t, \omega) = X_{1WV}(t, \omega) + X_{2WV}(t, \omega) + 2Re[X_{12WV}(t, \omega)] \tag{2.110}$$

where Re[·] denotes the real part of a complex value and

$$X_{12\mathrm{WV}}(t,\omega)=\int_{-\infty}^{\infty}x_1^*\left(t-\frac{\tau}{2}\right)x_2\left(t+\frac{\tau}{2}\right)\mathrm{e}^{-\mathrm{j}\omega\tau}\,\mathrm{d}\tau \tag{2.111}$$

It is seen that the distribution is related to the spectra of both auto- and cross-correlations. A pseudo Wigner–Ville distribution (PWVD) is defined by applying a window function, $w(\tau)$, centred at $\tau=0$ to the time-based correlations, i.e.

$$\breve{X}_{\mathrm{WV}}(t,\omega)=\int_{-\infty}^{\infty}x^*\left(t-\frac{\tau}{2}\right)x\left(t+\frac{\tau}{2}\right)w(\tau)\mathrm{e}^{-\mathrm{j}\omega\tau}\,\mathrm{d}\tau \tag{2.112}$$

In order to suppress the undesired cross-terms the two-dimensional WV distribution may be convolved with a TF-domain window. The window is a two-dimensional lowpass filter, which satisfies the time and frequency marginal (uncertainty) conditions, as described earlier. This can be performed as

$$C_x(t,\omega)=\frac{1}{2\pi}\int_{-\infty}^{\infty}\int_{-\infty}^{\infty}X_{\mathrm{WV}}(t',\omega')\Phi(t-t',\omega-\omega')\,\mathrm{d}t'\,\mathrm{d}\omega' \tag{2.113}$$

where

$$\Phi(t,\omega)=\frac{1}{2\pi}\int_{-\infty}^{\infty}\int_{-\infty}^{\infty}\phi(\tau,\nu)\mathrm{e}^{-\mathrm{j}\nu t}\mathrm{e}^{-\mathrm{j}\omega\tau}\,\mathrm{d}\nu\,\mathrm{d}\tau \tag{2.114}$$

and $\phi(.,.)$ is often selected from a set of well-known signals, the so-called *Cohen's class*. The most popular member of Cohen's class of functions is the bell-shaped function defined as

$$\phi(\tau,\nu)=\mathrm{e}^{-\nu^2\tau^2/(4\pi^2\sigma)},\quad \sigma>0 \tag{2.115}$$

A graphical illustration of such a function can be seen in Figure 2.17. In this case the distribution is referred to as a *Choi–Williams distribution*.

The application of a discrete time form of the Wigner–Ville distribution to BSS will be discussed later in this chapter and its application to seizure detection will be briefly explained in Chapter 4. To improve the distribution a signal-dependent kernel may also be used [57].

2.6 Coherency, Multivariate Autoregressive (MVAR) Modelling, and Directed Transfer Function (DTF)

In some applications such as in detection and classification of finger movement, it is very useful to establish how the associated movement signals propagate within the neural network of the brain. As will be shown in Chapter 7, there is a consistent movement of the source signals from the occipital to temporal regions. It is also clear that during the mental tasks different regions within the brain communicate with each other. The interaction and cross-talk among the EEG channels may be the only clue to understanding this process.

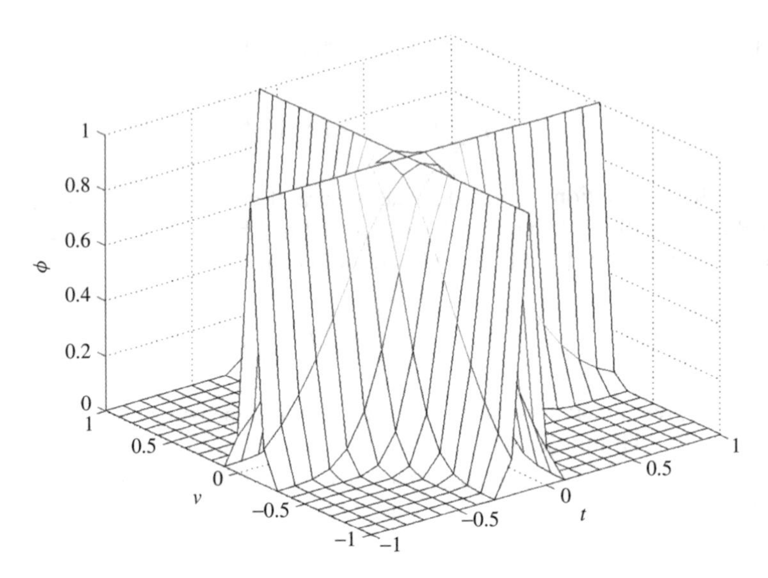

Figure 2.17 Illustration of $\phi(\tau, \nu)$ for the Choi–Williams distribution

This requires recognition of the transient periods of synchrony between various regions in the brain. These phenomena are not easy to observe by visual inspection of the EEGs. Therefore, some signal processing techniques have to be used in order to infer such causal relationships. One time series is said to be causal to another if the information contained in that time series enables the prediction of the other time series.

The spatial statistics of scalp EEGs are usually presented as coherence in individual frequency bands. These coherences result both from correlations among neocortical sources and volume conduction through the tissues of the head, i.e. brain, cerebrospinal fluid, skull, and scalp. Therefore, spectral coherence [58] is a common method for determining the synchrony in EEG activity. Coherency is given as

$$\mathrm{Coh}_{ij}^2(\omega) = \frac{E[C_{ij}(\omega)|^2]}{E[C_{ii}(\omega)]E[C_{jj}(\omega)]} \tag{2.116}$$

where $C_{ij}(\omega) = X_i(\omega)X_j^*(\omega)$ is the Fourier transform of the cross-correlation coefficients between channel i and channel j of the EEGs. Figure 2.18 shows an example of the cross-spectral coherence around one second prior to finger movement. A measure of this coherency, such as an average over a frequency band, is capable of detecting zero time lag synchronization and fixed time nonzero time lag synchronization, which may occur when there is a significant delay between the two neuronal population sites [59]. However, it does not provide any information on the directionality of the coupling between the two recording sites.

Granger causality (also called Wiener–Granger causality) [60] is another measure that attempts to extract and quantify the directionality from EEGs. Granger causality is based on bivariate AR estimates of the data. In a multichannel environment this causality is

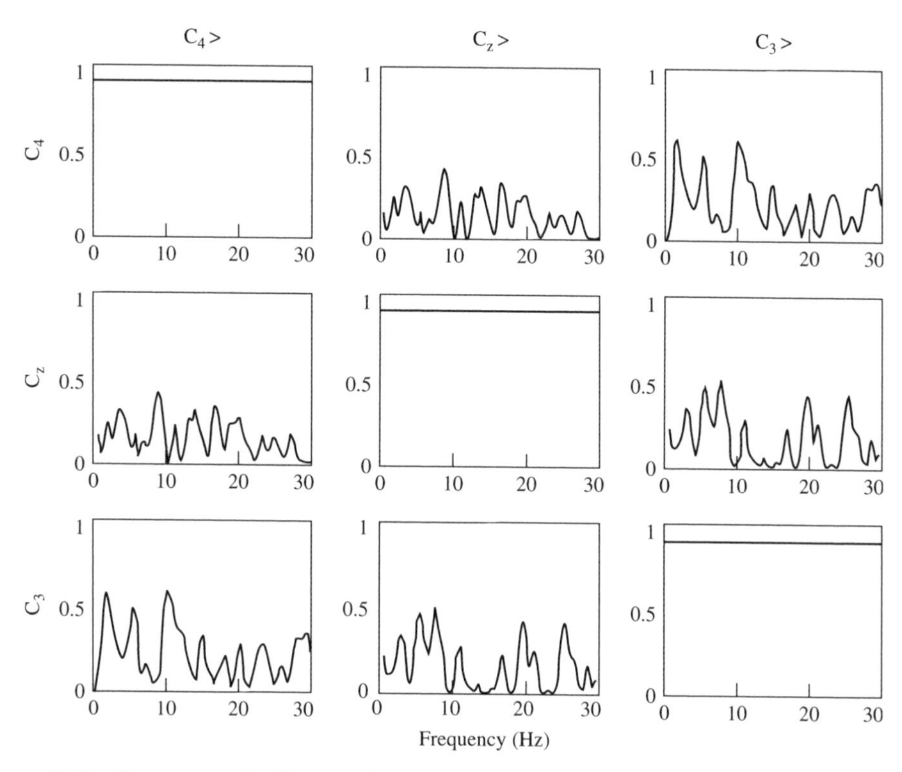

Figure 2.18 Cross-spectral coherence for a set of three electrode EEGs, one second before the right-finger movement. Each block refers to one electrode. By careful inspection of the figure, it is observed that the same waveform is transferred from C_z to C_3

calculated from pairwise combinations of electrodes. This method has been used to evaluate the directionality of the source movement from the local field potential in the visual system of cats [61].

For multivariate data in a multichannel recording, however, application of the Granger causality is not computationally efficient [61,62]. The directed transfer function (DTF) [63], as an extension of Granger causality, is obtained from multichannel data and can be used to detect and quantify the coupling directions. The advantage of the DTF over spectral coherence is that it can determine the directionality in the coupling when the frequency spectra of the two brain regions have overlapping spectra. The DTF has been adopted by some researchers for determining the directionality in the coupling [64,65] since it has been demonstrated that [66] there is a directed flow of information or cross-talk between the sensors around the sensory motor area before finger movement. The DTF is based on fitting the EEGs to an MVAR model. Assuming that $\boldsymbol{x}(n)$ is an M-channel EEG signal, it can be modelled in vector form as

$$\boldsymbol{x}(n) = -\sum_{k=1}^{p} \mathbf{L}_k \boldsymbol{x}(n-k) + \boldsymbol{v}(n) \tag{2.117}$$

where n is the discrete time index, p is the prediction order, $\boldsymbol{v}(n)$ is zero-mean noise, and $\mathbf{L}_k$ is generally an $M \times p$ matrix of prediction coefficients. A similar method to the Durbin algorithm for single channel signals, namely the Levinson–Wiggins–Robinson (LWR) algorithm is used to calculate the MVAR coefficients [20]. The Akaike AIC criterion [12] is also used for the estimation of prediction order p. By multiplying both sides of the above equation by $\boldsymbol{x}^{\mathrm{T}}(n-k)$ and performing the statistical expectation, the following Yule–Walker equation is obtained [67].

$$\sum_{k=0}^{p} \mathbf{L}_k \mathbf{R}(-k+p) = 0; \qquad \mathbf{L}_0 = \mathbf{I} \tag{2.118}$$

where $\mathbf{R}(q) = E[\boldsymbol{x}(n)\boldsymbol{x}^{\mathrm{T}}(n+q)]$ is the covariance matrix of $\boldsymbol{x}(n)$, and the cross-correlations of the signal and noise are zero since they are assumed to be uncorrelated. Similarly, the noise autocorrelation is zero for a nonzero shift since the noise samples are uncorrelated. The data segment is considered short enough for the signal to remain statistically stationary within that interval and long enough to enable accurate estimation of the prediction coefficients. Given the MVAR model coefficients, a multivariate spectrum can be achieved. Here it is assumed that the residual signal, $\boldsymbol{v}(n)$, is white noise. Therefore,

$$\mathbf{L}_f(\omega)\mathbf{X}(\omega) = \mathbf{V}(\omega) \tag{2.119}$$

where

$$\mathbf{L}_f(\omega) = \sum_{m=0}^{p} \mathbf{L}_m \mathrm{e}^{-\mathrm{j}\omega m} \tag{2.120}$$

and $\mathbf{L}(0) = \mathbf{I}$. Rearranging the above equation and replacing noise by $\sigma_v^2\mathbf{I}$ yields

$$\mathbf{X}(\omega) = \mathbf{L}_f^{-1}(\omega) \times \sigma_{\mathbf{V}}^2\mathbf{I} = \mathbf{H}(\omega) \tag{2.121}$$

which represents the model spectrum of the signals or the transfer matrix of the MVAR system. The DTF or causal relationship between channel i and channel j can be defined directly from the transform coefficients [62] given by

$$\Theta_{ij}^2(\omega) = |H_{ij}(\omega)|^2 \tag{2.122}$$

Electrode i is causal to j at frequency f if

$$\Theta_{ij}^2(\omega) > 0 \tag{2.123}$$

A time-varying DTF can also be generated (mainly to track the source signals) by calculating the DTF over short windows to achieve the short time DTF (SDTF) [62].

As an important feature in classification of left- and right-finger movements, or tracking the mental task related sources, the SDTF plays an important role. Some results of using the SDTF for detection and classification of finger movement are given in Chapter 7 in the context of brain–computer interfacing (BCI).

2.7 Chaos and Dynamical Analysis

As an effective tool for prediction and characterization of signals, *deterministic chaos* plays an important role. Although the EEG signals are considered chaotic, there are rules that do not in themselves involve any element of change and can be used in their characterization [68]. Mathematical research about chaos started before 1890 when people such as Andrey Kolmogorov or Henri Poincaré tried to establish whether planets would indefinitely remain in their orbits. In the 1960s Stephan Smale formulated a plan to classify all the typical kinds of dynamic behaviour. Many chaos-generating mechanisms have been created and used to identify the behaviour of the dynamics of the system. The Rossler system was designed to model a strange attractor using a simple stretch-and-fold mechanism. This was, however, inspired by the Lorenz attractor introduced more than a decade earlier [68].

To evaluate the city of chaotic a dynamical system different measures can be taken into account. A straightforward parameter is the attractor dimension. Different multidimensional attractors have been defined by a number of mathematicians. In many cases it is difficult to find the attractor dimension unless the parameters of the system can be approximated. However, later in this section it will be shown that the attraction dimension [69] can be simply achieved using the Lyapunov exponents.

2.7.1 Entropy

Entropy is a measure of uncertainty. The level of chaos may also be measured using entropy of the system. Higher entropy represents higher uncertainty and a more chaotic system. Entropy is given as

$$\text{Entropy of the signal } x(n) = \int_{\min(x)}^{\max(x)} p_x \log(1/p_x)\, \mathrm{d}x \tag{2.124}$$

where p_x is the probability density function (PDF) of signal $x(n)$. Generally, the distribution can be a joint PDF when the EEG channels are jointly processed. On the other hand, the PDF can be replaced by conditional PDF in places where the occurrence of the event is subject to another event. In this case, the entropy is called conditional entropy. Entropy is very sensitive to noise. Noise increases the uncertainty and noisy signals have higher entropy even if the original signal is ordered.

2.7.2 Kolmogorov Entropy

Also known as metric entropy, Kolmogorov entropy is an effective measure of chaos. To find the Kolmogorov entropy the phase space is divided into multidimensional hypercubes. Phase space is the space in which all possible states of a system are represented, each corresponding to one unique point in the phase space. In phase space, every degree of freedom or parameter of the system is represented as an axis of a multidimensional space. A phase space may contain many dimensions. The hypercube is a generalization of a 3-cube to n-dimensions, also called an n-cube or measure polytope. It is a regular polytope with mutually perpendicular sides, and is therefore an orthotope. Now, let $P_{i_0,\ldots,i_n}$ be the probability that a trajectory falls inside the hypercube, with i_0 at $t = 0$, i_1 at $t = T$,

i_2 at $t = 2T, \ldots$. Then define

$$K_n = -\sum_{i_0,\ldots,i_n} P_{i_0,\ldots,i_n} \ln P_{i_0,\ldots,i_n} \tag{2.125}$$

where $K_{n+1} - K_n$ is the information needed to predict which hypercube the trajectory will be in at $(n+1)T$, given trajectories up to nT. The Kolmogorov entropy is then defined as

$$K = \lim_{N\to\infty} \frac{1}{NT} \sum_{n=0}^{N-1} (K_{n+1} - K_n) \tag{2.126}$$

However, estimation of the above joint probabilities for large-dimensional data is computationally costly. On the other hand, in practice, long data sequences are normally required to perform a precise estimation of the Kolmogorov entropy.

2.7.3 Lyapunov Exponents

A chaotic model can be generated by a simple feedback system. Consider a quadratic iterator of the form $x(n) \to \alpha x(n)[1 - x(n)]$ with an initial value of x_0. This generates a time series such as that in Figure 2.19 (for $\alpha = 3.8$).

Although in the first 20 samples the time series seems to be random noise its semi-ordered alterations (cyclic behaviour) later show that some rules govern its chaotic behaviour. This time series is subject to two major parameters, α and x_0.

In order to adopt this model within a chaotic system a different initial value may be selected. Perturbation of an initial value generates an error E_0, which propagates during

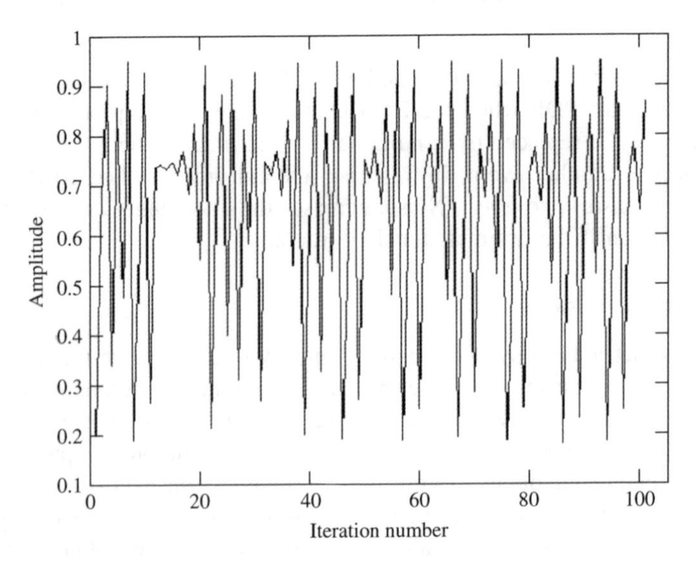

Figure 2.19 Generated chaotic signal using the model $x(n) \to \alpha x(n)[1 - x(n)]$ using $\alpha = 3.8$ and $x_0 = 0.2$

the signal evolution. After n samples the error changes to $E_n . E_n/E_0$ is a measure of how fast the error grows. The average growth of infinitesimally small errors in the initial point x_0 is quantified by Ljapunov (Lyapunov) exponents $\lambda(x_0)$. The total error amplification factor $|E_n/E_0|$, can be written in terms of sample error amplifications as

$$\left|\frac{E_n}{E_0}\right| = \left|\frac{E_n}{E_{n-1}}\right|\left|\frac{E_{n-1}}{E_{n-2}}\right|\cdots\left|\frac{E_1}{E_0}\right| \tag{2.127}$$

The average logarithm of this becomes

$$\frac{1}{n}\ln\left|\frac{E_n}{E_0}\right| = \frac{1}{n}\sum_{k=1}^{n}\ln\left|\frac{E_k}{E_{k-1}}\right| \tag{2.128}$$

Obviously, the problem is how to measure $|E_k/E_{k-1}|$. For the iterator $f(x(n))$ ($f(x(n)) = \alpha x(n)[1 - x(n)]$ in the above example) having a small perturbation ε at the initial point, the term in the above equation may be approximated as

$$\frac{1}{n}\ln\left|\frac{E_n}{E_0}\right| = \frac{1}{n}\sum_{k=1}^{n}\ln\left|\frac{\tilde{E}_k}{\varepsilon}\right| \tag{2.129}$$

where $\tilde{E}_k = f(x_{k-1} + \varepsilon) - f(x_{k-1})$. By replacing this in the above equation the Lyapunov exponent is approximated as

$$\lambda(x_0) = \lim_{n\to\infty}\frac{1}{n}\sum_{k=1}^{n}\ln|f'(x_{k-1})| \tag{2.130}$$

where $f'(x)$ represents differentiation of $f(x)$ with respect to x. This measure is very significant in separating unstable, unpredictable, or chaotic behaviour from predictable, stable, or ordered ones. If λ is positive the system is chaotic whereas it is negative for ordered systems.

Kaplan and Yorke [70] empirically concluded that it is possible to predict the dimension of a strange attractor from knowledge of the Lyapunov exponents of the corresponding transformation. This is termed the Kaplan–Yorke conjecture and has been investigated by many other researchers [71]. This is a very important conclusion since in many dynamical systems the various dimensions of the attractors are hard to compute, while the Lyapunov exponents are relatively easy to compute. This conjecture also claims that generally the information dimension D_{I} and Lyapunov dimension D_{L} respectively are defined as [68]

$$D_{\mathrm{I}} = \lim_{s\to 0}\frac{I(s)}{\log_2 1/s} \tag{2.131}$$

where s is the size of a segment of the attractor and $I(s)$ is the entropy of s, and

$$D_{\mathrm{L}} = m + \frac{1}{|\lambda_{m+1}|}\sum_{k=1}^{m}\lambda_k \tag{2.132}$$

where m is the maximum integer with $\gamma(m) = \lambda_1 + \cdots + \lambda_m \geq 0$, given that $\lambda_1 > \lambda_2 > \cdots > \lambda_m$.

2.7.4 Plotting the Attractor Dimensions from the Time Series

Very often it is necessary to visualize a phase space attractor and decide about the stability, chaocity, or randomness of a signal (time series). The attractors can be multidimensional. For a three-dimensional attractor a time delay T (a multiple of τ) can be chosen and the following sequence of vectors constructed.

$$\begin{matrix} [x(0) & x(T) & x(2T)] \\ [x(\tau) & x(\tau+T) & x(\tau+2T)] \\ [x(2\tau) & x(2\tau+T) & x(2\tau+2T)] \\ \vdots & \vdots & \vdots \\ [x(k\tau) & x(k\tau+T) & x(k\tau+2T)] \end{matrix}$$

By plotting these points in a three-dimensional coordinate space and linking the points together successively the attractor can be observed. Figure 2.20 shows the attractors for a sinusoidal and the above chaotic time series. Although the attractors can be defined for a

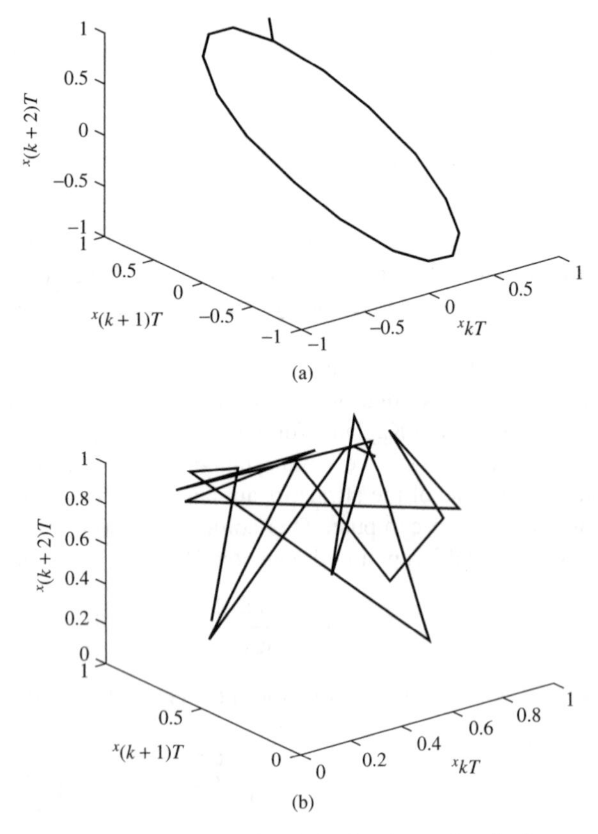

Figure 2.20 The attractors for (a) a sinusoid and (b) the above chaotic time sequence, both started from the same initial point

higher-dimensional space, visualization of the attractors is not possible when the number of dimensions increases above three.

2.7.5 Estimation of Lyapunov Exponents from the Time Series

Calculation of the Lyapunov exponents from the time series was first proposed by Wolf *et al.* [72]. In their method, initially a finite embedding sequence is constructed from the finite time series of $2N+1$ components as

$$x(0), x(\tau), x(2\tau), \ldots$$

This is the basic data (often called the reference trajectory or reference orbit) upon which the model builds. Generally, the start point is not given since there is no explicit governing equation that would generate the trajectory. From this sequence a point $x(k_0\tau)$ may be chosen that approximates the desired initial point $z_0(0)$. Considering Figure 2.21, these approximations should satisfy

$$|x(k_0\tau) - x(0)| < \delta \tag{2.133}$$

where δ is an *a priori* chosen tolerance. This point may be renamed as

$$z_0(0) = x(k_0\tau) \tag{2.134}$$

The successors of this point are known as

$$z_0(r\tau) = x((k_0 + r)\tau), \quad r = 1, 2, 3, \ldots \tag{2.135}$$

Now there are two trajectories to compare. The logarithmic error amplification factor for the first time interval becomes

$$l_0 = \frac{1}{\tau} \log \frac{|z_0(\tau) - z_0(0)|}{|x(\tau) - x(0)|} \tag{2.136}$$

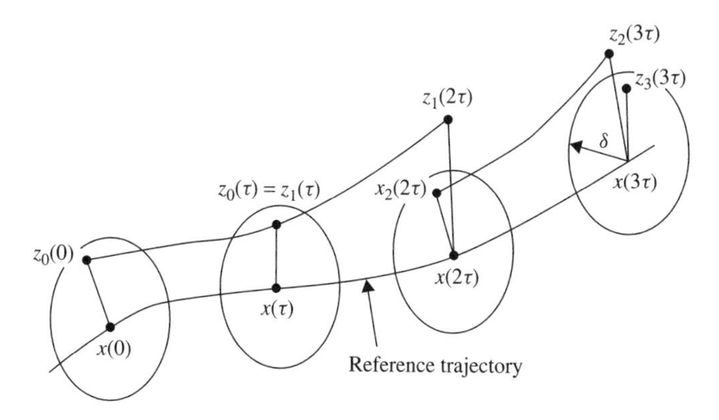

Figure 2.21 The reference and the model trajectories, evolution of the error, and start and end of the model trajectory segments. The model trajectory ends when its deviation from the reference trajectory is more than a threshold

This procedure is repeated for the next point $x(\tau)$ of the reference trajectory. For that point another point $z_1(\tau)$ needs to the formed from the trajectory, which represents an error with a direction close to the one obtained from $z_0(\tau)$ relative to $x(\tau)$. In the case where the previous trajectory is still close to the reference trajectory it may be possible simply to continue with that, thus setting $z_1(\tau) = z_0(\tau)$. This yields an error amplification factor l_1. Other factors, $l_2, \ldots, l_{m-1}$, can also be found by following the same procedure until the segment of the time series is exhausted. An approximation to the largest Lyapunov exponent for the current segment of the time series is obtained by averaging the logarithmic amplification factors over the whole reference trajectory:

$$\lambda = \frac{1}{m}\sum_{j=0}^{m-1} l_j \tag{2.137}$$

Instead of the above average, the maximum value of the error amplification factor may also be considered as the largest Lyapunov exponent. It is necessary to investigate the effect of noise as the data usually stem from a physical measurement and therefore contain noise. Hence, the perturbed points, $z_k(k\tau)$, should not be taken very close to each other, because then the noise would dominate the stretching effect on the chaotic attractor. On the other hand, the error should not be allowed to become too large in order to avoid nonlinear effects. Thus in practice some minimal error, δ_1, and a maximal error, δ_2, is prescribed and

$$\delta_1 < |x(k\tau) - z_k(k\tau)| < \delta_2 \tag{2.138}$$

is required.

2.7.5.1 Optimum Time Delay

In the above calculation it is important to find the *optimum time delay* τ. Very small time delays may result in near-linear reconstructions with high correlations between consecutive phase space points and very large delays might ignore any deterministic structure of the sequence. In an early proposal [73] the autocorrelation function is used to estimate the time delay. In this method τ is equivalent to the duration after which the autocorrelation reaches a minimum or drops to a small fraction of its initial value. In another attempt [74,75] it has been verified that the values of τ at which the mutual information has a local minimum are equivalent to the values of τ at which the logarithm of the correlation sum has a local minimum.

2.7.5.2 Optimum Embedding Dimension

To further optimize the measurement of Lyapunov exponents there is a need to specify the optimum value for m, named the *embedding dimension*. Before doing that some definitions have to be given as follows.

Fractal dimension is another statistic related to the dynamical measurement. The strange attractors are fractals and their fractal dimension D_f is simply related to the minimum number of dynamical variables needed to model the dynamics of the attractor. Conceptually, a simple way to measure D_f is to measure the *Kolmogorov capacity*. In this measurement a set is covered with small cells, depending on the dimensionality (i.e. squares for sets embedded in two dimensions, cubes for sets embedded in three dimensions, and so on), of size ε. If $M(\varepsilon)$ denotes the number of such cells within a set, the fractal dimension is defined as

$$D_f = \lim_{\varepsilon \to 0} \frac{\log[M(\varepsilon)]}{\log(1/\varepsilon)} \tag{2.139}$$

for a set of single points $D_f = 0$, for a straight line $D_f = 1$, and for a plane area $D_f = 2$. The fractal dimension, however, may not be an integer.

The *correlation dimension* is defined as

$$D_r = \lim_{r \to 0} \frac{\log C(r)}{\log r} \tag{2.140}$$

where

$$C(r) = \sum_{i=1}^{M(r)} p_i^2 \tag{2.141}$$

is the correlation sum and p_i the probability of cell i.

The optimal embedding dimension, m, as required for accurate estimation of the Lyapunov exponents, has to satisfy $m \geq 2D_f + 1$. D_f is, however, not often known *a priori*. The Grassberger–Procaccia algorithm can nonetheless be employed to measure the correlation dimension, C_r. The minimum embedding dimension of the attractor is $m+1$, where m is the embedding dimension above which the measured value of the correlation dimension C_r remains constant.

As another very important conclusion,

$$D_f = D_L = 1 + \frac{\lambda_1}{|\lambda_2|} \tag{2.142}$$

i.e. the fractal dimension D_f is equivalent to the Lyapunov dimension [68].

Chaos has been used as a measure in analysis of many types of signals and data. Its application to epileptic seizure prediction will be shown in Chapter 4.

2.7.6 Approximate Entropy

Approximate entropy (AE) is a statistic that can be estimated from the discrete-time sequences, especially for real-time applications [76,77]. This measure can quantify the complexity or irregularity of the system. The AE is less sensitive to noise and can be used for short-length data. In addition, it is resistant to short strong transient interferences (outliers) such as spikes [77].

Given the embedding dimension m, the m-vector $\boldsymbol{x}(i)$ is defined as

$$\boldsymbol{x}(i) = [x(i), x(i+1), \ldots, x(i+m-1)], \quad i = 1, \ldots, N-m+1 \tag{2.143}$$

where N is the number of data points. The distance between any two of the above vectors, $\boldsymbol{x}(i)$ and $\boldsymbol{x}(j)$, is defined as

$$d[\boldsymbol{x}(i), \boldsymbol{x}(j)] = \max_{k} |x(i+k) - x(j+k)| \tag{2.144}$$

where $|.|$ denotes the absolute value. Considering a threshold level of β, the number of times, $M^m(i)$, that the above distance satisfies $d[\boldsymbol{x}(i), \boldsymbol{x}(j)] \leq \beta$ is found. This is performed for all i. For the embedding dimension m,

$$\xi_\beta^m(i) = \frac{M^m(i)}{N-m+1} \quad \text{for } i = 1, \ldots, N-m+1 \tag{2.145}$$

Then, the average natural logarithm of $\xi_\beta^m(i)$ is found as

$$\psi_\beta^m = \frac{1}{N-m+1} \sum_{i=1}^{N-m+1} \ln \xi_\beta^m(i) \tag{2.146}$$

By repeating the same method for an embedding dimension of $m+1$, the AE will be given as

$$\mathrm{AE}(m, \beta) = \lim_{N \to \infty} (\psi_\beta^m - \psi_\beta^{m+1}) \tag{2.147}$$

In practice, however, N is limited and therefore the AE is calculated for N data samples. In this case the AE depends on m, β, and N, i.e.

$$\mathrm{AE}(m, \beta, N) = \psi_\beta^m - \psi_\beta^{m+1} \tag{2.148}$$

The embedding dimension can be found as previously mentioned. However, the threshold value has to be set correctly. In some applications the threshold value is taken as a value between 0.1 and 0.25 times the data standard deviation [76].

2.7.7 *Using the Prediction Order*

It is apparent that for signals with highly correlated time samples the prediction order of an AR or ARMA model is low and for noise-type signals where the correlation among the samples is low the order is high. This means that for the latter case a large number of previous samples is required to predict the current sample. A different criterion such as the Akaike information criterion (AIC) may be employed to find the prediction order from the time series. Figure 2.22 shows the prediction order automatically computed for overlapping segments of three sections of a time series in which the middle section is sinusoidal and the first and third sections are noise like signals.

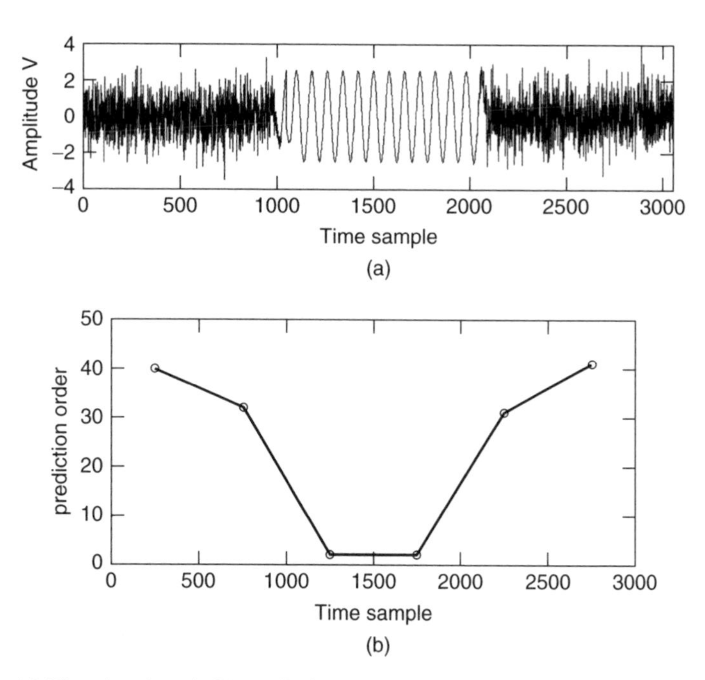

Figure 2.22 (a) The signal and (b) prediction order measured for overlapping segments of the signal

2.8 Filtering and Denoising

The EEG signals are subject to noise and artefacts. Electrocardiograms (ECGs), electrooc-clugrams (EOG), or eye blinks affect the EEG signals. Any multimodal recording such as EEG–fMRI significantly disturbs the EEG signals because of both magnetic fields and the change in the blood oxygen level and sensitivity of oxygen molecule to the magnetic field (ballistocardiogram). Artefact removal from EEGs will be explained in the related chapters. The noise in the EEGs, however, may be estimated and mitigated using adaptive and nonadaptive filtering techniques.

The EEG signals contain neuronal information below 100 Hz (in many applications the information lies below 30 Hz). Any frequency component above these frequencies can be simply removed by using lowpass filters. In the cases where the EEG data acquisition system is unable to cancel out the 50 Hz line frequency (due to a fault in grounding or imperfect balancing of the inputs to the differential amplifiers associated with the EEG system) a notch filter is used to remove it.

The nonlinearities in the recording system related to the frequency response of the amplifiers, if known, are compensated by using equalizing filters. However, the characteristics of the internal and external noises affecting the EEG signals are often unknown. The noise may be characterized if the signal and noise subspaces can be accurately separated. In Chapter 4 it is seen that the number of sources can be estimated. Using principal component analysis or independent component analysis it is possible to decompose the multichannel EEG observations to their constituent components, such as the

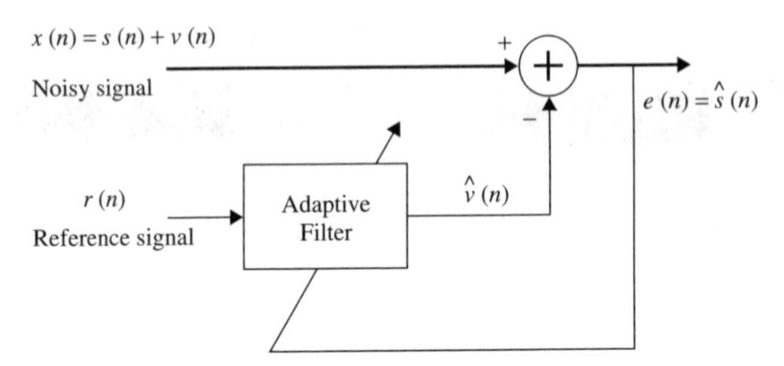

Figure 2.23 An adaptive noise canceller

neural activities and noise. Combining these two together, the estimated noise components can be extracted, characterized, and separated from the actual EEGs. These concepts are explained in the following sections and their applications to the artefact and noise removal will be given in the later chapters.

Adaptive noise cancellers used in communications, signal processing, and biomedical signal analysis can also be used to remove noise and artefacts from the EEG signals. An effective adaptive noise canceller, however, requires a reference signal. Figure 2.23 shows a general block diagram of an adaptive filter for noise cancellation. The reference signal carries significant information about the noise or artefact and its statistical properties. For example, in the removal of eye blinking artefacts (discussed in Chapter 7) a signature of the eye blink signal can be captured from the FP1 and FP2 EEG electrodes. In detection of the ERP signals, as another example, the reference signal can be obtained by averaging a number of ERP segments. There are many other examples such as ECG cancellation from EEGs and the removal of fMRI scanner artefacts from EEG–fMRI simultaneous recordings where the reference signals can be provided.

Adaptive Wiener filters are probably the most fundamental type of adaptive filters. In Figure 2.23 the optimal weights for the filter, $\boldsymbol{w}(n)$, are calculated such that $\hat{s}(n)$ is the best estimate of the actual signal $\boldsymbol{s}(n)$ in the mean-squared sense. The Wiener filter minimizes the mean-squared value of the error, defined as

$$e(n) = x(n) - \hat{v}(n) = x(n) - \boldsymbol{w}^{\mathrm{T}} r(n) \tag{2.149}$$

where $\boldsymbol{w}$ is the Wiener filter coefficient vector. Using the orthogonality principle [78] and assuming $x(n)$ and $r(n)$ are jointly statistically wide sense stationary, the final form of the mean-squared error will be

$$E[e(n)^2] = E[x(n)^2] - 2\boldsymbol{p}^{\mathrm{T}}\boldsymbol{w} + \boldsymbol{w}^{\mathrm{T}}\mathbf{R}\boldsymbol{w} \tag{2.150}$$

where $E(.)$ represents statistical expectation,

$$\boldsymbol{p} = E[x(n)\boldsymbol{r}(n)] \tag{2.151}$$

and

$$\mathbf{R} = E[\boldsymbol{r}(n)\boldsymbol{r}^{\mathrm{T}}(n)] \tag{2.152}$$

By taking the gradient with respect to $\boldsymbol{w}$ and equating it to zero,

$$\boldsymbol{w} = \mathbf{R}^{-1}\boldsymbol{p} \tag{2.153}$$

Since $\mathbf{R}$ and $\boldsymbol{p}$ are usually unknown, the above minimization is performed iteratively by substituting time averages for statistical averages. The adaptive filter in this case decorrelates the output signals. The general update equation is in the form of

$$\boldsymbol{w}(n+1) = \boldsymbol{w}(n) + \Delta\boldsymbol{w}(n) \tag{2.154}$$

where n is the iteration number, which typically corresponds to the discrete-time index. $\Delta\boldsymbol{w}(n)$ has to be computed such that $E[\boldsymbol{e}(n)]^2$ reaches a reasonable minimum. The simplest and most common way of calculating $\Delta\boldsymbol{w}(n)$ is by using the gradient descent or steepest descent algorithm [78]. In both cases, a criterion is defined as a function of the squared error (often called a performance index) such as $\eta\ (e(n)^2)$, that monotonically decreases after each iteration and converges to a global minimum. This requires

$$\eta(\boldsymbol{w} + \Delta\boldsymbol{w}) \leq \eta(\boldsymbol{w}) = \eta(\boldsymbol{e}(n)^2) \tag{2.155}$$

Assuming $\Delta\boldsymbol{w}$ to be very small, it is concluded that

$$\eta(\boldsymbol{w}) + \Delta\boldsymbol{w}^{\mathrm{T}}\nabla_{w}(\eta(\boldsymbol{w})) \leq \eta(\boldsymbol{w}) \tag{2.156}$$

where $\nabla_w(.)$ represents the gradient with respect to $\boldsymbol{w}$. This means that the above equation is satisfied by setting $\Delta\boldsymbol{w} = -\mu\nabla_w(.)$, where μ is the learning rate or convergence parameter. Hence, the general update equation takes the form

$$\boldsymbol{w}(n+1) = \boldsymbol{w}(n) - \mu\nabla_w(\eta(\boldsymbol{w}(n))) \tag{2.157}$$

Using the least mean square (LMS) approach, $\nabla_w(\eta(\boldsymbol{w}))$ is replaced by an instantaneous gradient of the squared error signal, i.e.

$$\nabla_w(\eta(\boldsymbol{w}(n))) \cong -2e(n)\boldsymbol{r}(n) \tag{2.158}$$

Therefore, the LMS-based update equation is

$$\boldsymbol{w}(n+1) = \boldsymbol{w}(n) + 2\ \mu e(n)\boldsymbol{r}(n) \tag{2.159}$$

Also, the convergence parameter, μ, must be positive and should satisfy

$$0 < \mu < \frac{1}{\lambda_{\max}} \tag{2.160}$$

where $\lambda_{\max}$ represents the maximum eigenvalue of the autocorrelation matrix $\mathbf{R}$. The LMS algorithm is the most simple and computationally efficient algorithm. However, the speed of convergence can be slow, especially for correlated signals. The recursive least-squares (RLS) algorithm attempts to provide a high-speed stable filter, but it is numerically unstable for real-time applications [79,80]. The performance index is defined as

$$\eta(\boldsymbol{w}) = \sum_{i=0}^{n} \gamma^{n-i} \mathrm{e}^2(i) \tag{2.161}$$

Then, by taking the derivative with respect to $\boldsymbol{w}$ gives

$$\nabla_w \eta(\boldsymbol{w}) = -2 \sum_{i=0}^{n} \gamma^{n-i} e(i)\boldsymbol{r}(i) \tag{2.162}$$

where $0 < \gamma \leq 1$ is the forgetting factor [79,80]. Replacing $e(n)$ in the above equation and writing it in vector form gives

$$\mathbf{R}(n)\boldsymbol{w}(n) = \boldsymbol{p}(n) \tag{2.163}$$

where

$$\mathbf{R}(n) = \sum_{i=0}^{n} \lambda^{n-i} \boldsymbol{r}(i)\boldsymbol{r}^{\mathrm{T}}(i) \tag{2.164}$$

and

$$\boldsymbol{p}(n) = \sum_{i=0}^{n} \lambda^{n-i} x(i)\boldsymbol{r}(i) \tag{2.165}$$

From this equation,

$$\boldsymbol{w}(n) = \mathbf{R}^{-1}(n)\boldsymbol{p}(n) \tag{2.166}$$

The RLS algorithm performs the above operation recursively such that **P** and **R** are estimated at the current time n as

$$\boldsymbol{p}(n) = \lambda \boldsymbol{p}(n-1) + x(n)\boldsymbol{r}(n) \tag{2.167}$$

$$\boldsymbol{R}(n) = \lambda \boldsymbol{R}(n-1) + \boldsymbol{r}(n)\boldsymbol{r}^{\mathrm{T}}(n) \tag{2.168}$$

In this case

$$\boldsymbol{r}(n) = \begin{bmatrix} r(n) \\ r(n-1) \\ \vdots \\ r(n-M) \end{bmatrix} \tag{2.169}$$

where M represents the finite impulse response (FIR) filter order. On the other hand,

$$\mathbf{R}^{-1}(n) = [\lambda\mathbf{R}^{-1}(n-1) + \boldsymbol{r}(n)\boldsymbol{r}^{\mathrm{T}}(n)]^{-1} \tag{2.170}$$

which can be simplified using the matrix inversion lemma [81]

$$\mathbf{R}^{-1}(n) = \frac{1}{\lambda}\left[\mathbf{R}^{-1}(n-1) - \frac{\mathbf{R}^{-1}(n-1)\boldsymbol{r}(n)\boldsymbol{r}^{\mathrm{T}}(n)\mathbf{R}^{-1}(n-1)}{\lambda + \boldsymbol{r}^{\mathrm{T}}(n)\mathbf{R}^{-1}(n-1)\boldsymbol{r}(n)}\right] \tag{2.171}$$

and finally the update equation can be written as

$$\boldsymbol{w}(n) = \boldsymbol{w}(n-1) + \mathbf{R}^{-1}(n)\boldsymbol{r}(n)\boldsymbol{g}(n) \tag{2.172}$$

where

$$\boldsymbol{g}(n) = x(n) - \boldsymbol{w}^{\mathrm{T}}(n-1)\boldsymbol{r}(n) \tag{2.173}$$

and the error $e(n)$ after each iteration is recalculated as

$$e(n) = x(n) - \boldsymbol{w}^{\mathrm{T}}(n)\boldsymbol{r}(n) \tag{2.174}$$

The second term on the right-hand side of the above equation is $\hat{\boldsymbol{v}}(n)$. The presence of $\mathbf{R}^{-1}(n)$ in Equation (2.172) is the major difference between the RLS and the LMS, but the RLS approach increases computation complexity by an order of magnitude.

2.9 Principal Component Analysis

All suboptimal transforms such as the DFT and DCT decompose the signals into a set of coefficients, which do not necessarily represent the constituent components of the signals. Moreover, as the transform kernel is independent of the data it is not efficient in terms of both decorrelation of the samples and energy compaction. Therefore, separation of the signal and noise components is generally not achievable using these suboptimal transforms.

Expansion of the data into a set of orthogonal components certainly achieves maximum decorrelation of the signals. This can enable separation of the data into the signal and noise subspaces.

For a single-channel EEG the Karhunen–Loève transform is used to decompose the ith channel signal into a set of weighted orthogonal basis functions:

$$x_i(n) = \sum_{k=1}^{N} w_{i,k}\varphi_k(n) \qquad \text{or} \qquad \boldsymbol{x}_{\mathrm{i}} = \boldsymbol{\Phi}\boldsymbol{w}_i \tag{2.175}$$

where $\boldsymbol{\Phi} = \{\varphi_k\}$ is the set of orthogonal basis functions. The weights $w_{i,k}$ are then calculated as

$$\boldsymbol{w}_i = \boldsymbol{\Phi}^{-1}\boldsymbol{x}_i \qquad \text{or} \qquad w_{i,k} = \sum_{n=0}^{N-1} \varphi_k^{-1}(n)x_i(n) \tag{2.176}$$

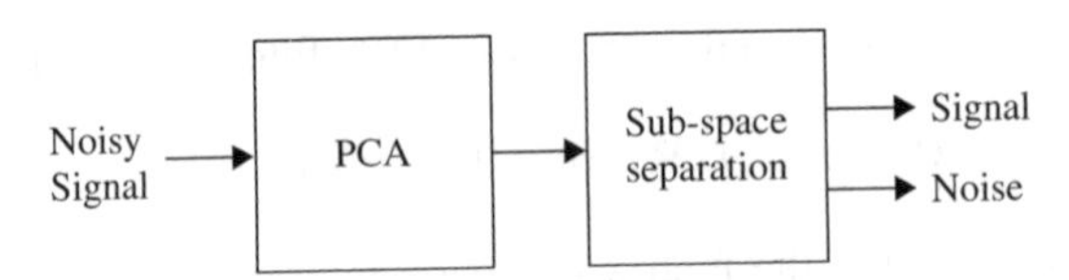

Figure 2.24 The general application of PCA

Often noise is added to the signal, i.e. $x_i(n) = s_i(n) + v_i(n)$, where $v_i(n)$ is additive noise. This degrades the decorrelation process. The weights are then estimated in order to minimize a function of the error between the signal and its expansion by the orthogonal basis, i.e. $\boldsymbol{e}_i = \boldsymbol{x}_i - \boldsymbol{\Phi}\boldsymbol{w}_i$. Minimization of the error in this case is generally carried out by solving the least-squares problem. In a typical application of PCA as depicted in Figure 2.24, the signal and noise subspaces are separated by means of some classification procedure.

2.9.1 Singular-Value Decomposition

The singular-value decomposition (SVD) is often used for solving the LS problem. This can be related to the decomposition of an $M \times M$ square autocorrelation matrix **R** into its eigenvalue matrix $\boldsymbol{\Lambda} = \text{diag}(\lambda_1, \lambda_2, \ldots, \lambda_M)$ and associated $M \times M$ orthogonal matrix of eigenvectors **V**, i.e. $\mathbf{R} = \mathbf{V}\boldsymbol{\Lambda}\mathbf{V}^{\text{H}}$, where $(.)^{\text{H}}$ denotes the Hermitian (conjugate transpose) operation. If **A** is an $M \times M$ data matrix such that $\mathbf{R} = \mathbf{A}^{\text{H}}\mathbf{A}$ then an $M \times M$ orthogonal (more generally unitary) matrix **U**, an $M \times M$ orthogonal matrix **V**, and an $M \times M$ diagonal matrix $\boldsymbol{\Sigma}$ exist with diagonal elements equal to $\lambda_i^{1/2}$, such that

$$\mathbf{A} = \mathbf{U}\boldsymbol{\Sigma}\,\mathbf{V}^{\text{H}} \tag{2.177}$$

Hence $\boldsymbol{\Sigma}^2 = \Lambda$. The columns of **U** are called left singular vectors and the rows of $\mathbf{V}^{\text{H}}$ are called right singular vectors. If **A** is a rectangular $N \times M$ matrix of rank k then **U** will be $N \times N$ and $\boldsymbol{\Sigma}$ will be

$$\boldsymbol{\Sigma} = \begin{bmatrix} \mathbf{S} & 0 \\ 0 & 0 \end{bmatrix} \tag{2.178}$$

where $S = \text{diag}(\sigma_1, \sigma_2, \ldots, \sigma_k)$, where $\sigma_i = \lambda_i^{1/2}$. For such a matrix the Moore–Penrose pseudoinverse is defined as an $M \times N$ matrix $\mathbf{A}^{\dagger}$ defined as

$$\mathbf{A}^{\dagger} = \mathbf{U}\boldsymbol{\Sigma}^{\dagger}\mathbf{V}^{\text{H}} \tag{2.179}$$

where $\boldsymbol{\Sigma}^{\dagger}$ is an $M \times N$ matrix defined as

$$\boldsymbol{\Sigma}^{\dagger} = \begin{bmatrix} \mathbf{S}^{-1} & 0 \\ 0 & 0 \end{bmatrix} \tag{2.180}$$

$\mathbf{A}^{\dagger}$ has a major role in the solutions of least-squares problems and $\mathbf{S}^{-1}$ is a $k \times k$ diagonal matrix with elements equal to the reciprocals of the assumed non zero singular values of $\mathbf{A}$, i.e.

$$\mathbf{S}^{-1} = \text{diag}\left(\frac{1}{\sigma_1}, \frac{1}{\sigma_2}, \ldots, \frac{1}{\sigma_k}\right) \tag{2.181}$$

In order to see the application of the SVD in solving the LS problem consider the error vector $\boldsymbol{e}$, defined as

$$\boldsymbol{e} = \boldsymbol{d} - \mathbf{A}\boldsymbol{h} \tag{2.182}$$

where $\boldsymbol{d}$ is the desired signal vector and $\mathbf{A}\boldsymbol{h}$ is the estimate $\hat{\boldsymbol{d}}$. To find $\boldsymbol{h}$, $\mathbf{A}$ is replaced with its SVD in the above equation, which thereby minimizes the squared Euclidean norm of the error vector, $||\boldsymbol{e}||^2$. By using the SVD, it is found that

$$\boldsymbol{e} = \boldsymbol{d} - \mathbf{U}\boldsymbol{\Sigma}\mathbf{V}^{\mathrm{H}}\boldsymbol{h} \tag{2.183}$$

or equivalently

$$\mathbf{U}^{\mathrm{H}}\boldsymbol{e} = \mathbf{U}^{\mathrm{H}}\boldsymbol{d} - \boldsymbol{\Sigma}\mathbf{V}^{\mathrm{H}}\boldsymbol{h} \tag{2.184}$$

Since $\mathbf{U}$ is a unitary matrix, $||\boldsymbol{e}||^2 = ||\mathbf{U}^{\mathrm{H}}\boldsymbol{e}||^2$. Hence, the vector $\boldsymbol{h}$ that minimizes $||\boldsymbol{e}||^2$ also minimizes $||\mathbf{U}^{\mathrm{H}}\boldsymbol{e}||^2$. Finally, the unique solution as an optimum $\boldsymbol{h}$ (coefficient vector) may be expressed as [82]:

$$\boldsymbol{h} = \sum_{i=1}^{k} \frac{\boldsymbol{u}_i^{\mathrm{H}}\boldsymbol{d}}{\sigma_i}\boldsymbol{v}_i \tag{2.185}$$

where k is the rank of $\mathbf{A}$. Alternatively, as the optimum least-squares coefficient vector

$$\boldsymbol{h} = (\mathbf{A}^{\mathrm{H}}\mathbf{A})^{-1}\mathbf{A}^{\mathrm{H}}\boldsymbol{d} \tag{2.186}$$

Performing a principal component analysis (PCA) is equivalent to performing an SVD on the covariance matrix. PCA uses the same concept as SVD and orthogonalization to decompose the data into constituent uncorrelated orthogonal components such that the autocorrelation matrix is diagonalized. Each eigenvector represents a principal component and the individual eigenvalues are numerically related to the variance they capture in the direction of the principal components. In this case the mean squared error (MSE) is simply the sum of the $N - K$ eigenvalues, i.e.

$$\text{MSE} = \sum_{k=N-K}^{N} \varphi_k^{\mathrm{T}} R_x \varphi_k = \sum_{k=N-K}^{N} \varphi_k^{\mathrm{T}}(\lambda_k \varphi_k) = \sum_{k=N-K}^{N} \lambda_k \tag{2.187}$$

PCA is widely used in data decomposition, classification, filtering, and whitening. In filtering applications the signal and noise subspaces are separated and the data are reconstructed from only the eigenvalues and eigenvectors of the actual signals. PCA is also

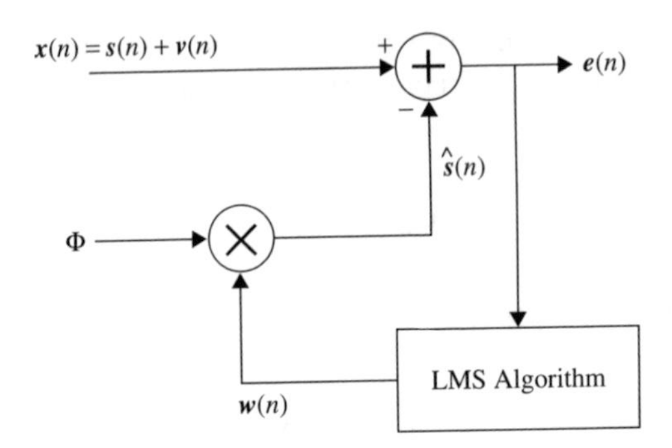

Figure 2.25 Adaptive estimation of the weight vector $\boldsymbol{w}(n)$

used for blind source separation of correlated mixtures if the original sources can be considered as statistically uncorrelated.

The PCA problem is then summarized as how to find the weights $\boldsymbol{w}$ in order to minimize the error given the observations only. The LMS algorithm is used here to minimize the MSE iteratively as

$$J_n = E[(\boldsymbol{x}(n) - \boldsymbol{\Phi}^{\mathrm{T}}(n)\boldsymbol{w}(n))^2] \tag{2.188}$$

The update rule for the weights is then

$$\boldsymbol{w}(n+1) = \boldsymbol{w}(n) + \mu \boldsymbol{e}(n)\boldsymbol{\Phi}(n) \tag{2.189}$$

where the error signal $\boldsymbol{e}(n) = \boldsymbol{x}(n) - \boldsymbol{\Phi}^{\mathrm{T}}(n)\boldsymbol{w}(n)$, $\mathbf{x}(n)$ is the noisy input, and n is the iteration index. The step size μ may be selected empirically or adaptively. These weights are then used to reconstruct the sources from the set of orthogonal basis functions. Figure 2.25 shows the overall system for adaptive estimation of the weight vector $\boldsymbol{w}$ using the LMS algorithm.

2.10 Independent Component Analysis

The concept of independent component analysis (ICA) lies in the fact that the signals may be decomposed into their constituent independent components. In places where the combined source signals can be assumed independent from each other this concept plays a crucial role in separation and denoising the signals.

A measure of independency may easily be described to evaluate the independence of the decomposed components. Generally, considering the multichannel signal as $\boldsymbol{y}(\mathrm{n})$ and the constituent signal components as $y_i(n)$, the $y_i(n)$ are independent if

$$p_{\mathbf{Y}}(\boldsymbol{y}(n)) = \prod_{i=1}^{m} p_y(y_i(n)) \quad \forall n \tag{2.190}$$

where $p(\mathbf{Y})$ is the joint probability distribution, $p_y(y_i(n))$ are the marginal distributions and m is the number of independent components.

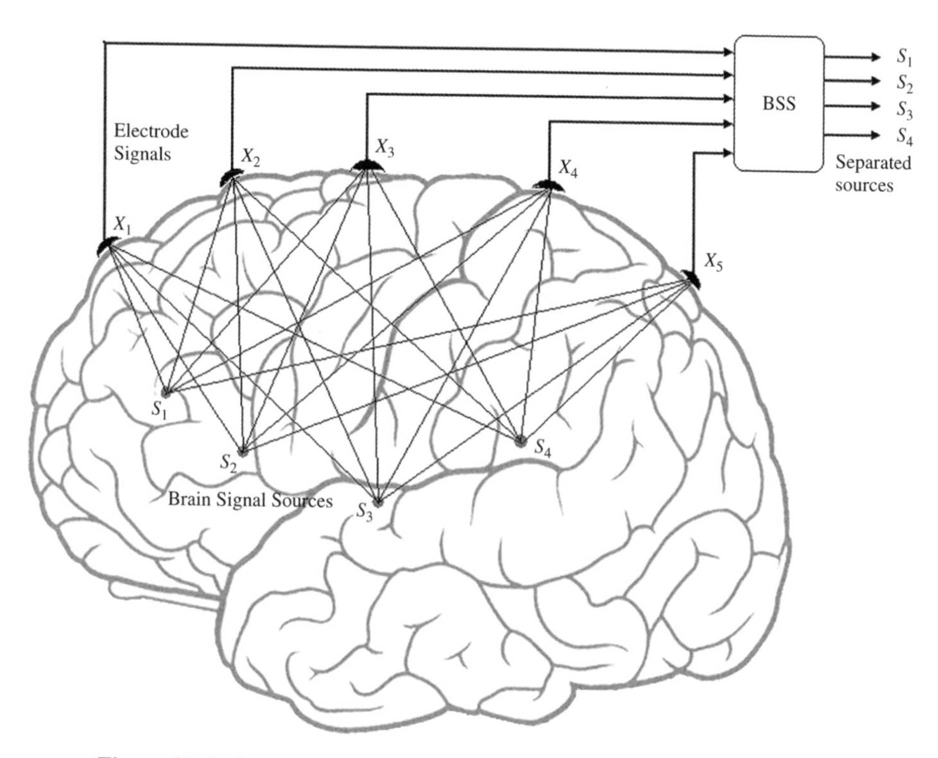

Figure 2.26 BSS concept; mixing and blind separation of the EEG signals

An important application of ICA is in blind source separation (BSS). BSS is an approach to estimate and recover the independent source signals using only the information of their mixtures observed at the recording channels. Due to its variety of applications BSS has attracted much attention recently. BSS of acoustic signals is often referred to as the 'cocktail party problem' [83], which means separation of individual sounds from a number of recordings in an uncontrolled environment such as a cocktail party. Figure 2.26 illustrates the BSS concept. As expected, ICA can be useful if the original sources are independent, i.e. $p(\mathbf{s}(n)) = \prod_{i=1}^{m} p_i(\mathbf{s}_i(n))$.

A perfect separation of the signals requires taking into account the structure of the mixing process. In a real-life application, however, this process is unknown, but some assumptions may be made about the source statistics.

Generally, the BSS algorithms do not make realistic assumptions about the environment in order to make the problem more tractable. There are typically three assumptions about the mixing medium. The most simple but widely used case is the instantaneous case, where the source signals arrive at the sensors at the same time. This has been considered for separation of biological signals such as the EEG, where the signals have narrow bandwidths and the sampling frequency is normally low. The BSS model in this case can be easily formulated as

$$\mathbf{x}(n) = \mathbf{H}\mathbf{s}(n) + \mathbf{v}(n) \tag{2.191}$$

where $m \times 1$ $\boldsymbol{s}(n)$, $n_e \times 1$ $\boldsymbol{x}(n)$, and $n_e \times 1$ $\boldsymbol{v}(n)$ denote respectively the vectors of source signals, observed signals, and noise at discrete time n. $\mathbf{H}$ is the mixing matrix of size $n_e \times m$. The separation is performed by means of a separating $m \times n_e$ matrix, $\mathbf{W}$, which uses only the information about $\boldsymbol{x}(n)$ to reconstruct the original source signals (or the independent components) as

$$\boldsymbol{y}(n) = \mathbf{W}\boldsymbol{x}(n) \tag{2.192}$$

In the context of EEG signal processing n_e denotes the number of electrodes. The early approaches in instantaneous BSS started from the work by Herault and Jutten [84] in 1986. In their approach, they considered non-Gaussian sources with similar number of independent sources and mixtures. They proposed a solution based on a recurrent artificial neural network for separation of the sources.

In acoustic applications, however, there are usually time lags between the arrival of the signals at the sensors. The signals also may arrive through multiple paths. This type of mixing model is called a convolutive model. One example is in places where the acoustic properties of the environment vary, such as a room environment surrounded by walls. Based on these assumptions the convolutive mixing model can be classified into two more types: anechoic and echoic. In both cases the vector representations of mixing and separating processes are changed to $\boldsymbol{x}(n) = \mathbf{H}(n)^{*}\boldsymbol{s}(n) + \boldsymbol{v}(n)$ and $\boldsymbol{y}(n) = \mathbf{W}(n)^{*}\boldsymbol{x}(n)$ respectively, where $*$ denotes the convolution operation.

In an anechoic model, however, the expansion of the mixing process may be given as

$$x_i(n) = \sum_{j=1}^{M} h_{ij} s_j(n - \delta_{ij}) + v_i(n), \quad \text{for } i = 1, \ldots, N \tag{2.193}$$

where the attenuation, h_{ij}, and delay, δ_{ij}, of source j to sensor i would be determined by the physical position of the source relative to the sensors. Then the unmixing process will be given as

$$y_j(m) = \sum_{i=1}^{N} w_{ji} x_i(m - \delta_{ji}), \quad \text{for } j = 1, \ldots, M \tag{2.194}$$

where the w_{ji}s are the elements of W. In an echoic mixing environment it is expected that the signals from the same sources reach to the sensors through multiple paths. Therefore the expansion of the mixing and separating models will be changed to

$$x_i(n) = \sum_{j=1}^{M} \sum_{k=1}^{K} h_{ij}^{k} s_j(n - \delta_{ij}^{k}) + v_i(n), \quad \text{for } i = 1, \ldots, N \tag{2.195}$$

where K denotes the number of paths and $v_i(n)$ is the accumulated noise at sensor i. The unmixing process will be formulated similarly to the anechoic one. Obviously, for a known number of sources an accurate result may be expected if the number of paths is known.

The aim of BSS using ICA is to estimate an unmixing matrix $\mathbf{W}$ such that $\mathbf{Y} = \mathbf{WX}$ best approximates the independent sources S, where $\mathbf{Y}$ and $\mathbf{X}$ are respectively matrices with columns $\boldsymbol{y}(n) = [y_1(n), y_2(n), \ldots, y_m(n)]^{\mathrm{T}}$ and $\boldsymbol{x}(n) = [x_1(n), x_2(n), \ldots x_{ne}(n)]^{\mathrm{T}}$. In any case, the unmixing matrix for the instantaneous case is expected to be equal to the inverse of the mixing matrix, i.e. $\mathbf{W} = \mathbf{H}^{-1}$. However, in all ICAs algorithms based upon restoring independence, the separation is subject to permutation and scaling ambiguities in the output independent components, i.e. $\mathbf{W} = \mathbf{PDH}^{-1}$, where $\mathbf{P}$ and $\mathbf{D}$ are the permutation and scaling matrices respectively.

There are three major approaches in using ICA for BSS:

1. Factorizing the joint PDF of the reconstructed signals into its marginal PDFs. Under the assumption that the source signals are stationary and non-Gaussian, the independence of the reconstructed signals can be measured by a statistical distance between the joint distribution and the product of its marginal PDFs. Kullback–Laibler (KL) divergence (distance) is an example. For nonstationary cases and for the short-length data, there will be poor estimation of the PDFs. Therefore, in such cases, this approach may not lead to good results. On the other hand, such methods are not robust for noisy data since in this situation the PDF of the signal will be distorted.
2. Decorrelating the reconstructed signals through time, i.e. diagonalizing the covariance matrices at every time instant. If the signals are mutually independent, the off-diagonal elements of the covariance matrix vanish, although the reverse of this statement is not always true. If the signals are nonstationary the time-varying covariance structure can be used to estimate the unmixing matrix. An advantage of this method is that it only uses second-order statistics, which implies that it is likely to perform better in noisy and short data length conditions than higher-order statistics.
3. Eliminating the temporal cross-correlation functions of the reconstructed signals as much as possible. In order to perform this, the correlation matrix of observations can be diagonalized at different time lags simultaneously. Here, second-order statistics are also normally used. As another advantage, it can be applied in the presence of white noise since such noise can be avoided by using the cross-correlation only for $\tau \neq 0$. Such a method is appropriate for stationary and weakly stationary sources (i.e. when the stationarity condition holds within a short segment of data).

It has been shown [85] that mutual information (MI) is a measure of independence and that maximizing the non-Gaussianity of the source signals is equivalent to minimizing the mutual information between them.

In the majority of cases the number of sources is known. This assumption avoids any ambiguity caused by false estimation of the number of sources. In exactly determined cases the number of sources is equal to the number of mixtures. In overdetermined situations, however, the number of mixtures is more than the number of sources.

There have been many attempts to apply BSS to EEG signals [86–96] for separation of normal brain rhythms, event-related signals, or mental or physical movement-related sources. If the number of sources is unknown, a criterion has to be established to estimate the number of sources beforehand. This process is a difficult task, especially when noise is involved. In those cases where the number of sources is more than the number of mixtures (known as underdetermined systems), the above BSS schemes cannot be applied simply because the unmixing matrix will not be invertible, and generally the original sources cannot be extracted. However, when the signals are sparse other methods based on clustering may be utilized.

A signal is said to be sparse when it has many zero or at least approximately zero samples. Separation of the mixtures of such signals is potentially possible in the situation where at each sample instant the number of nonzero sources is not more than the number of sensors. The mixtures of sparse signals can also be instantaneous or convolutive. However, as will be briefly described later, the solution for only a simple case of a small number of idealized sources has been given in the literature.

In the context of EEG analysis, although the number of signals mixed at the electrodes seems to be limited, the number of sources corresponding to the neurons firing at a time can be enormous. However, if the objective is to study a certain rhythm in the brain the problem can be transformed to the time–frequency domain or even to the space–time–frequency domain. In such domains the sources may be considered disjoint and generally sparse. Also it is said that in the brain neurons encode data in a sparse way if their firing pattern is characterized by a long period of inactivity [97,98].

2.10.1 *Instantaneous BSS*

This is the most commonly used scheme for processing of the EEGs. The early work by Jutten and Herault led to a simple but fundamental adaptive algorithm [99]. Linsker [100] proposed unsupervised learning rules based on information theory that maximize the average mutual information between the inputs and outputs of an artificial neural network. Comon [85] performed minimization of mutual information to make the outputs independent. The Infomax algorithm [101] was developed by Bell and Sejnowski, which in spirit is similar to the Linsker method. Infomax uses an elegant stochastic gradient learning rule that was proposed by Amari *et al.* [102]. Non-Gaussianity of the sources was first exploited by Hyvarinen and Oja [103] in developing their fast ICA (fICA) algorithm. fICA is actually a blind source extraction algorithm, which extracts the sources one-by-one based on their kurtosis; the signals with transient peaks have high kurtosis. Later it was demonstrated that the Infomax algorithm and maximum likelihood estimation are in fact equivalent [104,105].

Based on the Infomax algorithm [101] for signals with positive kurtosis such as simultaneous EEG-fMRI and speech signals, minimizing the mutual information between the source estimates and maximizing the entropy of the source estimates are equivalent. Therefore, a stochastic gradient ascent algorithm can be used to iteratively find the unmixing matrix by maximization of the entropy. The Infomax algorithm finds a $\mathbf{W}$ that minimizes the following cost function:

$$J(\mathbf{W}) = I(z, \boldsymbol{x}) = H(z) - H(z|\boldsymbol{x}) \tag{2.196}$$

where $H(z)$ is the entropy of the output, $H(z|\boldsymbol{x})$ is the entropy of the output subject to a known input, and $z = f(\boldsymbol{y})$ is a nonlinear activation function applied element wise to $\boldsymbol{y}$, the estimated sources. $I(z,\boldsymbol{x})$ is the mutual information between the input and output of the constructed adaptive neural network (ANN). $H(z|\boldsymbol{x})$ is independent of $\mathbf{W}$; therefore, the gradient of J is only proportional to the gradient of $H(z)$. Correspondingly, the natural gradient [107] of J denoted as $\nabla_{\mathbf{W}} J$ will be

$$\nabla_{\mathbf{W}} J = \nabla_{\mathbf{W}} I(z, \boldsymbol{x})\mathbf{W}^{\mathrm{T}}\mathbf{W} = \nabla_{\mathbf{W}} I(z, \boldsymbol{x})\mathbf{W}^{\mathrm{T}}\mathbf{W} \tag{2.197}$$

in which the time index n is dropped for convenience of presentation. Then, the sequential adaptation rule for the unmixing matrix $\mathbf{W}$ becomes

$$\mathbf{W}(n+1) = \mathbf{W}(n) + \mu[\mathbf{I} - 2f(\mathbf{y}(n))\boldsymbol{y}^{\mathrm{T}}(n)]\mathbf{W}(n) \tag{2.198}$$

where $f(\mathbf{y}(n)) = \{1 + \exp[-\boldsymbol{y}(n)]\}^{-1}$, assuming the outputs are super-Gaussian and μ is the learning rate, which is either a small constant or gradually changes following the speed of convergence.

Joint approximate diagonalization of eigenmatrices (JADE) is another well-known BSS algorithm [106] based on higher-order statistics (HOS). The JADE algorithm effectively diagonalizes the fourth-order cumulant of the estimated sources. This procedure uses certain matrices $\mathbf{Q}_z(\mathbf{M})$ formed by the inner product of the fourth-order cumulant tensor of the outputs with an arbitrary matrix $\mathbf{M}$, i.e.

$$\{\mathbf{Q}_z(\mathbf{M})\}_{ij} = \sum_{k=1}^{n_e}\sum_{l=1}^{n_e} \mathrm{Cum}(z_i, z_j^*, z_k, z_l^*)m_{lk} \tag{2.199}$$

where the (l, k)th component of the matrix $\mathbf{M}$ is written as m_{lk}, $\mathbf{Z} = \mathbf{CY}$, and $*$ denotes complex conjugate. The matrix $\mathbf{Q}_z(\mathbf{M})$ has the important property that it is diagonalized by the correct rotation matrix $\mathbf{U}$, i.e. $\mathbf{U}^{\mathrm{H}}\mathbf{Q}\mathbf{U} = \mathbf{\Lambda}_{\mathbf{M}}$, and $\mathbf{\Lambda}_{\mathbf{M}}$ is a diagonal matrix whose diagonal elements depend on the particular matrix $\mathbf{M}$ as well as $\mathbf{Z}$. By using Equation (2.199), for a set of different matrices $\mathbf{M}$, a set of cumulant matrices $\mathbf{Q}_z(\mathbf{M})$ can be calculated. The desired rotation matrix $\mathbf{U}$ then jointly diagonalizes these matrices. In practice, only approximate joint diagonalization is possible [106], i.e. the problem can be stated as minimization of

$$J(\mathbf{u}) = \sum_{j=1}^{n_e}\sum_{i=1}^{n_e} \mathrm{off}\,\{\boldsymbol{u}^{\mathrm{H}}\mathbf{Q}_{ij}\mathbf{u}\}$$

where

$$\mathrm{off}\,(\mathbf{M}) = \sum_{i \neq j} |m_{ij}|^2 \tag{2.200}$$

EEG signals are, however, nonstationary. Nonstationarity of the signals has been exploited in developing an effective BSS algorithm based on second-order statistics called SOBI (second-order blind identification) [107]. In this algorithm separation is performed at a number of discrete time lags simultaneously. At each lag the algorithm unitarily diagonalizes the whitened data covariance matrix. It also mitigates the effect of noise on the observation by using a whitening matrix calculation, which can improve robustness to noise. Unitary diagonalization can be explained as follows. If $\mathbf{V}$ is a whitening matrix and $\mathbf{X}$ is the observation matrix, the covariance matrix of the whitened observation is $\mathbf{C_X} = \mathrm{E}[\mathbf{VXX}^{\mathrm{H}}\mathbf{V}^{\mathrm{H}}] = \mathbf{VR_XV}^{\mathrm{H}} = \mathbf{VHR_SH}^{\mathrm{H}}\mathbf{V}^{\mathrm{H}} = \mathbf{I}$, where $\mathbf{R_X}$ and $\mathbf{R_S}$ denote respectively the covariance matrices of the observed data and the original sources. It is assumed that $\mathbf{R_S} = \mathbf{I}$, i.e. the sources have unit variance and are uncorrelated, so $\mathbf{VH}$ is a unitary matrix. Therefore $\mathbf{H}$ can be factored as $\mathbf{H} = \mathbf{V}^{-1}\mathbf{U}$, where $\mathbf{U} = \mathbf{VH}$. The joint approximate diagonalization for a number of time lags can be obtained efficiently using a generalization of the Jacobi technique for the exact diagonalization of a single Hermitian matrix. The SOBI algorithm is implemented through the following steps as given in Reference [106]:

(a) The sample covariance matrix $\hat{\boldsymbol{R}}(0)$ is estimated from T data samples. The m largest eigenvalues and their corresponding eigenvectors of $\hat{\boldsymbol{R}}(0)$ are denoted as $\lambda_1, \lambda_2, \ldots, \lambda_m$ and $\boldsymbol{h}_1, \boldsymbol{h}_2, \ldots, \boldsymbol{h}_m$ respectively.
(b) Under the white noise assumption, an estimate $\hat{\sigma}^2$ of the noise variance is the average of the $n_{\mathrm{e}}-m$ smallest eigenvalues of $\hat{\boldsymbol{R}}(0)$. The whitened signals are $z(n) = [z_1(n), z_2(n), \ldots, z_{n_{\mathrm{e}}}(n)]^{\mathrm{T}}$, computed by $z_i(n) = (\lambda_i - \hat{\sigma}^2)^{-1/2}\mathbf{h}_i^{\mathrm{H}}\mathbf{x}(n)$ for $1 \le i \le n_{\mathrm{e}}$. This is equivalent to forming a whitening matrix as $\hat{W} = \left[(\lambda_1 - \hat{\sigma}^2)^{-1/2}\boldsymbol{h}_1, \ldots, (\lambda_{n_{\mathrm{e}}} - \hat{\sigma}^2)^{-1/2}\boldsymbol{h}_{n_{\mathrm{e}}}\right]^{\mathrm{H}}$.
(c) Form sample estimates $\underline{\hat{\mathbf{R}}}(\tau)$ by computing the sample covariance matrices of $z(t)$ for a fixed set of time lags $\tau \in \{\tau_j | j = 1, \ldots, K\}$.
(d) A unitary matrix $\hat{U}$ is then obtained as a joint diagonalizer of the set $\{\underline{\hat{\mathbf{R}}}(\tau_j) | j = 1, \ldots, K\}$.
(e) The source signals are estimated as $\hat{s}(t) = \hat{\boldsymbol{U}}^{\mathrm{H}}\hat{\boldsymbol{W}}\boldsymbol{x}(t)$ or the mixing matrix $\mathbf{A}$ is estimated as $\hat{\boldsymbol{A}} = \hat{\boldsymbol{W}}^{\dagger}\hat{\boldsymbol{U}}$, where the superscript † denotes the Moore–Penrose pseudoinverse.

The FICA algorithm [103] is another very popular BSS technique which extracts the signals one by one based on their kurtosis. In fact, the algorithm uses an independence criterion that exploits non-Gaussianity of the estimated sources. In some places where the objective is to remove the spiky artefacts, such as the removal of the fMRI artefact from the simultaneous EEG-fMRI recordings, application of an iterative fICA followed by deflation of the artefact component gives excellent results [108]. A typical signal of this type is given in Figure 2.27.

Practically, fICA maximizes the negentropy, which represents the distance between a distribution and a Gaussian distribution having the same mean and variance, i.e.

$$\mathrm{Neg}(y) \propto \{E[f(y)] - E[f(y_{\mathrm{Gaussian}})]\}^2 \tag{2.201}$$

where f is a score function [109] and Neg stands for negentropy. This, as mentioned previously, is equivalent to maximizing the kurtosis. Therefore, the cost function can be simply defined as

$$J(\mathbf{W}) = -\frac{1}{4}|k_4(\mathbf{y})| = -\frac{\beta}{4}k_4(\mathbf{y}) \tag{2.202}$$

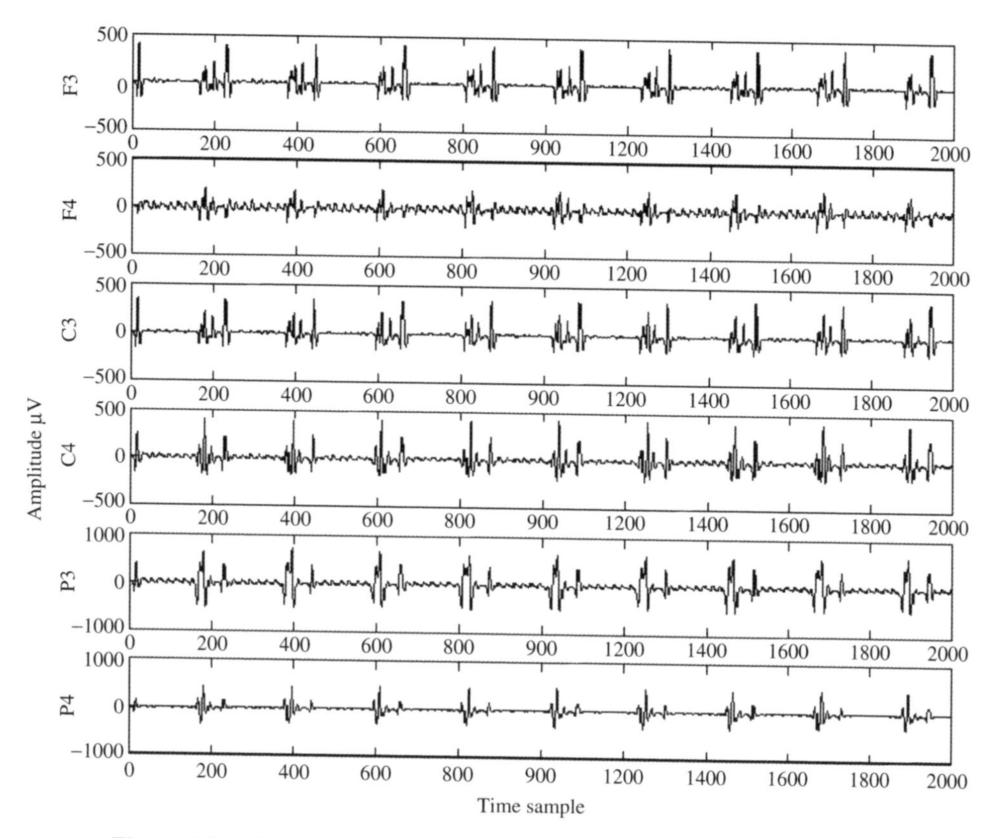

Figure 2.27 A sample of an EEG signal simultaneously recorded with fMRI

where $k_4(\mathbf{y})$ is the kurtosis, and β is the sign of the kurtosis. Applying the standard gradient decent approach to minimize the cost function gives

$$\mathbf{W}(n+1) = \mathbf{W}(n) - \mu \left.\frac{\partial J(\mathbf{W})}{\partial \mathbf{W}}\right|_{\mathbf{W}=\mathbf{W}(n)} \tag{2.203}$$

where

$$-\mu \left.\frac{\partial J(\mathbf{W})}{\partial \mathbf{W}}\right|_{\mathbf{W}=\mathbf{W}(n)} = \mu(n)\varphi(\mathbf{y}(n))\boldsymbol{x}(n) \tag{2.204}$$

Here $\mu(n)$ is a learning rate,

$$\varphi(y_i) = \beta \frac{\hat{m}_4(y_i)}{\hat{m}_2^3(y_i)} \left[\frac{\hat{m}_2(y_i)}{\hat{m}_4(y_i)} y_i^3 - y_i \right] \tag{2.205}$$

and $\hat{m}_q(y_i) = \hat{E}[y_i^q(n)]$, which is an estimate of the qth-order moment of the actual sources. Since fICA extracts the sources one-by-one a deflation process is followed to exclude the extracted source from the mixtures. The process reconstructs the mixtures iteratively by

$$\boldsymbol{x}_{j+1} = \boldsymbol{x}_j - \tilde{\boldsymbol{w}}_j \boldsymbol{y}_j, \quad j = 1, 2, \ldots \tag{2.206}$$

where $\tilde{\boldsymbol{w}}_j$ is estimated by minimization of the following cost function:

$$J(\tilde{\boldsymbol{w}}_j) = \frac{1}{2} E\left[\sum_{p=1}^{n_r} \boldsymbol{x}_{j+1,p}^2\right] \tag{2.207}$$

where n_r is the number of remaining mixtures.

Figure 2.28 shows the results after application of fICA to remove the scanner artefact from the EEGs. In addition to the separation of EEGs using fICA, very good results have been reported after application of fICA to separation of temporomanibular joint sounds [110].

In a time–frequency (TF) approach, which assumes that the sources are approximately cyclostationary and nonstationary, the auto-terms and cross-terms of the covariance matrix of the mixtures are first separated and BSS is applied to both terms [111,112]. In this approach, the spatial time–frequency distribution (STFD) of the mixed signals is

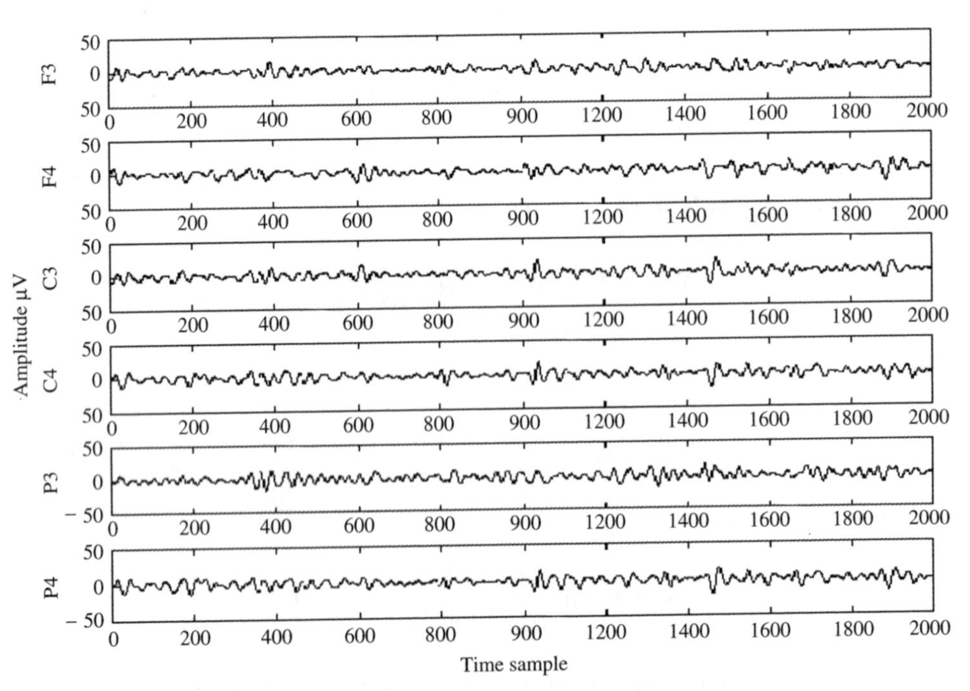

Figure 2.28 The EEG signals after removal of the scanner artefact.

defined as

$$\mathbf{D}_{xx}(n,\omega)=\frac{1}{2\pi}\sum_{u=\tau/2}^{N-\tau/2}\sum_{\tau=0}^{N/2-1}\phi(n-u,\tau)\mathrm{e}^{-\mathrm{i}\omega\tau}E\left[\boldsymbol{x}\left(u+\frac{\tau}{2}\right)\boldsymbol{x}\left(u-\frac{\tau}{2}\right)\right] \tag{2.208}$$

where $\phi(.)$ is the discretized kernel function defining a distribution from Cohen's class of TF distributions [113] and $\boldsymbol{x}(.)$ is an N sample observation of the signals, which is normally contaminated by noise. Assuming $\boldsymbol{x}(t)=\mathbf{A}\boldsymbol{s}(t)+\boldsymbol{v}(t)$, using the above equation it is found that

$$\mathbf{D}_{xx}(n,\omega)=\mathbf{A}\mathbf{D}_{ss}(n,\omega)\mathbf{A}^{\mathrm{H}}+\sigma^2\mathbf{I} \tag{2.209}$$

where $\mathbf{D}_{ss}(.)$ is the STFD of the source signals and σ^2 is the noise variance and depends on both noise power and the kernel function. From this equation it is clear that both $\mathbf{D}_{xx}$ and $\mathbf{D}_{ss}$ exhibit the same eigenstructure. The covariance matrix of the source signals is then replaced by the source STFD matrix composed of auto- and cross-source time–frequency distributions (TFDs) respectively, on the diagonal and off-diagonal entries.

Defining a whitening matrix $\mathbf{W}$ such that $\mathbf{U}=\mathbf{WA}$ is unitary, a whitened and noise-compensated STFD matrix is defined as

$$\begin{aligned}\tilde{\boldsymbol{D}}_{xx}(n,\omega)&=\mathbf{W}(\mathbf{D}_{xx}(n,\omega)-\sigma^2\mathbf{I})\mathbf{W}^{\mathrm{H}}\\&=\mathbf{U}\mathbf{D}_{ss}(n,\omega)\mathbf{U}^{\mathrm{H}}\end{aligned} \tag{2.210}$$

$\mathbf{W}$ and σ^2 can be estimated from the sample covariance matrix and $\mathbf{D}_{xx}$ is estimated based on the discrete-time formulation of the TFDs. From Equation (2.191) it is known that the sensor STFD matrix exhibits the same eigenstructure as the data covariance matrix commonly used for cyclic data [111]. The covariance matrix of the source signals is replaced by a source STFD matrix composed of the auto- and cross-source TFDs on the diagonal and off-diagonal entries respectively. The peaks occur in mutually exclusive locations on the TF plane. The kernel function can be defined in such a way as to maximize disjointness of the points in the TF plane. By estimation of the STFD in Equation (2.192) at appropriate TFD points, it is possible to recover the source signals by estimating a unitary transformation $\hat{\boldsymbol{U}}$, via optimization of a joint diagonal and off-diagonal criterion, to have

$$\hat{\boldsymbol{s}}(n)=\hat{\boldsymbol{U}}^{\mathrm{H}}\mathbf{W}\boldsymbol{x}(n)\quad\text{for } n=1,\ldots,N-1 \tag{2.211}$$

In order to define and extract the peaks of $\mathbf{D}_{xx}$ a suitable clustering approach has a to be followed. This algorithm has potential application for estimating the EEG sources since in most normal cases the sources are cyclic or quasicyclic.

2.10.2 Convolutive BSS

In many practical situations the signals reach the sensors with different time delays. The corresponding delay between source j and sensor i, in terms of number of samples, is directly proportional to the sampling frequency and conversely to the speed of sound, i.e. $\delta_{ij}\propto d_{ij}f_s/c$, where d_{ij}, f_s, and c are the distance between source j and sensor i, the sampling frequency, and the speed of sound respectively. For speech and music in air, as an example the following could be taken: d_{ij} in terms of metres, f_s between 8 and 44 kHz, and $c = 330$ m/s. Also, in an acoustic environment the sound signals can reach the sensors through multipaths after reflections by obstacles (such as walls). The above two cases have been addressed as anechoic and echoic BSS models respectively and formulated at the beginning of this section. The solution to echoic cases is obviously more difficult and normally involves some approximations to the actual system. As an example, in the previously mentioned cocktail party problem the source signals propagate through a dynamic medium with many parasitic effects, such as multiple echoes and reverberation. Therefore, the received signals are to a first approximation a weighted sum of mixed and delayed components. In other words, the received signals at each microphone are the convolutive mixtures of speech signals.

Unfortunately, most of the proposed BSS approaches to instantaneous mixtures fail or are limited in separation of convolutive mixtures, generally due to:

(a) noise;
(b) possibly a smaller number of sensors than the number of source signals (from the sources directly and through multipaths);
(c) nonstationarity of the signals;
(d) time delays, which make the overall mixing not instantaneous.

Convolutive BSS has recently been a focus of research in the acoustic signal processing community. Two major approaches have been followed for both anechoic and echoic cases. The first approach is to solve the problem in the time domain. In such methods, in order to have accurate results both the weights of the unmixing matrix and the delays have to be estimated. However, in the second approach, the problem can be transformed into the frequency domain as $\boldsymbol{h}(n)^*\boldsymbol{s}(n)\xrightarrow{\mathbf{F}}\mathbf{H}(\omega)\cdot\mathbf{S}(\omega)$ and instantaneous BSS applied to each frequency bin mixed signal. The separated signals at different frequency bins are then combined and transformed to the time domain to reconstruct the estimated sources. The short-term discrete Fourier transform is often used for this purpose. Figure 2.29 clearly represents the frequency-domain BSS of convolutive mixtures. However, the inherent permutation problem of BSS severely deteriorates the results since the order of the separated sources in different frequency bins can vary from segment to segment of the signals.

An early work in convolutive BSS by Platt and Faggin [114], who applied the adaptive noise cancellation network to the BSS model of Herault and Jutten [115], which has delays in the feedback path, was based on the minimum output power principle. This scheme exploits the fact that the signal corrupted by noise has more power than the clean signal. The feedback path cancels out the interferences as the result of delayed versions of the other sources. This circuit was also used later to extend the Infomax BSS to convolutive cases [116]. The combined network maximizes the entropy at the output of the network with respect to the weights and delays. Torkkola [117] extended this algorithm to the echoic cases. In order to achieve a reasonable convergence, some prior knowledge of the recording situation is necessary.

In another work an extension of the SOBI algorithm has been used for anechoic BSS [118]. The problem has been transformed to the frequency domain and joint diagonalization of spectral matrices has been utilized to estimate the mixing coefficients as well as the

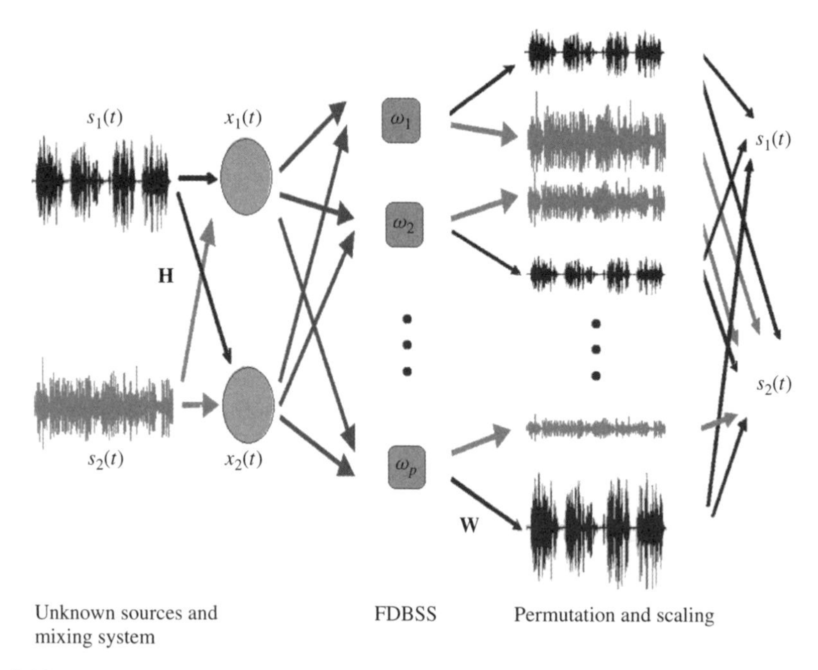

Figure 2.29 A schematic diagram of the frequency-domain BSS of convolutive mixtures for a simple two-source model

delays [119]. In attempts by Parra *et al.* [120], Ikram and Morgan [121], and Cherkani and Deville [122], second-order statistics have been used to ensure that the estimated sources, $\mathbf{Y}(\omega, m)$, are uncorrelated at each frequency bin. $\mathbf{W}(\omega)$ is estimated in such a way that it diagonalizes the covariance matrices $\mathbf{R}_{\mathbf{Y}}(\omega, k)$ simultaneously for all time blocks k, $k = 0, 1, \ldots, K-1$, i.e.

$$\begin{aligned}\mathbf{R}_{\mathbf{Y}}(\omega, k) &= \mathbf{W}(\omega)\mathbf{R}_{\mathbf{X}}(\omega, k)\mathbf{W}^{\mathrm{H}}(\omega) \\ &= \mathbf{W}(\omega)\mathbf{H}(\omega)\mathbf{\Lambda}_{\mathrm{S}}(\omega, k)\mathbf{H}^{\mathrm{H}}(\omega)\mathbf{W}^{\mathrm{H}}(\omega) \\ &= \mathbf{\Lambda}_{\mathrm{c}}(\omega, k)\end{aligned} \tag{2.212}$$

where $\mathbf{\Lambda}_{\mathbf{S}}(\omega, k)$ is the covariance matrix of the source signals, which changes with k, $\mathbf{\Lambda}_{\mathrm{c}}(\omega, k)$ is an arbitrary diagonal matrix, and $\mathbf{R}_{\mathbf{X}}(\omega, k)$ is the covariance matrix of $\mathbf{X}(\omega)$, estimated by

$$\hat{\boldsymbol{R}}_X(\omega, k) = \frac{1}{N}\sum_{n=0}^{N-1}\mathbf{X}(\omega, NK+n)\mathbf{X}^{\mathrm{H}}(\omega, NK+n) \tag{2.213}$$

where N is the number of mixtures; the unmixing filter $\mathbf{W}(\omega)$ for each frequency bin ω that simultaneously satisfies the K decorrelation equations can then be obtained using an overdetermined least-squares solution. Since the output covariance matrix $\mathbf{R}_{\mathbf{Y}}(\omega, k)$ has to be diagonalized the update equation for estimation of the unmixing matrix $\mathbf{W}$ can be found by minimizing the off-diagonal elements of $\mathbf{R_Y}(\omega, k)$, which leads to

$$\mathbf{W}_{\rho\upsilon+1}(\omega) = \mathbf{W}_{\rho}(\omega) - \mu(\omega)\frac{\partial}{\partial \mathbf{W}_{\rho}^{\mathrm{H}}(\omega)}\{\|\mathbf{V}_{\rho}(\omega, k)\|^2\} \tag{2.214}$$

where ρ is the iteration index, $\|\cdot\|^2$ is the squared Frobenius norm,

$$\mu(\omega) = \frac{\alpha}{\sum_k \|\mathbf{R_X}(\omega, k)\|^2} \tag{2.215}$$

and

$$\mathbf{V}(\omega, k) = \mathbf{W}(\omega)\mathbf{R_X}(\omega, k)\mathbf{W}^{\mathrm{H}}(\omega) - \mathrm{diag}[\mathbf{W}(\omega)\mathbf{R_X}(\omega, k)\mathbf{W}^{\mathrm{H}}(\omega)] \tag{2.216}$$

and α is a constant, which is adjusted practically.

In these methods a number of solutions for mitigating the permutation ambiguity have been suggested. Smaragdis [123] reformulated the Infomax algorithm for the complex domain and used it to solve the BSS in the frequency domain. Murata *et al.* [124] also formulated the problem of BSS in each frequency bin using a simultaneous diagonalization method similar to the SOBI method. To mitigate the permutation problem a method based on the temporal structure of signals, which exploits the nonstationarity of speech was introduced. The method exploits the correlations between the frequency bins of the spectrum of the signals.

However, for the EEG mixing model the f_s is normally low (since the bandwidth <100 Hz) and the propagation velocity is equivalent to that of electromagnetic waves (300,000 km/s). Therefore, the delay is almost zero and the mixing model can always be considered to be instantaneous. The main drawbacks for the application of BSS to separation of EEG signals is due to the:

(a) noisy environment;
(b) unknown number of sources;
(c) nonstationarity of the sources;
(d) movement of the ERP sources.

Although many attempts have been made to solve the above problems more efforts are required to provide robust solutions for different applications.

2.10.3 Sparse Component Analysis

In places where the sources are sparse, i.e. at each time instant, the number of non-zero values are less or equal to the number of mixtures. The columns of the mixing matrix may be calculated individually, which makes the solution to the underdetermined case possible. The problem can be stated as a clustering problem since the lines in the scatter plot can be separated based on their directionalities by means of clustering [125,126]. The same idea has been followed more comprehensively by Li *et al.* [127]. In their method, however, the separation has been performed in two different stages. First, the unknown mixing matrix is estimated using the k-means clustering method. Then, the source matrix is estimated using a standard linear programming algorithm. The line orientation of a dataset may be thought of as the direction of its greatest variance. One way is to perform eigenvector decomposition on the covariance matrix of the data, the resultant principal eigenvector, i.e. the eigenvector corresponding with the largest eigenvalue, indicates the direction of the data. There are many cases for which the sources are disjoint in other domains rather than the time domain. In these cases the sparse component analysis can be performed in those domains more efficiently. One such approach, called DUET [128], transforms the anechoic convolutive observations into the time–frequency domain using a short-time discrete Fourier transform, and the relative attenuation and delay values between the two observations are calculated from the ratio of corresponding time–frequency points. The regions of significant amplitudes (atoms) are then considered to be the source components in the time–frequency domain.

For instantaneous cases, in separation of sparse sources the common approach used by most researchers is to attempt to maximize the sparsity of the extracted signals in the output of the separator. The columns of the mixing matrix **A** assign each observed data point to only one source based on some measure of proximity to those columns [129]; i.e. at each instant only one source is considered active. Therefore the mixing system can be presented as

$$x_i(n) = \sum_{j=1}^{M} a_{ji}s_j(n), \quad i = 1, \ldots, N \tag{2.217}$$

where in an ideal case $a_{ji} = 0$ for $i \neq j$. Minimization of the L_1-norm is one of the most logical methods for estimation of the sources. L_1-norm minimization is a piecewise linear operation that partially assigns the energy of $\boldsymbol{x}(n)$ to the M columns of $\mathbf{A}$ that form a cone around $\boldsymbol{x}(n)$ in $\Re^M$ space. The remaining $N-M$ columns are assigned zero coefficients; therefore the L_1-norm minimization can be manifested as

$$\min ||\boldsymbol{s}(n)||_1 \text{ subject to } \mathbf{A}\boldsymbol{s}(n) = \boldsymbol{x}(n) \tag{2.218}$$

A detailed discussion of signal recovery using L_1-norm minimization is presented by Takigawa *et al.* [130], but it is worth highlighting its potential advantages for short datasets. As mentioned above, it is important to choose a domain in which the signals are more sparse. Para-factor (PARAFAC) analysis is an effective tool in detection and classification of sources in a multidimensional space. In a very recent approach it has been considered that the brain signal sources in the space–time–frequency domain are disjoint. Therefore clustering the observation points in the space–time–frequency domain can be effectively used for separation of brain sources [131]. The outcome is highly valuable for detection of μ rhythms corresponding to left and right finger movements in the context of brain–computer interfacing (BCI). The details of PARAFAC and its application to BCI can be found in Chapter 7 of this book.

2.10.4 Nonlinear BSS

Consider the cases where the parameters of the mixing system change because of changes in the mixing environment or change in the statistics of the sources. For example, if the images of both sides of a semi-transparent paper are photocopied the results will be two mixtures of the original sources. However, since the minimum observable grey level is black (or zero) and the maximum is white (say 1), the sum of the grey levels cannot go beyond these limits. This represents a nonlinear mixing system. As another example, think of the joint sounds heard from surface electrodes from over the skin. The mixing medium involves acoustic parameters of the body tissues. However, the tissues are not rigid. In such cases, if the tissues vibrate due to the sound energy then the mixing system will be a nonlinear system. The mixing and unmixing can generally be modelled respectively as

$$\boldsymbol{x}(n) = f(\mathbf{A}\boldsymbol{s}(n) + \boldsymbol{n}(n)) \tag{2.219}$$

$$\boldsymbol{y}(\mathrm{t}) = g(\mathbf{W}\boldsymbol{x}(n)) \tag{2.220}$$

where $f(.)$ and $g(.)$ represent respectively the nonlinearities in the mixing and unmixing processes. There have been some attempts to solve nonlinear BSS problems, especially for separation of image mixtures [132,133]. In one attempt [132] the mixing system has been modelled as a radial basis function (RBF) neural network. The parameters of this network are then computed iteratively. However, in these methods an assumption is often made about the mixing model. Unfortunately, none of these methods currently give satisfactory results.

2.10.5 Constrained BSS

The optimization problem underlying the solution to the BSS problem may be subject to fulfilment of a number of conditions. These may be based as *a priori* knowledge of the sources or the mixing system. Any constraint on the estimated sources or the mixing system (or unmixing system) can lead to a more accurate estimation of the sources. Statistical [104] as well as geometrical constraints [105] have very recently been used in developing new BSS algorithms. In most of the cases the constrained problem is converted to an unconstrained one by means of a regularization parameter such as a Lagrange multiplier or more generally a nonlinear penalty function, as used in Reference [104].

Incorporating nonlinear penalty functions [118] into a joint diagonalization problem not only exploits nonstationarity of the signals but also ensures fast convergence of the update equation. A general formulation for the cost function of such a system can be in the form of

$$\mathbf{J}(\mathbf{W}) = \mathbf{J}_{\mathrm{m}}(\mathbf{W}) + \kappa\varphi(\mathbf{J}_{\mathrm{c}}(\mathbf{W})) \tag{2.221}$$

where $\mathbf{J}_{\mathrm{m}}(\mathbf{W})$ and $\mathbf{J}_{\mathrm{c}}(\mathbf{W})$ are respectively the main and the constraint cost functions, $\varphi(\cdot)$ is the nonlinear penalty function, and κ is the penalty parameter.

Constrained BSS has a very high potential in incorporating clinical information into the main optimization formulation. As a new application of constrained BSS, an effective algorithm has been developed for removing the eye-blinking artefacts from EEGs. A similar method to the joint diagonalization of correlation matrices by using gradient methods [134] has been developed [88], which exploits the temporal structure of the underlying EEG sources. The algorithm is an extension of SOBI, with the aim of iteratively performing the joint diagonalization of multiple time lagged covariance matrices of the estimated sources and exploiting the statistical structure of the eye-blinking signal as a constraint. The estimated source covariance matrix is given by

$$\mathbf{R}_{\mathbf{Y}}(k) = \mathbf{W}\mathbf{R}_{\mathbf{X}}(k)\mathbf{W}^{\mathrm{T}} \tag{2.222}$$

where $\mathbf{R_X}(k) = E\{\boldsymbol{x}(n)\mathbf{x}^{\mathrm{T}}(n-k)\}$ is the covariance matrix of the electrode data. Following the same procedure as in Reference [135], the least-squares (LS) estimate of W is found from

$$J_{\mathrm{m}}(\mathbf{W}) = \arg\min_{\mathbf{W}} \sum_{k=1}^{T_{\mathrm{B}}} \|E(k)\|_{\mathrm{F}}^2 \tag{2.223}$$

where $\|\cdot\|_{\mathrm{F}}^2$ is the squared Frobenius norm and $E(k)$ is the error to be minimized between the covariances of the source signals, $\mathbf{R}_S(k)$ and the estimated sources, $\mathbf{R}_Y(k)$. The corresponding cost function has been defined, based on minimizing the off-diagonal elements for each time block, i.e.

$$J(\mathbf{W}) = J_{\mathrm{m}}(\mathbf{W}) + \Lambda J_{\mathrm{c}}(\mathbf{W}) \tag{2.224}$$

where

$$J_{\mathrm{m}}(\mathbf{W}) = \sum_{k=1}^{T_{\mathrm{B}}} \|\mathbf{R_Y}(k) - \mathrm{diag}(\mathbf{R_Y}(k)\|_{\mathrm{F}}^2 \tag{2.225}$$

and

$$J_{\mathrm{c}}(\mathbf{W}) = F(E[\boldsymbol{g}(n)\mathbf{y}^{\mathrm{T}}(n)]) \tag{2.226}$$

is a second-order constraint term. $F(.)$ is a nonlinear function approximating the cumulative density function (CDF) of the data and $\Lambda = \{\lambda_{ij}\}(i, j = 1, \ldots, N)$ is the weighted factor which is governed by the correlation (matrix) between the EOG and EEG signals ($\mathbf{R_{GY}}$), defined as $\Lambda = \kappa\ \mathrm{diag}(\mathbf{R_{GY}})$, where κ is an adjustable constant. Then a gradient approach [112] is followed to minimize the cost function. The incremental update equation is

$$\mathbf{W}(n+1) = \mathbf{W}(n) - \mu\frac{\partial J(\mathbf{W})}{\partial \mathbf{W}} \tag{2.227}$$

which concludes the algorithm.

Blind source separation has been widely used for processing EEG signals. Although the main assumptions about the source signals, such as uncorrelatedness or independency of such signals, have not yet been verified, the empirical results illustrate the effectiveness of such methods. EEG signals are noisy and nonstationary signals, which are normally affected by one or more types of internal artefacts. The most efficient approaches are those that consider all different domain statistics of the signals and take the nature of the artefacts into account. In addition, a major challenge is in how to incorporate and exploit the physiological properties of the signals and characteristics of the actual sources into the BSS algorithm. Some examples have been given here; more will be presented in the other chapters of this book.

In the case of brain signals the independency or uncorrelatedness conditions for the sources may not be satisfied. This, however, may be acceptable for abnormal sources, movement-related sources, or ERP signals. Transforming the problem into a domain such as the space–time–frequency domain, where the sources can be considered disjoint, may be a good solution.

2.11 Application of Constrained BSS: Example

In practice, the natural signals such as EEG source signals are not always independent. A topographic ICA method proposed in Reference [136] incorporates the dependency among the nearby sources in not only grouping the independent components related to nearby sources but also separating the sources originating from different regions of the brain. In this ICA model it is proposed that the residual dependency structure of the independent components (ICs), defined as dependencies that cannot be cancelled by ICA, could be used to establish a topographic order between the components. Based on this model, if the topography is defined by a lattice or grid, the dependency of the components is a function of the distance of the components on that grid. Therefore, the generative model, which implies correlation of energies for components that are close in the topographic grid, is defined. The main assumption is that the nearby sources are correlated and those far from each other are independent.

To develop such an algorithm a neighbourhood relation is initially defined as

$$h(i,j) = \begin{cases} 1, & \text{if } |i-j| \le m \\ 0, & \text{otherwise} \end{cases} \tag{2.228}$$

where the constant m specifies the width of the neighbourhood. Such a function is therefore a matrix of hyperparameters. This function can be incorporated into the main cost function of BSS. The update rule is then given as [136]

$$\boldsymbol{w}_i \propto E[z(\boldsymbol{w}_i^{\mathrm{T}} z)r_i] \tag{2.229}$$

where z_i is the whitened mixed signals and

$$r_i = \sum_{k=1}^{N} h(i,k) g\left(\sum_{j=1}^{N} h(k,j)(\boldsymbol{w}_j^{\mathrm{T}} z)^2\right) \tag{2.230}$$

The function g is the derivative of a nonlinear function such as those defined in Reference [136]. It is seen that the vectors $\boldsymbol{w}_i$ are constrained to some topographic boundary defined by $h(i,j)$. Finally, the orthogonalization and normalization can be accomplished, for example, by the classical method involving matrix square roots:

$$\mathbf{W} \leftarrow (\mathbf{W}\mathbf{W}^{\mathrm{T}})^{-1/2}\mathbf{W} \tag{2.231}$$

where $\mathbf{W}$ is the matrix of the vectors $\boldsymbol{w}_i$, i.e. $\mathbf{W} = [\boldsymbol{w}_1, \boldsymbol{w}_2, \ldots, \boldsymbol{w}_N]^{\mathrm{T}}$. The original mixing matrix $\mathbf{A}$ can be computed by inverting the whitening process as $\mathbf{A} = (\mathbf{WV})^{-1}$, where $\mathbf{V}$ is the whitening matrix.

In a yet unpublished work by Jing and Sanei this algorithm has been modified for separation of seizure signals, by (a) iteratively finding the best neighbourhood m and (b) constraining the desired estimated source to be within a specific frequency band and originating from certain brain zones (confirmed clinically). Figure 2.30 illustrates the independent components of a set of EEG signals from an epileptic patient, using the above constrained topographic ICA method. In Figure 2.31 the corresponding topographic maps

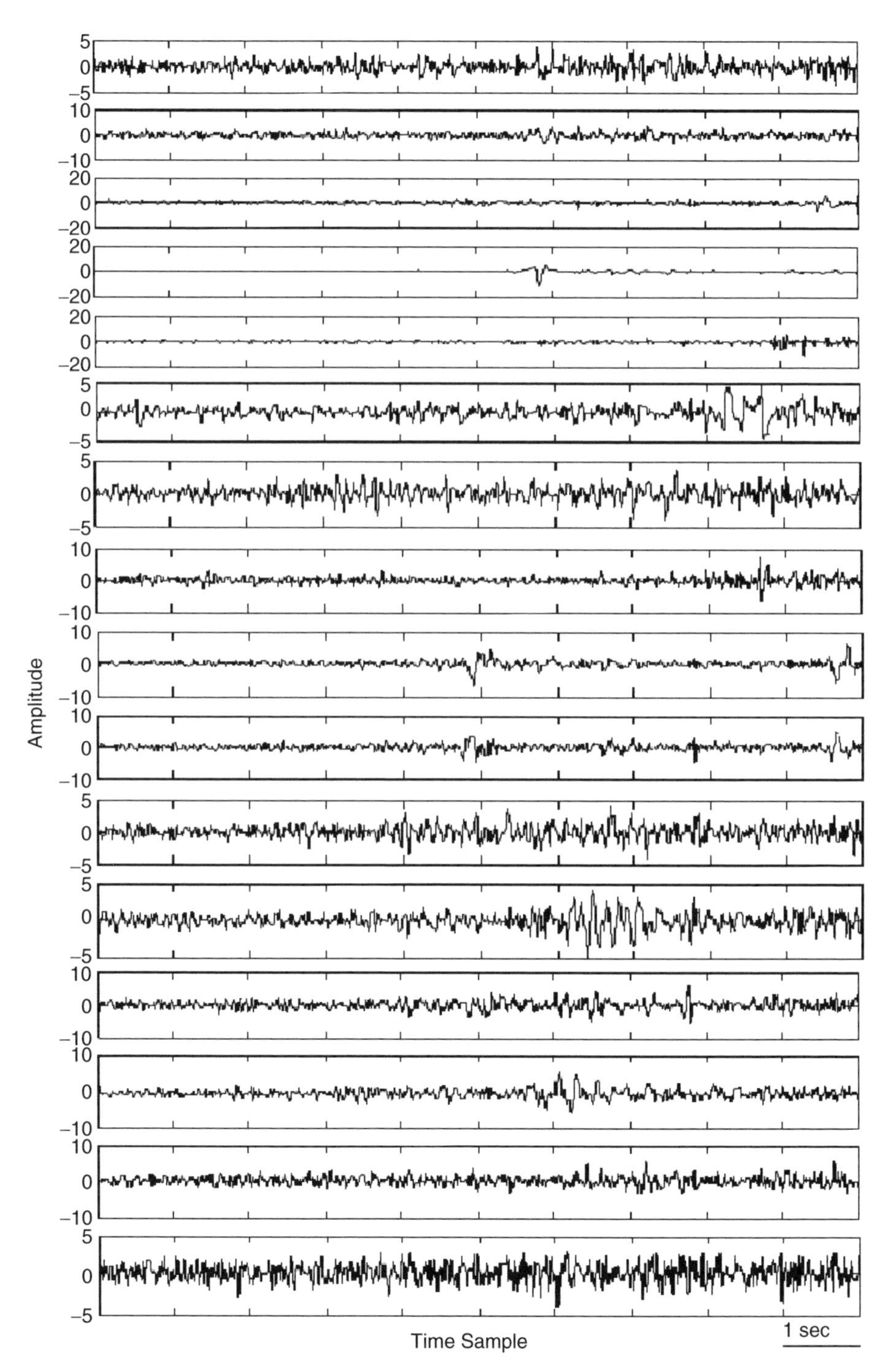

Figure 2.30 The estimated independent components of a set of EEG signals, acquired from 16 electrodes, using constrained topographic ICA. It is seen that similar ICS are gouped together

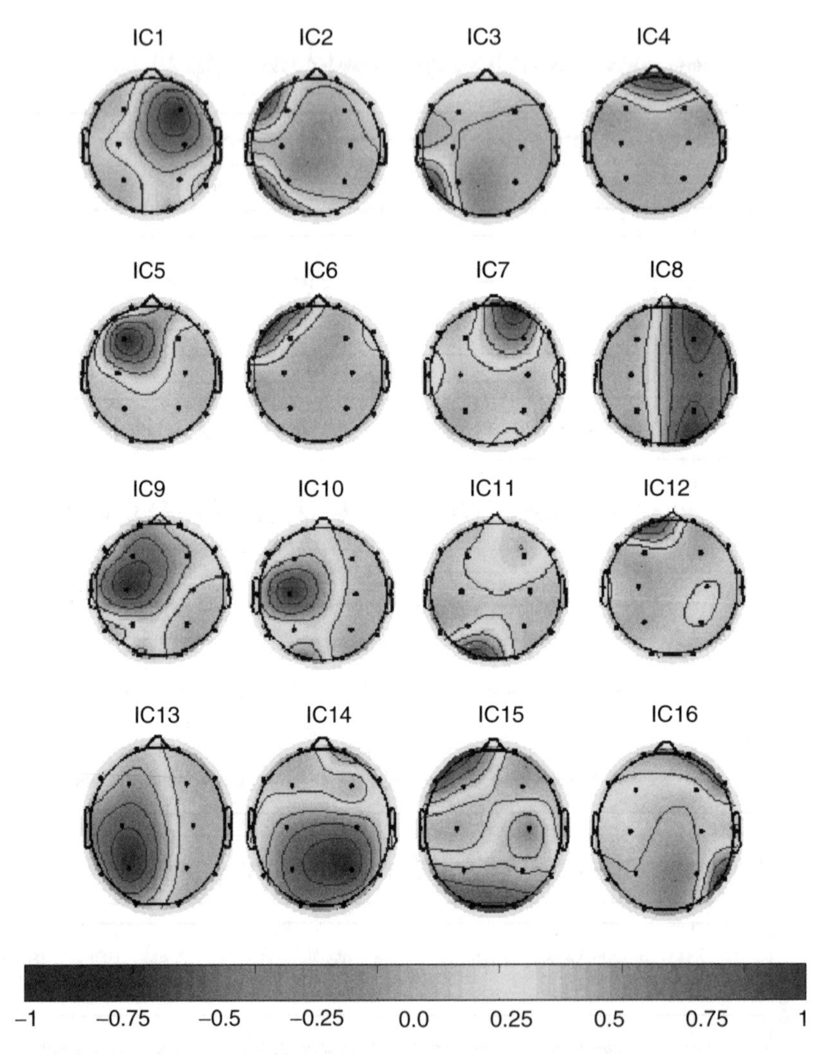

Figure 2.31 The topographic maps, each illustrating an IC. It is clear that the sources are geometrically localized

for all independent components (i.e. backprojection of each IC to the scalp using the inverse of estimated unmixing matrix) are shown. From these figures the sixth IC from the top clearly shows the seizure component. Consequently, the corresponding topograph shows the location of a seizure over the left temporal electrodes.

2.12 Signal Parameter Estimation

In many applications such as modelling, denoising, or prediction, some parameters of signal models or distributions have often to be estimated. For example, in AR modelling the prediction coefficients may be recursively computed using the least mean square

(LMS) algorithm as

$$a_p(k+1) = a_p(k) - \mu e(k)x(k) \tag{2.232}$$

where p is the prediction order, $e(.)$ is the error (residual) signal, and k is the iteration number.

The same algorithm can be used for estimation of the intersection point of a number of spheres. The spheres can be those centred at each EEG electrode and their radius proportional to the inverse of the correlations between each estimated independent component (source) and the scalp EEG signals. The intersection point is then related to the location of the source. This problem will be explained in Chapters 3 and 5.

2.13 Classification Algorithms

In the context of biomedical signal processing, especially with application to EEG signals, the classification of the data in feature spaces is often required. For example, the strength, locations, and latencies of P300 subcomponents may be classified to not only detect whether the subject has Alzheimer's disease but also to determine the stage of the disease. As another example, to detect whether there is a left or right finger movement in the BCI area the time, frequency, and spatial features need to be classified. Also, in blind source separation using the Cohen class Wigner–Ville distribution method, the status of the auto-terms and cross-terms has to be estimated in order to separate the sources. This has to be carried out by means of some clustering techniques such as k-means followed by evaluation of the clusters.

The objective of classification is to draw a boundary between two or more classes and to label them based on their measured features. In a multidimensional feature space this boundary takes the form of a separating hyperplane. The art of the work here is to find the best hyperplane that has a maximum distance from all the classes.

There have been several clustering and classification techniques developed within the last forty years. Among them artificial neural networks (ANNs), linear discriminant analysis (LDA), hidden Markov modelling (HMM), k-means clustering, fuzzy logic, and support vector machines (SVMs) have been very popular. These techniques have been developed and are well explained in the literature [137]. The explanation for all these methods is beyond the objective of this chapter. However, here a summary of an SVM is provided since it has been applied to EEG signals for the removal of the eye-blinking artefact [138], detection of epileptic seizures [139], detection of evoked potentials (EPs), classification of left and right finger movements in BCI [140], and many other issues related to EEGs [141].

Unlike many mathematical problems in which some form of explicit formula based on a number of inputs results in an output, in certain forms of classification of data there will be no model or formula of this kind. In such cases the system should be trained to be able to recognize the inputs. Many classification algorithms do not perform efficiently when:

(a) the number of features is high;
(b) there is a limited time for performing the classification;
(c) there is a nonuniform weighting among the features;

(d) there is a nonlinear map between the inputs and the outputs;
(e) the distribution of the data is not known;
(f) the convergence is not convex (monotonic), so it may fall into a local minimum.

There are two types of machine learning algorithms for classification of data: supervised learning and unsupervised learning. In the former case the target is known and the classifier is trained to minimize a difference between the actual output and the target values. A good example of such classifiers is the multilayered perceptron (MLP). In unsupervised learning, however, the classifier clusters the data into the groups having farthest distances from each other. A popular example for these classifiers is the k-means algorithm.

2.13.1 Support Vector Machines

Among all supervised classifiers, the SVM is the one that performs well in the above situations [142–147]. The concept of SVM was initiated in 1979 by Vapnik [147]. To understand the concept of the SVM consider a binary classification for the simple case of a two-dimensional feature space of linearly separable training samples (Figure 2.32) $S = \{(\boldsymbol{x}_1, y_1), (\boldsymbol{x}_2, y_2), \ldots, (\boldsymbol{x}_m, y_m)\}$ where $\boldsymbol{x} \in R^d$ is the input vector and $\boldsymbol{y} \in \{-1, 1\}$ is the class label. A discriminating function could be defined as

$$f(\boldsymbol{x}) = \mathrm{sgn}(\langle \boldsymbol{w}, \boldsymbol{x} \rangle + b) = \begin{cases} +1 & \text{if } x \text{ belongs to the first class } \bullet \\ -1 & \text{if } x \text{ belongs to the second class } \circ \end{cases} \tag{2.233}$$

In this formulation $\boldsymbol{w}$ determines the orientation of a discriminant plane (or hyperplane). Clearly, there is an infinite number of possible planes that could correctly classify the training data. One can be as shown in Figure 2.32.

An optimal classifier finds the hyperplane for which the best generalizing hyperplane is equidistant or farthest from each set of points. The set of input vectors is said to be optimally separated by the hyperplane if they are separated without error and the distance between the closest vector and the hyperplane is maximal. In that case there will be only one hyperplane to achieve optimal separation. This can be similar to the one shown in Figure 2.33.

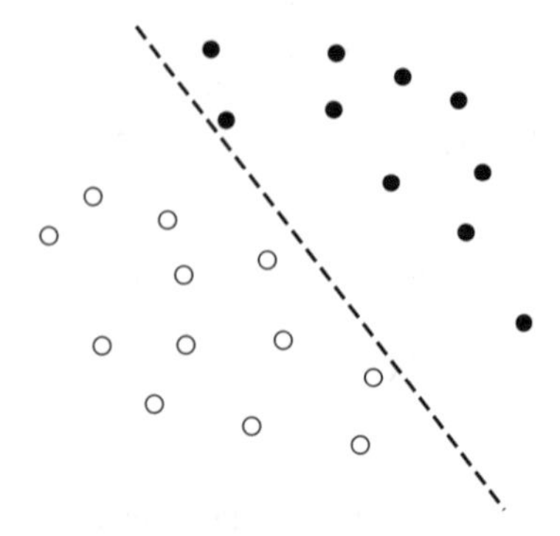

Figure 2.32 A two-dimensional separable dataset and a separating hyperplane

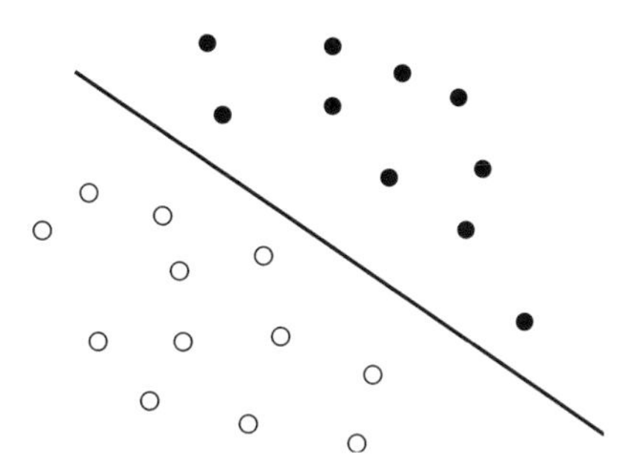

Figure 2.33 An optimal separating hyperplane

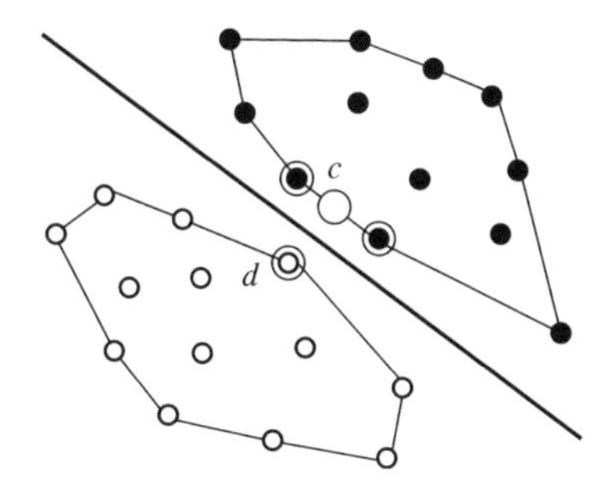

Figure 2.34 Graphical determination of the convex hulls, support vectors, and the separating hyperplane

One way to find the separating hyperplane in a separable case is by constructing the so-called *convex hulls* of each dataset as in Figure 2.34. The encompassed regions are the convex hulls for the datasets. By examining the hulls it is possible then to determine the closest two points lying on the hulls of each class (note that these do not necessarily coincide with actual data points). By constructing a plane that is perpendicular and equivalent to these two points an optimal hyperplane should result and the classifier should be robust in some sense.

Notice in Figure 2.35 that three data points have been identified with circles. These are the only data points required to determine the optimal hyperplane, and are commonly referred to as the *support vectors* (SVs). In places where the data are multi dimensional and the number of points is high the graphical solution to find the hyperplane will no longer be practical. A mathematical solution will then be necessary.

To formulate an SVM, start with the simplest case: linear machines trained on separable data (it will be seen that in the analysis for the general case, nonlinear machines trained on nonseparable data result in a very similar quadratic programming problem). Again label the training data $\{\boldsymbol{x}_i, y_i\}, i = 1, \ldots, m, y_i \in \{-1, 1\}, \boldsymbol{x}_i \in R^d$.

Suppose that a hyperplane separates the positive from the negative examples. The points $\boldsymbol{x}$ which lie on the hyperplane satisfy $\langle \boldsymbol{w},\boldsymbol{x} \rangle + b = 0$, where $\boldsymbol{w}$ is normal to the hyperplane, $|b|/\|\boldsymbol{w}\|_2$ is the perpendicular distance from the hyperplane to the origin, and $\|\boldsymbol{w}\|_2$ is the Euclidean norm of $\boldsymbol{w}$. Define the 'margin' of a separating hyperplane as in Figure 2.36. For

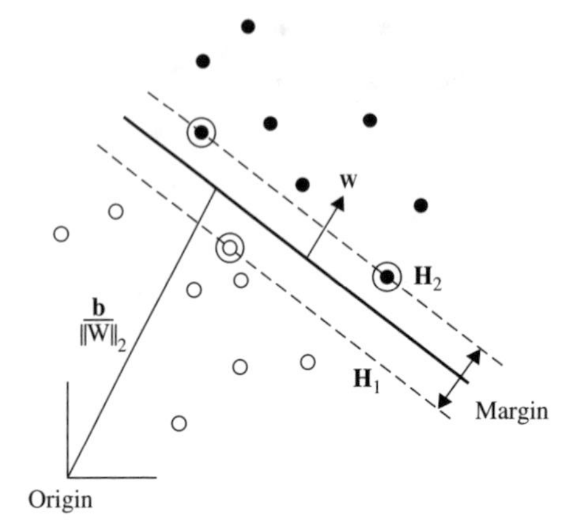

Figure 2.35 Linear separating hyperplane for the separable case, the support vectors are circled

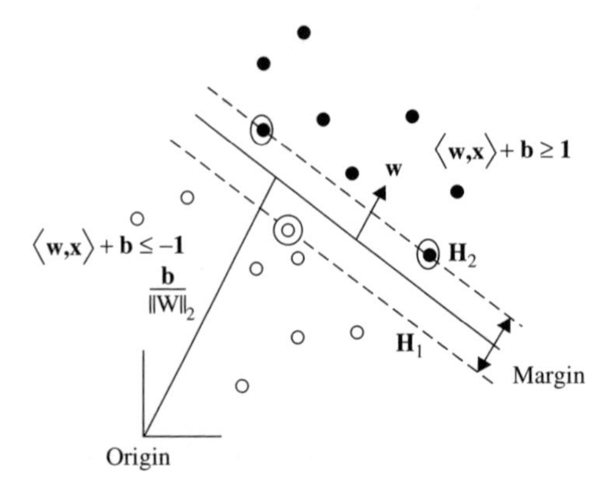

Figure 2.36 The constraints for the SVM

the linearly separable case, the support vector algorithm simply looks for the separating hyperplane with the largest margin. The approach here is to reduce the problem to a convex optimization by minimizing a quadratic function under linear inequality constraints. First it should be noted that in the definition of linear classifiers there is an inherent degree of freedom, in that the function can be scaled arbitrarily. This allows the margins to be set to be equal to unity for simplicity (hyperplanes with a functional margin of unity are sometimes referred to as canonical hyperplanes) and subsequently to minimize the norm of the weight vector. To find the plane farthest from both classes of data, the margins between the supporting canonical hyperplanes for each class are simply maximized. The support planes are pushed apart until they meet the closest data points, which are then deemed to be the support vectors (circled in Figure 2.36). Therefore, since

$$\begin{aligned} \langle \boldsymbol{x}_i, \boldsymbol{w} \rangle + b &\geq +1 \quad \text{for } y_i = +1 \\ \langle \boldsymbol{x}_i, \boldsymbol{w} \rangle + b &\leq -1 \quad \text{for } y_i = -1 \end{aligned} \tag{2.234}$$

which can be combined into one set of inequalities as $y_i(\langle \boldsymbol{x}_i, \boldsymbol{w}\rangle + b) - 1 \geq 0 \forall i$, the margin between these supporting planes (H_1 and H_2) can be shown to be $\gamma = 2/\|\boldsymbol{w}\|_2$. To maximize this margin, the following is therefore required:

$$\begin{aligned} &\text{Minimize } \langle \boldsymbol{w}, \boldsymbol{w}\rangle \\ &\text{subject to } y_i(\langle \mathbf{x}_i . \boldsymbol{w}\rangle + b) - 1 \geq 0, \quad i = 1, \ldots, m. \end{aligned} \tag{2.235}$$

To perform this constrained optimization problem the constraint can be incorporated into the main cost (risk) function by using Lagrange multipliers. This leads to minimization of an unconstrained empirical risk function (Lagrangian) which consequently results in a set of conditions called the Kuhn–Tucker (KT) conditions.

In order to perform Lagrangian optimization the so-called *primal form* must be constructed:

$$L(\boldsymbol{w}, b, \boldsymbol{\alpha}) = \frac{1}{2}\langle \boldsymbol{w}, \boldsymbol{w}\rangle - \sum_{i=1}^{m} \alpha_i [y_i(\langle \boldsymbol{x}_i, \boldsymbol{w}\rangle + b) - 1] \tag{2.236}$$

where the $\alpha_i, i = 1, \ldots, m$, are the Lagrangian multipliers. Thus, the Lagrangian primal has to be minimized with respect to $\boldsymbol{w}$, b and maximized with respect to $\alpha_i \geq 0$. Constructing the classical Lagrangian dual form facilitates this solution. This is achieved by setting the derivatives of the primal to zero and resubstituting them back into the primal. Hence,

$$\frac{\partial L(\boldsymbol{w}, b, \boldsymbol{\alpha})}{\partial \boldsymbol{w}} = \boldsymbol{w} - \sum_{i=1}^{m} y_i \alpha_i \boldsymbol{x}_i = 0 \tag{2.237}$$

Thus

$$\boldsymbol{w} = \sum_{i=1}^{m} y_i \alpha_i \boldsymbol{x}_i \tag{2.238}$$

and

$$\frac{\partial L(\boldsymbol{w}, b, \boldsymbol{\alpha})}{\partial b} = \sum_{i=1}^{m} y_i \alpha_i = 0 \tag{2.239}$$

By replacing these into the primal form the dual form is obtained as

$$L(\boldsymbol{w}, b, \boldsymbol{\alpha}) = \frac{1}{2}\sum_{j=1}^{m}\sum_{i=1}^{m} y_i y_j \alpha_i \alpha_j \langle \boldsymbol{x}_i, \boldsymbol{x}_j\rangle - \sum_{i=1}^{m} y_i y_j \alpha_i \alpha_j \langle \boldsymbol{x}_i, \boldsymbol{x}_j\rangle + \sum_{i=1}^{m} \alpha_i \tag{2.240}$$

which is reduced to

$$L(\boldsymbol{w}, b, \boldsymbol{\alpha}) = \sum_{i=1}^{m} \alpha_i - \frac{1}{2}\sum_{i=1}^{m} y_i y_j \alpha_i \alpha_j \langle \boldsymbol{x}_i, \boldsymbol{x}_j\rangle \tag{2.241}$$

considering that $\sum_{i=1}^{m} y_i \alpha_i = 0$ and $\alpha_i \geq 0$.

These equations can be solved mathematically (with the aid of a computer) using quadratic programming (QP) algorithms. There are many algorithms available within numerous publicly viewable websites [148,149].

However, in many practical situations the datasets are not separable (i.e. they have overlaps in the feature space). Therefore the maximum margin classifier described above will no longer be applicable. Obviously, it may be possible to define a complicated nonlinear hyperplane to separate the datasets perfectly but, as seen later this causes the overfitting problem which reduces the robustness of the classifier.

As can be seen in Figure 2.37, the convex hulls overlap and the datasets are no longer linearly separable. The ideal solution where no points are misclassified and no points lie within the margin is no longer feasible. This means that the constraints need to be relaxed to allow for the minimum amount of misclassification. In this case, the points that subsequently fall on the wrong side of the margin are considered to be errors. They are, however, apportioned a lower influence (according to a preset *slack* variable) on the location of the hyperplane and therefore are considered to be support vectors (see Figure 2.38). The classifier obtained in this way is called a *soft margin classifier* (see Figure 2.39).

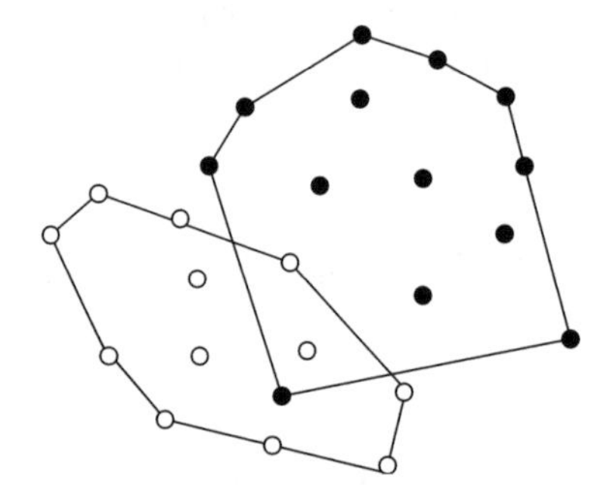

Figure 2.37 Encompassed regions for the nonseparable case

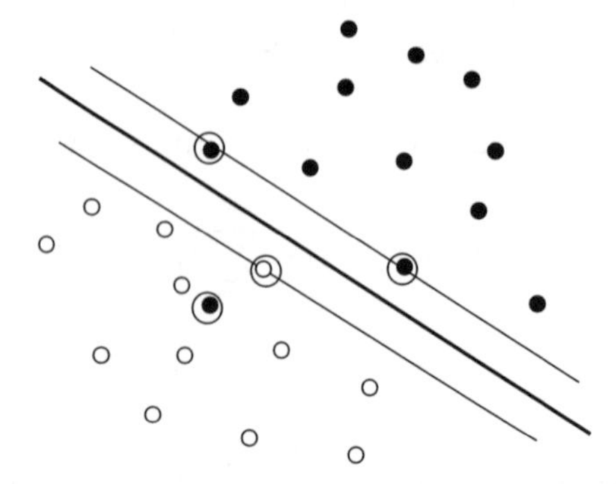

Figure 2.38 Support vectors in a nonseparable case with a linear hyperplane

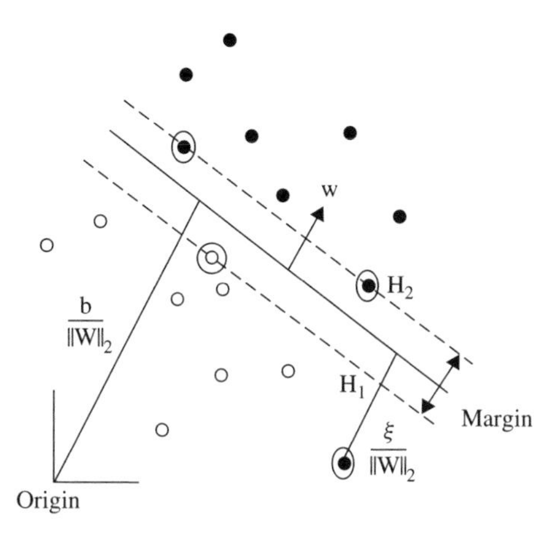

Figure 2.39 Soft margin and the concept of the slack parameter

In order to optimize the soft margin classifier, the margin constraints must be allowed to be violated according to a preset *slack* variable ξ_i in the constraints, which then become

$$\begin{aligned} &\langle \boldsymbol{x}_i, \boldsymbol{w}\rangle + b \geq +1 - \xi_i \quad \text{for } y_i = +1 \\ &\langle \boldsymbol{x}_i, \boldsymbol{w}\rangle + b \leq -1 + \xi_i \quad \text{for } y_i = -1 \\ &\text{and } \xi_i \geq 0 \qquad\qquad \forall i \end{aligned} \tag{2.242}$$

Thus for an error to occur, the corresponding ξ_i must exceed unity, so $\sum_i \xi_i$ is an upper bound on the number of training errors. Hence a natural way to assign an extra cost for errors is to change the objective function to

$$\begin{aligned} &\text{Minimize}\langle \boldsymbol{w},\boldsymbol{w}\rangle + C\sum_{i=1}^{m}\xi_i \\ &\text{subject to} y_i(\langle \boldsymbol{x}_i, \boldsymbol{w}\rangle + b) \geq 1 - \xi_i \quad \text{and } \xi_i \geq 0 \; i = 1, \ldots, m \end{aligned} \tag{2.243}$$

The primal form will then be

$$L(\boldsymbol{w}, b, \xi, \boldsymbol{\alpha}, \boldsymbol{r}) = \frac{1}{2}\langle \boldsymbol{w},\boldsymbol{w}\rangle - C\sum_{i=1}^{m}\xi_i - \sum_{i=1}^{m}\alpha_i[y_i(\langle \boldsymbol{w},\boldsymbol{x}_i\rangle + b) - 1 + \xi_i] - \sum_{i=1}^{m} r_i\xi_i \tag{2.244}$$

Hence,

$$\frac{\partial L(\boldsymbol{w}, b, \xi, \boldsymbol{\alpha}, \boldsymbol{r})}{\partial \boldsymbol{w}} = \boldsymbol{w} - \sum_{i=1}^{m} y_i\alpha_i\boldsymbol{x}_i = 0 \tag{2.245}$$

Thus again

$$w = \sum_{i=1}^{m} y_i \alpha_i x_i \tag{2.246}$$

and

$$\frac{\partial L(\boldsymbol{w}, b, \xi, \boldsymbol{\alpha}, \boldsymbol{r})}{\partial \xi} = C - \alpha_i - r_i = 0 \tag{2.247}$$

so that

$$\alpha_i + r_i = C \tag{2.248}$$

and

$$\frac{\partial L(\boldsymbol{w}, b, \boldsymbol{\alpha})}{\partial b} = \sum_{i=1}^{m} y_i \alpha_i = 0 \tag{2.249}$$

By replacing these into the primal form the dual form is obtained as

$$L(\boldsymbol{w}, b, \xi_i, \boldsymbol{\alpha}, \boldsymbol{r}) = \sum_{i=1}^{m} \alpha_i - \frac{1}{2} \sum_{i=1}^{m} y_i y_j \alpha_i \alpha_j \langle \boldsymbol{x}_i, \boldsymbol{x}_j \rangle \tag{2.250}$$

by again considering that $\sum_{i=1}^{m} y_i \alpha_i = 0$ and $\alpha_i \geq 0$. This is similar to the maximal marginal classifier. The only difference is the new constraints of $\alpha_i + r_i = C$, where $r_i \geq 0$ and hence $0 \leq \alpha_i \leq C$. This implies that the value C sets an upper limit on the Lagrangian optimization variables α_i. This is sometimes referred to as the box constraint. The value of C offers a trade-off between accuracy of data fit and regularization. A small value of C (i.e. <1) significantly limits the influence of error points (or outliers), whereas if C is chosen to be very large (or infinite) then the soft margin approach (as in Figure 2.39) becomes identical to the maximal margin classifier. Therefore in the use of the soft margin classifier, the choice of the value of C will depend heavily on the data. Appropriate selection of C is of great importance and is an area of research. One way to set C is gradually to increase C from max (α_i) for $\forall i$ and then find the value for which the error (outliers, cross-validation, or number of misclassified points) is minimum. Finally, C can be found empirically [150].

There will be no change in formulation of the SVM for the multidimensional cases. Only the dimension of the hyperplane changes depending on the number of feature types.

In many nonseparable cases use of a nonlinear function may help to make the datasets separable. As can be seen in Figure 2.40, the datasets are separable if a nonlinear hyperplane is used. *Kernel mapping* offers an alternative solution by nonlinearly projecting the data into a (usually) higher-dimensional feature space to allow the separation of such cases.

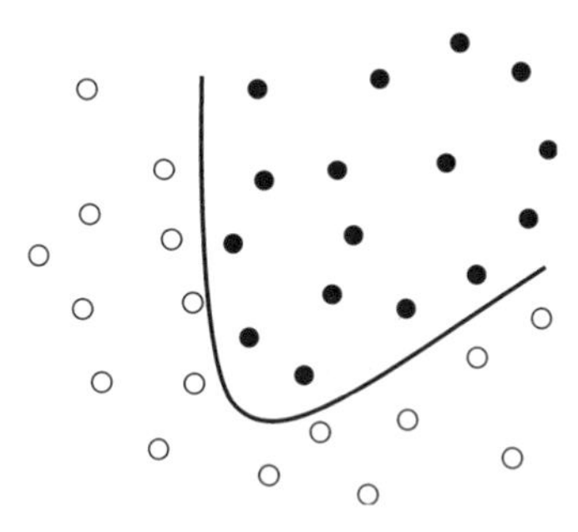

Figure 2.40 Nonlinear discriminant hyperplane

The key success of kernel mapping is that special types of mapping that obey Mercer's theorem, sometimes called reproducing kernel Hilbert spaces (RKHSs) [147], offer an implicit mapping into the feature space:

$$K(\boldsymbol{x}, z) = \langle \varphi(\boldsymbol{x}), \varphi(z) \rangle \tag{2.251}$$

This means that the explicit mapping need not be known or calculated; rather the inner product itself is sufficient to provide the mapping. This simplifies the computational burden dramatically and in combination with the inherent generality of SVMs largely mitigates the dimensionality problem. Further, this means that the input feature inner product can simply be substituted with the appropriate kernel function to obtain the mapping while having no effect on the Lagrangian optimization theory. Hence,

$$L(\boldsymbol{w}, b, \xi_i, \boldsymbol{\alpha}, \boldsymbol{r}) = \sum_{i=1}^{m} \alpha_i - \frac{1}{2} \sum_{i=1}^{m} y_i y_j \alpha_i \alpha_j K(\boldsymbol{x}_i, \boldsymbol{x}_j) \tag{2.252}$$

The relevant classifier function then becomes

$$f(\boldsymbol{x}) = \text{sgn}\left[\sum_{i=1}^{n\text{SVs}} y_i \alpha_i K(\boldsymbol{x}_i, \boldsymbol{x}_j) + b \right] \tag{2.253}$$

In this way all the benefits of the original linear SVM method are maintained. A highly nonlinear classification function, such as a polynomial or a radial basis function or even a sigmoidal neural network, can be trained using a robust and efficient algorithm that does not suffer from local minima. The use of kernel functions transforms a simple linear classifier into a powerful and general nonlinear classifier [150].

Some examples of popular RKHS functions used in SVMs are given below:

Polynomial
$$K(u, v) = (\langle u, v \rangle + c)^d \tag{2.254}$$

Gaussian radial basis function
$$K(u, v) = \exp\left(-\frac{\|u - v\|_2^2}{2\sigma^2} \right) \tag{2.255}$$

Exponential radial basis function $$K(u, v) = \exp\left(-\frac{\|u - v\|_2^2}{2\sigma^2}\right) \quad (2.256)$$

Multilayer perceptron $$K(u, v) = \tanh(\rho(\langle u, v\rangle + c)) \quad (2.257)$$

Potentially, it is possible to fit a hyperplane using an appropriate kernel to the data in order to avoid overlapping the sets (or nonseparable cases) and therefore produce a classifier with no error on the training set. This, however, is unlikely to generalize well. More specifically, the main problem with this is that the system may no longer be robust since a testing or new input can be easily misclassified.

Another issue related to the application of SVMs is the cross-validation problem. The distribution of the output of the classifier (without the hard limiter 'sign' in Equation (2.253)) for a number of inputs of the same class may be measured. The probability distributions of the results (which are centred at -1 for class '-1' and at '$+1$' for class '$+1$') are plotted in the same figure. Less overlap between the distributions represents a better performance of the classifier. The choice of the kernel influences the performance of the classifier with respect to the cross validation concept.

SVMs may be slightly modified to enable classification of multiclass data [151]. Moreover, some research has been undertaken to speed up the training step of the SVMs [152].

2.13.2 The k-Means Algorithm

The k-means algorithm [153] is an effective and generally a simple clustering tool that has been widely used for many applications such as in those given in References [126] and [154]. This algorithm divides a set of features (such as points in Figure 2.41) into k clusters.

The algorithm is initialized by setting 'k' to be the assumed number of clusters. Then the centre for each cluster k is identified by selecting k representative data points. The next

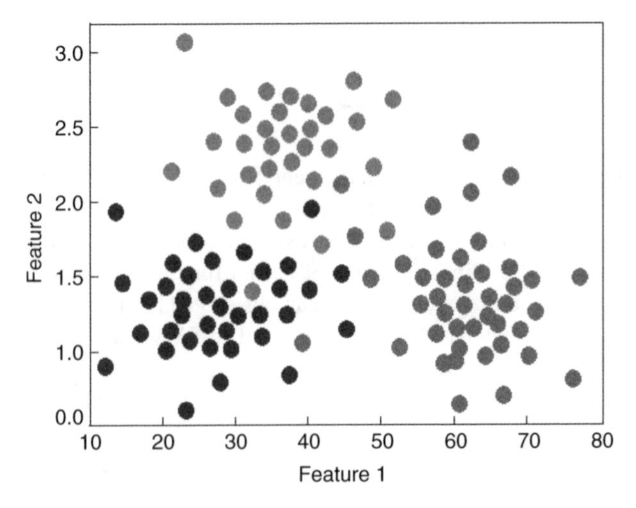

Figure 2.41 A two-dimensional feature space with three clusters, each with a different colour

step in the k-means clustering algorithm after initialization is to assign the remaining data points to the closest cluster centre. Mathematically, this means that each data point needs to be compared with every existing cluster centre and the minimum distance found. This is performed most often in the form of error checking (which will be discussed shortly). However, before this, new cluster centres are calculated. This is essentially the remaining step in k-means clustering: once clusters have been established (i.e. each data point is assigned to its closest cluster centre), the geometric centre of each cluster is recalculated.

The Euclidian distance of each data point within a cluster to its centre can be calculated. It can be repeated for all other clusters, whose resulting sums can themselves be summed together. The final sum is known as the *sum of within-cluster sum of squares*. Consider the within-cluster variation (*sum of squares* for cluster c) error as ε_c:

$$\varepsilon_c = \sum_{i=1}^{n_c} d_i^2 = \sum_{i=1}^{n_c} \|x_i^c - \overline{x}_c\|_2^2 \quad \forall c \tag{2.258}$$

where d_i^2 is the squared Euclidean distance between data point i and its designated cluster centre $\overline{x}_c$, n_c is the total number of data points (features) in cluster c, and x_i^c is an individual data point in cluster c. The cluster centre (mean of data points in cluster c) can be defined as

$$\overline{x}_c = \frac{1}{n_c}\sum_{i=1}^{n_c} x_i^c \tag{2.259}$$

and the total error is

$$E_k = \sum_{c=1}^{k} \varepsilon_c \tag{2.260}$$

The overall k-means algorithm may be summarized as:

1. Initialization
   (a) Define the number of clusters (k).
   (b) Designate a cluster centre (a vector quantity that is of the same dimensionality of the data) for each cluster, typically chosen from the available data points.
2. Assign each remaining data point to the closest cluster centre. That data point is now a member of that cluster.
3. Calculate the new cluster centre (the geometric average of all the members of a certain cluster).
4. Calculate the sum of within-cluster sum of squares. If this value has not significantly changed over a certain number of iterations, stop the iterations. Otherwise, go back to Step 2.

Therefore, an optimum clustering depends on an accurate estimation of the number of clusters. A common problem in k-means partitioning is that if the initial partitions are not chosen carefully enough the computation will run the chance of converging to a *local* minimum rather than the *global* minimum solution. The initialization step is therefore very important.

One way to combat this problem is to run the algorithm several times with different initializations. If the results converge to the same partition then it is likely that a global minimum has been reached. This, however, has the drawback of being very time consuming and computationally expensive. Another solution is to change the number of partitions (i.e. number of clusters) dynamically as the iterations progress. The ISODATA (iterative self-organizing data analysis technique algorithm) is an improvement on the original k-means algorithm that does exactly this. ISODATA introduces a number of additional parameters that allow it to progressively check within- and between-cluster similarities so that the clusters can dynamically split and merge.

Another approach for solving this problem is to use so-called gap statistics [155]. In this approach the number of clusters are iteratively estimated. The steps of this algorithm are:

1. For a varying number of clusters $k = 1, 2, \ldots, K$, compute the error measurement E_k using Equation (2.238).
2. Generate a number B of reference datasets. Cluster each one with the k-means algorithm and compute the dispersion measures, $\breve{E}_{kb}$, $b = 1, 2, \ldots, B$. The gap statistics are then estimated using

$$G_k = \frac{1}{B}\sum_{b=1}^{B}\log(\breve{E}_{kb}) - \log(E_k) \tag{2.261}$$

where the dispersion measure $\breve{E}_{kb}$ is the E_k of the reference dataset B.
3. To account for the sample error in approximating an ensemble average with B reference distributions, the standard deviation is computed as

$$S_k = \left[\frac{1}{B}\sum_{b=1}^{B}[\log(\breve{E}_{kb}) - \overline{E}_b]^2\right]^{1/2} \tag{2.262}$$

where

$$\overline{E}_b = \frac{1}{B}\sum_{b=1}^{B}\log(\breve{E}_{kb}) \tag{2.263}$$

4. By defining $\breve{S}_k = S_k\,(1 + 1/B)^{1/2}$, the number of clusters is estimated as the smallest k such that $G_k \geq G_{k+1} - \breve{S}_{k+1}$.
5. With the number of clusters identified, use the $k-$means algorithm to partition the feature space into k subsets (clusters).

The above clustering method has several advantages since it can estimate the number of clusters within the feature space. It is also a multiclass clustering system and unlike SVM can provide the boundary between the clusters.

2.14 Matching Pursuits

EEG signals are often combinations of rhythmical and transient features. These features may best be explored in the time–frequency (TF) domain. The matching pursuit (MP) algorithm [156] is often used instead of popular TF approaches such as the STFT and the WT because of its higher temporal–spatial resolution in the TF space [4], local adaptivity to transient structure, and its computational compatibility to the EEG data structure despite its computational complexity.

Here the formulation of MP is given for continuous-time signals and the dictionary. A similar presentation can be given for discrete signals simply by changing t to n. The definition of matching pursuits is straightforward; given a set of functions $D(t) = \{g_1(t), g_2(t), \ldots, g_K(t)\}$, called the dictionary of MP, where $\|g_i\| = 1$, m signals from D can be found to best approximate a signal $f(t)$. The approximation error is obtained as

$$\varepsilon = \left\| f(t) - \sum_{i=1}^{m} w_i g_{\gamma_i}(t) \right\|_2 \tag{2.264}$$

where $\|\cdot\|_2$ denotes the Euclidean norm, w_i are the weights, and $\{\gamma_i\}_{i=1,\ldots,m}$ represents the indices of the selected functions g_{γ_i}. The MP algorithm provides a suboptimal iterative solution for the above expansion. The MP algorithm performs the following steps. In the first step the waveform $g_{\gamma_0}(t)$ that best matches the signal $f(t)$ is chosen. Then, in the consecutive steps, the waveform g_{γ_i} is matched to the signal $\tilde{f}_k$, which is the residual from the previous iteration:

$$\begin{aligned} \tilde{f}_0(t) &= f(t) \\ \tilde{f}_{k+1}(t) &= \tilde{f}_k(t) - \langle \tilde{f}_k(t), g_{\gamma_k}(t)\rangle g_{\gamma_k}(t) \quad \text{for } k = 1, \ldots, m \\ g_{\gamma_n}(t) &= \arg\max_{g_{\gamma_i} \in D} |\langle \tilde{f}_n(t), g_{\gamma_i}(t)\rangle| \end{aligned} \tag{2.265}$$

where $\langle f(t), g(t)\rangle$ represents the cross correlation of $f(t)$ and $g(t)$. The orthogonality of $\tilde{f}_{k+1}(t)$ and $g_{\gamma_k}(t)$ at each step implies energy conservation, i.e.

$$\|f(t)\|^2 = \sum_{k=0}^{m-1} |\langle \tilde{f}_k(t), g_{\gamma_k}(t)\rangle|^2 + \|\tilde{f}_m(t)\|^2 \tag{2.266}$$

where $\tilde{f}_{k+1}(t)$ in the above procedure converges to $f(t)$ if the complete dictionary ($m = D$) is used. In that case

$$f(t) = \sum_{k=0}^{\infty} \langle \tilde{f}_k(t), g_{\gamma_k}(t)\rangle g_{\gamma_n}(t) \tag{2.267}$$

From this equation it is possible to derive a TF distribution of the signal's energy $F(t, \omega)$ that is free of cross-terms (i.e. the sum is 100 % correlated with the data $f(t)$) by adding

Wigner distributions of selected functions

$$F(t,\omega) = \sum_{n=0}^{m} |\langle \tilde{f}_n(t), g_{\gamma_n}(t)\rangle|^2 G_{\gamma_n}(t,\omega) \quad \forall t,\ \forall \omega \tag{2.268}$$

where $G_{\gamma_k}(t,\omega)$ is the Wigner time–frequency distribution of the kth selected function. A combination of MP and the WT has also been proposed [157].

Gabor functions (sine-modulated Gaussian functions), are often used as the dictionary functions and provide optimal joint TF localization. A real Gabor function may be expressed as [158]

$$g_\gamma(t) = K(\gamma)\mathrm{e}^{-\pi[(t-\tau)/\sigma]^2} \sin\left[\frac{\omega}{N}(t-\tau)+\theta\right] \tag{2.269}$$

where $K(\gamma)$ is the normalization factor, i.e. it makes $\|g_\gamma(t)\| = 1$, N is the length of the signals, and $\gamma = \{\tau, \omega, \sigma, \theta\}$ are the parameters of the functions (time–frequency atoms) that form the dictionary. In the original MP algorithm proposed by Mallat and Zhang [156] the parameters of the dictionary are selected from dyadic sequences of integers and their sampling interval is governed by another integer parameter (octave) j. The parameter σ, the width of the signal in the time domain, is set to $2^j, 0 \leq j \leq L$ (signal size $N = 2^L$). The time–frequency coordinates of τ and ω are sampled for each octave j with interval $\sigma = 2^j$. In the case of oversampling by l it is sampled with interval 2^{j-1}.

Analysing sleep EEG data by means of the MP algorithm has been attempted [4]. In this approach a statistical bias of the decomposition, resulting from the structure of the applied dictionary, has been considered. In the proposed stochastic dictionaries the parameters of the waveforms within the dictionary are randomized before each decomposition. The MP algorithm was modified for this purpose and tuned for maximum time–frequency resolution.

The above method was also applied to analysis of single-trial event-related potentials, in particular ERD (event-related desynchronization)/ERS related to a voluntary movement. The main idea was based upon averaging energy distributions of single EEG trials in the time–frequency plane. Consistent results, essential for the brain–computer interfacing (BCI) problem, have been reported.

Several other applications of MP for analysis of the EEG signals have been reported [50,157,159]. It is a powerful method for detection of the features localized in the time–frequency domain [160] and transient signals [159]. This includes ERP detection, detection and classification of movement-related potentials, seizure detection [50], and identification of gamma bursts.

2.15 Summary and Conclusions

In this chapter several concepts in the processing of EEG signals, including signal modelling, signal segmentation, signal transforms, multivariate modelling and direct transfer functions, chaos and dynamic analysis, independent component analysis and blind source separation, classification and clustering, and matching pursuits, have been reviewed. It is very difficult to bring all the methods and algorithms used in the processing of EEG signals into a single chapter or even a book. In reality, algorithms are developed generally based on the specific requirements of certain applications. Therefore, this chapter is not expected to cover all the aspects of digital signal processing applied to EEGs.

However, to the best knowledge of the authors, the sections included cover the important fundamental signal processing techniques required by the EEG research community. This chapter also provides certain key references for further reading in the field of signal processing for the analysis of the EEG signals. With this information the readers will be better able to digest the contents of the later chapters of this book.

References

[1] Lebedev, M. A., and Nicolelis, M. A., 'Brain–machine interfaces: past, present and future', *Trends. Neurosci.*, **29**, 2006, 536–546.

[2] Lopes da Silva, F., 'Functional localization of brain sources using EEG and/or MEG data: volume conductor and source models', *J. Magnetic Resonance Imaging*, **22**(10), 2004, 1533–1538.

[3] Malinovska, U., Durka, P. J., Blinowska, K. J., Szelenberger, W., and Wakarow, A., 'Micro- and macrostructure of sleep EEG, a universal, adaptive time–frequency parametrization', *IEEE Engng. in Medicine and Biology Mag.*, **25**(4), July/August 2006, pp. 26–31.

[4] Durka, P. J., Dobieslaw, I., and Blinowska, K. J., 'Stochastic time–frequency dictionaries for matching pursuit', *IEEE Trans. Signal Process.*, **49**(3), March 2001.

[5] Strang, G., *Linear Algebra and Its Applications*, 4th edn., Belmont, California, USA, 2006.

[6] Benedek, G., and Villars, F., *Physics, with Illustrative Examples from Medicine and Biology*, Springer-Verlag, New York, 2000.

[7] Hille, B., *Ionic Channels of Excitable Membranes*, Sinauer, Sunderland, Massachussetts, 1992.

[8] Hodgkin, A., and Huxley, A., 'A quantitative description of membrane current and its application to conduction and excitation in nerve', *J. Physiol. (Lond.)*, **117**, 1952, 500–544.

[9] Simulator for Neural Networks and Action Potentials (SNNAP) Tutorial, The University of Texas-Houston Medical School, 2003, http://snnap.uth.tmc.edu.

[10] Ziv, I., Baxter, D. A., and Byrne, J. H., 'Simulator for neural networks and action potentials: description and application', *J. Neurophysiol.*, **71**, 1994, 294–308.

[11] Gerstner, W., and Kistler, W. M., *Spiking Neuron Models*, 1st edn. Cambridge University Press, Cambridge, August 2002.

[12] Akaike, H., 'A new look at statistical model order identification', *IEEE Trans. Autom. Control*, **19**, 1974, 716–723.

[13] Kay, S. M., *Modern Spectral Estimation: Theory and Application*, Prentice-Hall, Englewood Cliffs, New Jersey, 1988.

[14] Guegen, C., and Scharf, L., 'Exact maximum likelihood identification of ARMA models: a signal processing perspective', *Signal Processing Theory Applications*, Eds M. Kunt and F. de Coulon, North-Holland, Lausanne, Switzerland, 1980, pp. 759–769.

[15] Akay, M., *Biomedical Signal Processing*, Academic Press, New York, 2001.

[16] Kay, S. M., *Modern Spectral Estimation, Theory and Application*, Prentice-Hall, Englewood Cliffs, New Jersey, 1988.

[17] Akaike, H., 'A New Look at statistical model identification', *IEEE Trans. Autom. Control*, **19**, 1974, 716–723.

[18] Durbin, J., 'Efficient estimation of parameters in moving average models', *Biometrika*, **46**, 1959, 306–316.

[19] Trench, W. F., 'An algorithm for the inversion of finite Toelpitz matrices', *J. Soc. Ind. Appl. Math.*, **12**, 1964, 515–522.

[20] Morf, M., Vieria, A., Lee, D., and Kailath, T., 'Recursive multichannel maximum entropy spectral estimation', *IEEE Trans. Geosci. Electronics*, **16**, 1978, 85–94.

[21] Spreckelesen, M., and Bromm, B., 'Estimation of single-evoked cerebral potentials by means of parametric modelling and Kalman filtering', *IEEE Trans. Biomed. Engng.*, **33**, 1988, 691–700.

[22] Demiralp, T., and Ademoglu, A., 'Modeling of evoked potentials as decaying sinusoidal oscillations by Prony's method', in Proceeding of IEEE EMBS, Paris, 1992.

[23] De Prony, B. G. R., 'Essai experimental et analytique: sur les lois de la dilatabilite de fluids elastiques et sur celles de la force expansive de la vapeur de l'eau et de la vapeur de l'alkool, a differentes temperatures', *J. E. Polytech.*, **1**(2), 1795, 24–76.

[24] Marple, S. L., *Digital Spectral Analysis with Applications*, Prentice-Hall, Englewood Cliffs, New Jersey, 1987.

[25] Lawson, C. L., and Hanson, R. J., *Solving Least Squares Problems*, Prentice-Hall, Englewood Cliffs, New Jersey, 1974.

[26] Demiralp, T., and Ademoglu, A., 'Modelling of evoked potentials as decaying sinusoidal oscillations by the Prony method', in Proceeding of IEEE EMBS, Paris, 1992.

[27] Bouattoura, D., Gaillard, P., Villon, P., and Langevin, F., 'Multilead evoked potentials modelling based on the Prony's method', Proceeding of IEEE TECON on *Digital Signal Processing Applications*, 1996, pp. 565–568.

[28] Dacorogna, M., Muller, U., Olsen, R. B., and Pictet, O., 'Modelling short-term volatility with GARCH and HARCH models', in *Nonlinear Modelling of High Frequency Financial Time Series, Econometrics*, Eds L. Christian Dunis and B. Zhou, John Wiley & Sons, Ltd, Chichester, 1998.

[29] McLeod, A. J., and Li, W. K., 'Diagnostics checking ARMA time series models using squared residual autocorrelations', *J. Time Series Analysis*, **4**, 1983, 269–273.

[30] Brock, W. A., Hsieh, D. A., and LeBaron, B., *Nonlinear Dynamics, Chaos, and Instability: Statistical Theory and Economic Evidence*, MIT press, Cambridge, Massachussetts, 1992.

[31] Hsieh, D. A., 'Testing for nonlinear dependence in daily foreign exchange rates', *J. Business*, **62**, 1989, 339–368.

[32] Engle, R. F., Lilien, D. M., and Robin, R. P., 'Estimating time-varying risk premia in the term structure: the ARCH-M model', *Econometrica*, **55**, 1987, 391–407.

[33] Nelson, D. B., 'Stationarity and persistence in the GARCH(1,1) model', *J. Econometrics*, **45**, 1990, 7–35.

[34] Glosten, L. R., Jagannathan, R., and Runkle, D., 'On the relation between the expected value and the volatility of the nominal excess return on stocks', *J. Finance*, **2**, 1995, 225–251.

[35] Zakoian, J. M., 'Threshold heteroskedastic models', *J. Economic Dynamics and Control*, **18**, 1994, 931–955.

[36] Ding, Z., Engle, R. F., and Granger, C. W. J., 'A long memory property of stock market returns and a new model', *J. Empirical Finance*, **1**, 1993, 83–106.

[37] Sentana, E., 'Quadratic ARCH models: a potential reinterpretation of ARCH models as second-order Taylor approximations', London School of Economics, 1991.

[38] Galka, A., Yamashita, O., and Ozaki, T., 'GARCH modelling of covariance in dynamical estimation of inverse solutions', *Phys. Lett. A*, **333**, 2004, 261–268.

[39] Tikhonov, A. N, editor, *Ill-Posed Problems in Natural Sciences*, The Netherlands, 1992, 155–165.

[40] Pascual-Marqui, R. D., Esslen, M., Kochi, K., and Lehmann, D., 'Functional imaging with low resolution brain electromagnetic tomography (LORETA): a review', *Methods and Findings in Expl. Clin. Pharmacol.*, **24C**, 2002, 91–95.

[41] Lagerlund, T. D., Sharbrough, F. W., and Busacker, N. E., 'Spatial filtering of multichannel electroencephalographic recordings through principal component analysis by singular value decomposition', *J. Clin. Neurophysiol.*, **14**(1), 1997, 73–82.

[42] Da Silva, F. H., Hoeks, A., Smits, H., and Zetterberg, L. H., 'Model of brain rhythmic activity: the alpha-rhythm of the thalamus', *Kybernetic*, **15**, 1974, 27–37.

[43] Lopes da Silva, F. H., van Rotterdam, A., Barts, P., van Heusden, E., Burr, W., 'Models of neuronal populations: the basic mechanisms of rhythmicity', *Prog. Brain Res.*, **45**, 281–308, 1976.

[44] Wilson, H. R., and Cowan, J. D., 'Excitatory and inhibitory interaction in localized populations of model neurons', *J. Biophys.*, **12**, 1972, 1–23.

[45] Zetterberg, L. H., 'Stochastic activity in a population of neurons–a system analysis approach', Report of the Institute of Medical Physics, TNO, Utrecht, Vol. 1, 1973, 53.

[46] Hyvarinen, A., Kahunen, J., and Oja, E., *Independent Component Analysis*, John Wiley & Sons, Ltd, Chichester, 2001.

[47] Cover, T. M., and Thomas, J. A., *Elements of Information Theory*, John Wiley & Sons, Ltd, Chichester, 2001.

[48] Bro, R., Multi-way analysis in the food industry: models, algorithms, and applications', PhD thesis, University of Amsterdam (NL) and Royal Veterinary and Agricultural University, 1998; MATLAB toolbox available on-line at http://www.models.kvl.dk/users/rasmus/.

[49] Harris, F. J., *Multirate Signal Processing for Communication Systems*, Prentice-Hall, Englewood Cliffs, New Jersey, 2004.

[50] Franaszczuk, P. J., Bergey, G. K., and Durka, P. J., 'Time–frequency analysis of mesial temporal lobe seizures using the matching pursuit algorithm', *Soc. Neurosci. Abstr.*, **22**, 1996, 184.

[51] Murenzi, R., Combes, J. M., Grossman, A., and Tchmitchian, P. (Eds), *Wavelets*, Springer-Verlag, Berlin, Heidelberg, New York, 1988.

[52] Vaidyanathan, P. P., *Multirate Systems and Filter Banks*, Prentice-Hall, Englewood Cliffs, New Jersey, 1993

[53] Holschneider, M., Kronland-Martinet, R., Morlet, J., and Tchamitchian, Ph., 'A real-time algorithm for signal analysis with the help of the wavelet transform', in *Wavelets: Time–Frequency Methods and Phase Space*, Eds J. M. Combes, A. Grossman, and Ph. Tchamitchian, Springer-Verlag, Berlin, 1989, pp. 286–297.

[54] Chui, C. K., *An Introduction to Wavelets*, Academic Press, New York, 1992.

[55] Stein, E. M., 'On the functions of Littlewood–Paley, Lusin and Marcinkiewicz', *Trans. Am. Math. Soc.*, **88**, 1958, 430–466.

[56] Vetterli, M., and J. Kovačevic, *Wavelets and Subband Coding*, Prentice-Hall, Englewood Cliffs, New Jersey, 1995.

[57] Glassman, E. L., 'A wavelet-like filter based on neuron action potentials for analysis of human scalp electroencephalographs', *IEEE Trans. Biomed. Engng.*, **52**(11), 2005, 1851–1862.

[58] Gerloff, G., Richard, J., Hadley, J., Schulman, A. E., Honda, M., and Hallett, M., 'Functional coupling and regional activation of human cortical motor areas during simple, internally paced and externally paced finger movements', *Brain*, **121**(8), 1998, 1513–1531.

[59] Sharott, A., Magill, P. J., Bolam, J. P., and Brown, P., 'Directional analysis of coherent oscillatory field potentials in cerebral cortex and basal ganglia of the rat', *J. Physiol.*, **562**(3), 2005, 951–963.

[60] C. W. J. Granger, 'Investigating causal relations in econometric models and cross-spectral methods', *Econometrica*, **37**, 1969, 424–438.

[61] Bernosconi, C., and P. König, 'On the directionality of cortical interactions studied by spectral analysis of electrophysiological recordings', *Biol. Cybern.*, **81**(3), 1999, 199–210.

[62] Kaminski, M., Ding, M., Truccolo, W., and Bressler, S., 'Evaluating causal relations in neural systems: Granger causality, directed transfer function, and statistical assessment of significance', *Biol. Cybern.*, **85**, 2001, 145–157.

[63] Kaminski, M., and Blinowska, K., 'A new method of the description of information flow in the brain structures', *Biol. Cybern.*, **65**, 1991, 203–210.

[64] Jing, H., and Takigawa, M., 'Observation of EEG coherence after repetitive transcranial magnetic stimulation', *Clin. Neurophysiol.*, **111**, 2000, 1620–1631.

[65] Kuś, R., Kaminski, M., and Blinowska, K., 'Determination of EEG activity propagation: pair-wise versus multichannel estimate', *IEEE Trans. Biomed. Engng.*, **51**(9), 2004, 1501–1510.

[66] Ginter Jr, J., Kaminski, M., Blinowska, K., and Durka, P., 'Phase and amplitude analysis in time–frequency–space; application to voluntary finger movement', *J. Neurosci. Meth.*, **110**, 2001, 113–124.

[67] Ding, M., Bressler, S. L., Yang, W., and Liang, H., 'Short-window spectral analysis of cortical event-related potentials by adaptive multivariate autoregressive modelling: data preprocessing, model validation, and variability assessment', *Biol. Cybern.*, **83**, 2000, 35–45.

[68] Peitgen, H.-O., Lurgens, H., and Saupe, D., *Chaos and Fractals*, Springer-Verlag, New York, 1992.

[69] Grassberger, P., and Procaccia, I., 'Characterization of strange attractors', *Phys. Rev. Lett.*, **50**, 1983, 346–349.

[70] Kaplan, L., and Yorke, J. A., 'Chaotic behaviour of multidimensional difference equations', in *Functional Differential Equations and Approximation of Fixed Points*, Ed. H.-O. Peitgen and H.-O. Walther, Springer-Verlag, New York, 1979.
[71] Russell, D. A., Hanson, J. D., and Ott, E., 'Dimension of strange attractors', *Phys. Rev. Lett.* **45**, 1980, 1175–1179.
[72] Wolf, A., Swift, J. B., Swinny, H. L., and Vastano, J. A., 'Determining Lyapunov exponents from a time series', *Physical*, **16D**, 1985, 285–317.
[73] Albano, A. M., Muench, J., Schwartz, C., Mees, A. I., and Rapp, P. E., 'Singular value decomposition and the Grassberger–Procaccia algorithm', *Phys. Rev. A*, **38**, 1988, 3017.
[74] King, G. P., Jones, R., and Broomhead, D. S., Phase portraits from a time series: a singular system approach', *Nucl. Phys. B*, **2**, 1987, 379.
[75] Fraser, A. M., and Swinney, H., 'Independent coordinates for strange attractors from mutual information', *Phys. Rev. A*, **33**, 1986, 1134–1139.
[76] Pincus, S. M., 'Approximate entropy as a measure of system complexity', *Proc. Natl Acad. Sci., USA*, **88**, 1991, 2297–2301.
[77] Fusheng, Y., Bo, H., and Qingyu, T., 'Approximate entropy and its application in biosignal analysis', in *Nonlinear Biomedical Signal Processing*, Vol. II, Ed. Metin Akay, IEEE Press, New York, 2001, pp. 72–91.
[78] Widrow, B., Glover, J. R., McCool Jr, J., Kaunitz, J. M., Williams, J., Hearn, C. S., and Zeidler, R. H., 'Adaptive noise cancelling principles and applications', *Proc. IEEE*, **63**(12), 1975, 1692–1716.
[79] Satorius, E. H., and Shensa, M. J., 'Recursive lattice filters: a brief overview', in Proceeding of 19th IEEE Conference on *Decision Control*', 1980, pp. 955–959.
[80] Lee, D., Morf, M., and Friedlander, B., 'Recursive square-root ladder estimation algorithms, *IEEE Trans. Accous., Speech, Signal Proces.*, **29**, 1981, 627–641.
[81] Lawsen, C. L., and Hansen, R. J., *Solving the Least-Squares Problem*, Prentice-Hall, Englewood Cliffs, New Jersey, 1974.
[82] Proakis, J. G., Rader, C. M., Ling, F., Moonen, M., Proudler, I. K., and Nikias, C. L., *Algorithms for Statistical Signal Processing*, Prentice-Hall, Englewood Cliffs, New Jersey, 2001.
[83] Cherry, C. E., 'Some experiments in the recognition of speech, with one and two ears', *J. Acoust. Soc. Am.*, **25**, 1953, 975–979.
[84] Herault, J., and Jutten, C., 'Space or time adaptive signal processing by neural models', Proceeding of the American Institute of Physics (AIP) Conference on *Neural Networks for Computing*, 1986 pp. 206–211.
[85] Comon, P., 'Independent component analysis: a new concept', *Signal Process.*, **36**, 1994, 287–314.
[86] Jung, T. P., Makeig, S., Westereld, M., Townsend, J., Courchesne, E., and Sejnowski, T. J., 'Analyzing and visualizing single-trial event-related potentials', *Advances in Neural Information Processing Systems* Vol. 8, MIT Press, Cambridge, Massachussetts, 1999.
[87] Jung, T. P., Makeig, S., Humphries, C., Lee, T. W., McKeown, M. J., Iragui, V., and Sejnowski, T. J., 'Removing electroencephalographic artefacts by blind source separation', *Psychophysiology*, **37**, 2000, 163–178.
[88] Shoker, L., Sanei, S., and Chambers, J., 'Artifact removal from electroencephalograms using a hybrid BSS-SVM algorithm', *IEEE Signal Process. Lett.*, **12**(10), October 2005.
[89] Shoker, L., Sanei, S., and Chambers, J., 'A hybrid algorithm for the removal of eye blinking artefacts from electroencephalograms', in Proceedings of the IEEE Statistical Signal Processing Workshop, France, SSP2005, February 2005.
[90] Corsini, J., Shoker, L., Sanei, S., and Alarcon, G., 'Epileptic seizure prediction from scalp EEG incorporating BSS', *IEEE Trans. Biomed. Engng.*, **53**(5), 790–799, May 2006.
[91] Latif, M. A., Sanei, S., and Chambers, J., 'Localization of abnormal EEG sources blind source separation partially constrained by the locations of known sources', *IEEE Signal Process. Lett.*, **13**(3), March 2006, 117–120.
[92] Spyrou, L., Jing, M., Sanei, S., and Sumich, A., 'Separation and localisation of P300 sources and the subcomponents using constrained blind source separation', *EURASIP J. Adv. Signal Process.*, Article ID 82912, November, 2006, 10 pages.
[93] Sanei, S., 'Texture segmentation using semi-supervised support vector machines', *Int. J. Computat. Intell. Applics*, **4**(2), 2004, 131–142.

[94] Tang, A. C., Sutherland, T., and Wang, Y., 'Contrasting single trial ERPs between experimental manipulations: improving differentiability by blind source separation', *Neuroimage*, **29**, 2006, 335–346.

[95] Makeig, S., Bell, A. J., Jung, T., and Sejnowski, T. J., 'Independent component analysis of electroencephalographic data', *Advances in Neural Information Processing Systems* Vol. 8, MIT Press, Cambridge, Massachussetts, 1996, pp. 145–151.

[96] Makeig, S., Jung, T. P., Bell, A. J., Ghahremani, D., and Sejnowski, T. J., 'Blind separation of auditory event-related brain responses into independent components', *Proc. Natl Academy of Sci., USA*, **94**, September, 1997 10979–10984.

[97] P. Földiák, and Young, M., 'Sparse coding in the primate cortex', in *The Handbook of Brain Theory and Neural Networks*, Eds Michael A. Arbib, 2nd edn., MIT Press, Cambridge, Massachussetts, 1995, pp. 895–898.

[98] Einhauser, W., Kayser, C., Konig, P., and Kording, K. P., 'Learning the invariance properties of complex cells from their responses to natural stimuli', *Eur. J. Neurosci.*, **15**, 2002, 475–486.

[99] Jutten, C., and Herault, J., 'Blind separation of sources, Part I: an adaptive algorithm based on neuromimetic architecture', *Signal Processing*, **24**, 1991, 1–10.

[100] Linsker, R., 'An application of the principle of maximum information preservation to linear systems', *Advances in Neural Information Processing Systems*, Morgan Kaufmann, Palo Alto, California, 1989, pp. 186–194.

[101] Bell, A. J., and Sejnowski, T. J., 'An information-maximization approach to blind separation, and blind deconvolution', *Neural Computation*, **7**(6), 1995, 1129–1159.

[102] Amari, S., Cichocki, A., and Yang, H. H., 'A new learning algorithm for blind signal separation', in *Advances in Neural Information Processing Systems*, Vol. 8, MIT Press, Cambridge, Massachussetts, 1996.

[103] Hyvarinen, A., and Oja, E., 'A fast-fixed point algorithm for independent component analysis', *Neural Computation*, **9**(7), 1997, 1483–1492.

[104] Parra, L., and Spence, C., 'Convolutive blind separation of non-stationary sources', *IEEE Trans. Speech Audio Process.*, **8**(3), 2000, 320–327.

[105] Cardoso, J.-F., 'Infomax and maximum likelihood for blind source separation', *IEEE Signal Process. Lett.*, **4**(4), 1997, 112–114.

[106] Cardoso, J., 'Source separation using higher order moments', in Proceedings of the International Conference on *Acoustics, Speech and Signal Processing* (*ICASSP*), Glasgow, Scotland, 1989, pp. 2109–2112.

[107] Belouchrani, A., Abed-Meraim, K., J.-F. Cardoso, and Moulines, E., 'A blind source separation technique using second order statistics', *IEEE Trans. Signal Process.*, **45**(2), 1997, 434–444.

[108] Jing, M., and Sanei, S., 'Scanner artifact removal in simultaneous EEG–fMRI for epileptic seizure prediction', in *IEEE 18th International Conference on Pattern Recognition, (ICPR)*, Vol. 3, Hong Kong, August 2006, pp. 722–725.

[109] Mathis, H., and Douglas, S. C., 'On the existence of universal nonlinearities for blind source separation', *IEEE Trans. Signal Process.*, **50**, May 2004, 1007–1016.

[110] Cheong, C., Sanei, S., and Chambers, J., 'A filtering approach to underdetermined BSS with application to temporomandibular disorders', in Proceedings of IEEE ICASSP, France, 2006.

[111] Belouchrani, A., Abed-Mariam, K., Amin, M. G., and Zoubir, A. M., 'Blind source separation of nonstationary signals', *IEEE Signal Process. Lett.*, **11**(7), 2004, 605–608.

[112] Cirillo, L., and Zoubir, A., 'On blind separation of nonstationary signals', in Proceedings of the 8th Symposium on *Signal Processing and Its Applications* (*ISSPA*), Sydney, Australia, 2005.

[113] Cohen, L., *Time–Frequency Analysis*, Prentice-Hall, Englewood Cliffs, New Jersey, 1995.

[114] Platt, C., and Fagin, F., 'Networks for the separation of sources that are superimposed and delayed', *Advances in Neural Information Processing Systems.*, Vol. 4, Morgan Kaufmann, San Mateo, 1992, pp. 730–737.

[115] Herault, J., and Jutten, C., 'Space or time adaptive signal processing by neural models', *Proc. AIP Conf., Neural Network for Computing*, American Institute of Physics, New York, 1986, pp. 206–211.

[116] K. Torkkola, 'Blind separation of delayed sources based on information maximization', in Proceeding of ICASSP, Atlanta, Georgia, 1996, pp. 3509–3512.

[117] Torkkola, K., 'Blind separation of convolved sources based on information maximization', in Proceeding of the IEEE Workshop on *Neural Networks and Signal Processing (NNSP)*, Kyoto, Japan, 1996, 315–323.

[118] Wang, W., Sanei, S., and Chambers, J. A., 'Penalty function based joint diagonalization approach for convolutive blind separation of nonstationary sources', *IEEE Trans. Signal Process.*, **53**(5), 2005, 1654–1669.

[119] Yeredor, A., 'Blind source separation with pure delay mixtures', in Proceedings of Conference on *Independent Component Analysis and Blind Source Separation*, San Diego, California, 2001.

[120] Parra, L., Spence, C., Sajda, P., Ziehe, A., and Muller, K. R., 'Unmixing hyperspectral data', in *Advances in Neural Information Process.*, **13**, MIT Press, Cambridge, Massachussetts, 2000, pp. 942–948.

[121] Ikram, M., and Morgan, D., 'Exploring permutation inconstancy in blind separation of speech signals in a reverberant environment', in proceedings of the IEEE conference on *acoustic, speech, and signal processing* (*ICASSP 2000*), Turkey, 2000.

[122] Cherkani, N., and Deville, Y., 'Self adaptive separation of convolutively mixed signals with a recursive structure, Part 1: stability analysis and optimisation of asymptotic behaviour', *Signal Process.*, **73**(3), 1999, 225–254.

[123] Smaragdis, P., 'Blind separation of convolved mixtures in the frequency domain', *Neurocomputing*, **22**, 1998, 21–34.

[124] Murata, N., Ikeda, S., and Ziehe, A., 'An approach to blind source separation based on temporal structure of speech signals', *Neurocomputing*, **41**, 2001, 1–4.

[125] Zibulevsky, M., 'Relative Newton method for quasi-ML blind source separation', http://ie.technion.ac.il/mcib, 2002.

[126] Luo, Y., Chambers, J., Lambotharan, S., and Proudler, I., 'Exploitation of source non-stationarity in underdetermined blind source separation with advanced clustering techniques', *IEEE Trans. Signal Process.*, **54**(6), 2006, 2198–2212.

[127] Li, Y., Amari, S., Cichocki, A., Ho, D. W. C., and Shengli, X., 'Underdetermined blind source separation based on sparse representation', *IEEE Trans Signal Process.*, **54**(2), 2006, 423–437.

[128] Jurjine, A., Rickard, S., Yilmaz, O., 'Blind separation of disjoint orthogonal signals: demixing *N* sources from 2 mixtures', in Proceedings of the IEEE Conference on *Acoustic, Speech, and Signal Processing* (*ICASSP 2000*), Turkey Vol. 5, 2000, pp. 2985–2988.

[129] Vielva, L., Erdogmus, D., Pantaleon, C., Santamaria, I., Pereda, J., and Principle, J. C., 'Underdetermined blind source separation in a time-varying environment', in *Proceedings of the IEEE Conference on Acoustic, Speech, and Signal Processing (ICASSP 2000), Turkey*, Vol. 3, 2002, pp. 3049–3052.

[130] Takigawa, I., Kudo, M., Nakamura, A., and Toyama, J., 'On the minimum l_1-norm signal recovery in underdetermined source separation.' in Proceedings of the 5th International Conferences on *Independent Component Analysis*, Granada, Spain, 2004, pp. 22–24.

[131] Nazarpour, K., Sanei, S., Shoker, L., and Chambers, J., 'Parallel space–time–frequency decomposition of EEG signals for brain computer interfacing', in Proceedings of EUSIPCO 2006, Florence, Italy, 2006.

[132] Jutten, C., and Karhunen, J., 'Advances in blind source separation (BSS) and independent component analysis (ICA) for nonlinear mixtures', *Int. J. Neural Systems*, **14**(5), 2004, 267–292.

[133] Almeida, L. B., 'Nonlinear Source Separation', in *Synthesis Lectures on Signal Processing*, Ed. J. Moura, Morgan & Claypool, California, USA, 2006.

[134] Joho, M., and Mathis, H., 'Joint diagonalization of correlation matrices by using gradient methods with application to blind signal processing', in Proceedings of the 1st Annual Conference *Sensor Array and Multichannel Signal Processing* (*SAM2002*), Rosslyn, VA, USA, 2002, 273–277.

[135] Parra, L., and Spence, C., 'Convolutive blind separation of non-stationary sources', *IEEE Trans. Speech and Audio Process.*, **8**, 2000, 320–327.

[136] Hyvärinen, A., Hoyer, P. O., and Inkl, M., 'Topographic independent component analysis', *Neural Computation*, **13**, 2001, 1527–1558.

[137] Vapnik, V., *Statistical Learning Theory*, John Wiley & Sons, Inc., New York, 1998.

[138] Shoker, L., Sanei, S., and Chambers, J., 'Artifact removal from electroencephalograms using a hybrid BSS-SVM algorithm', *IEEE Signal Process. Lett.*, **12**(10), October. 2005.

[139] Shoker, L., Sanei, S., Wang, W., and Chambers, J., 'Removal of eye blinking artifact from EEG incorporating a new constrained BSS algorithm', *IEE J. Med. Biolog. Engng. and Computing*, **43**, 2004, 290–295.

[140] Gonzalez, B., and Sanei, S., and Chambers, J., 'Support vector machines for seizure detection', in Proceedings of the IEEE ISSPIT2003, Germany, December 2003, pp. 126–129.

[141] Shoker, L., Sanei, S., and Sumich, A., 'Distinguishing between left and right finger movement from EEG using SVM', in Proceedings of the IEEE EMBS, Shanghai, China, September 2005, pp. 5420–5423.

[142] Bennet, K. P., and Campbell, C., 'Support vector machines: hype or hallelujah?', *SIGKDD Explorations*, **2**(2), 2000, 1–13; http://www.rpi.edu/~bennek.

[143] Christianini, N., and Shawe-Taylor, J., *An Introduction to Support Vector Machines*, Cambridge University Press, Cambridge, 2000.

[144] DeCoste, D., and Scholkopf, B., *'Training invariant support vector machines', Machine Learning*, Kluwer Press, Dordrecht, The Netherlands, 2001.

[145] Burges, C., 'A tutorial on support vector machines for pattern recognition', *Data Mining and Knowledge Discovery*, **2**, 1998, 121–167.

[146] Gunn, S., 'Support vector machines for classification and regression', Technical Reports, Department of Electronics and Computer Science, Southampton University, 1998.

[147] Vapnik, V., *The Nature of Statistical Learning Theory*, Springer, New York, 1995.

[148] http://www.support-vector.net.

[149] http://www.kernel-machines.org.

[150] Chapelle, O., and Vapnik, V., 'Choosing multiple parameters for support vector machines', *Machine Learning*, **46**, 2002, 131–159.

[151] Weston, J., and Watkins, C., 'Support vector machines for multi-class pattern recognition', in Proceedings of the Seventh European Symposium on *Artificial Neural Networks*, Bruges, Belgium, 1999.

[152] Platt, J., 'Sequential minimal optimisation: a fast algorithm for training support vector machines', Technical Report MSR-TR-98-14, Microsoft Research, 1998.

[153] Hartigan, J., and Wong, M., 'A k-mean clustering algorithm', *Appl. Statistics.*, **28**, 1979, 100–108.

[154] Fellous, J.-M., Tiesinga, P. H. E., Thomas, P. J., and Sejnowski, T. J., 'Discovering spike patterns in neural responses', *J. Neurosci.*, **24**(12), 2004, 2989–3001.

[155] Hastie, T., Tibshirani, R., and Walter, G., 'Estimating the number of clusters in a dataset via the gap statistic', Technical Report 208, Stanford University, Stanford, California, 2000.

[156] Mallat, S., and Zhang, Z., 'Matching pursuit with time–frequency dictionaries', *IEEE Trans. Signal Process.*, **41**, 1993, 3397–3415.

[157] Blinowska, K. J., and Durka, P. J., 'The application of wavelet transform and matching pursuit to the time-varying EEG signals', in *Intelligent Engineering Systems Through Artificial Neural Networks*, Eds C. H. Dagli and B. R. Fernandez Vol. 4, ASME Press, New York, 1994, pp. 535–540.

[158] Gabor, D., 'Theory of communication', *J. IEE*, **93**, 1946, 429–457.

[159] Durka, P. J., and Blinowska, K. J., 'Analysis of EEG transients by means of matching pursuit', *Epilepsia*, **37**, 1996, 386–399.

[160] Durka, P. J., Durka, P. J., Ircha, D., Neuper, C., and Pfurtscheller, G., 'Time–frequency microstructure of event-related EEG desynchronization and synchronization', *Med. & Biolog. Engng. Computing*, **39**(3), May 2001.

3

Event-Related Potentials

Event-related potentials (ERPs) were first explained in 1964 [1,2], and have remained as a useful diagnostic tool, in both psychiatry and neurology. In addition, they have been widely used in brain–computer interfacing (BCI).

ERPs are those EEGs that directly measure the electrical response of the cortex to sensory, affective, or cognitive events. They are voltage fluctuations in the EEG induced within the brain, as a sum of a large number of action potentials (APs) that are time locked to sensory, motor, or cognitive events. They are typically generated in response to peripheral or external stimulations, and appear as somatosensory, visual, and auditory brain potentials, or as slowly evolving brain activity observed before voluntary movements or during anticipation of conditional stimulation.

ERPs are quite small (1–30 μV) relative to the background EEG activity. Therefore, they often need the use of a signal-averaging procedure for their elucidation. In addition, although evaluation of the ERP peaks does not result in a reliable diagnosis the application of ERP in psychiatry has been very common and widely followed.

The ERP waveform can be quantitatively characterized across three main dimensions: amplitude, latency, and scalp distribution [3]. In addition, an ERP signal may also be analysed with respect to the relative latencies between its subcomponents. The amplitude provides an index of the extent of neural activity (and how it responds functionally to experimental variables), the latency (i.e. the time point at which peak amplitude occurs) reveals the timing of this activation, and the scalp distribution provides the pattern of the voltage gradient of a component over the scalp at any time instant.

The ERP signals are either positive, represented by the letter P, such as P300, or negative, represented by the letter N, such as N100 and N400. The digits indicate the time in terms of milliseconds after the stimuli (audio, visual, or somatosensory). The amplitude and latency of the components occurring within 100 ms after stimulus onset are labelled oxogenous, and are influenced by physical attributes of stimuli such as intensity, modality, and presentation rate. On the other hand, endogenous components such as P300 are nonobligatory responses to stimuli, and vary in amplitude, latency, and scalp distribution with strategies, expectancies, and other mental activities triggered by the event eliciting the ERP. These components are not influenced by the physical attributes of the stimuli.

An analysis of event-related potentials, such as visual event potentials (VEPs), within the EEG signals is important in the clinical diagnosis of many psychiatric diseases such

EEG Signal Processing S. Sanei and J. Chambers

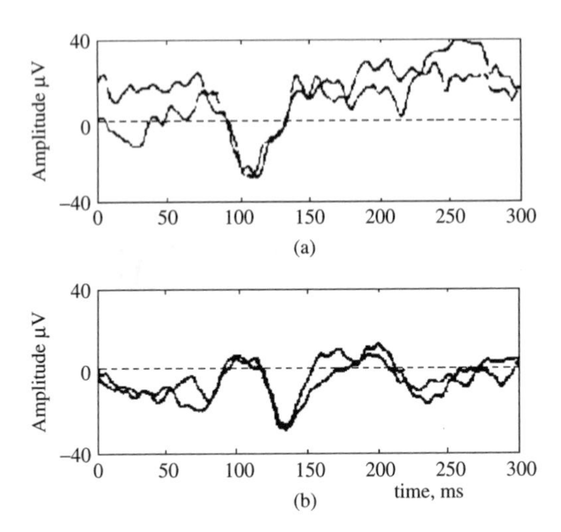

Figure 3.1 Four P100 components: (a) two normal P100 and (b) two abnormal P100 components. In (a) the P100 peak latency is at approximately 106 ms, whereas in (b) the P100 peak latency is at approximately 135 ms

as dementia. Alzheimer's disease, the most common cause of dementia, is a degenerative disease of the cerebral cortex and subcortical structures. The relative severity of the pathological changes in the associated cortex accounts for the clinical finding of diminished visual interpretation skills with normal visual acuity. Impairment of visuocognitive skills often happens with this disease. This means the patient may have difficulties with many complex visual tasks, such as tracing a target figure embedded in a more complex figure or identifying single letters that are presented briefly and followed by a pattern-making stimulus. A specific indicator of dementia is the information obtained by using a VEP. Several studies have confirmed that flash VEPs are abnormal while pattern VEPs are normal in patients with Alzheimer's disease. The most consistent abnormality in the flash VEP is the increase in the latency of the P100 component and an increase in its amplitude [4,5]. Other ERP components such as N130, P165, and N220 also have longer latencies in patients with Alzheimer's disease [6,7]. Figure 3.1 illustrates the shape of two normal and two abnormal P100 components. It is seen that the peak of the normal P100s (normally shown in reversed polarity) has a latency of approximately 106 ms, whereas the abnormal P100 peak has a latency of approximately 135 ms and has lower amplitude.

Although determination of the locations of the ERP sources within the brain is a difficult task the scalp distribution of an ERP component can often provide very useful and complementary information to that derived from amplitude and latency. Generally, two types of topographic maps can be generated: raw voltage (or surface potentials) and current source density (CSD), both derived from the electrode potentials.

The scalp-recorded ERP voltage activity reflects the summation of both cortical and subcortical neural activity within each time window. On the other hand, CSD maps reflect primary cortical surface activity [8,9]. CSD is obtained by spatially filtering the subcortical areas as well as cortical areas distal to the recording electrodes to remove the volume-conducted activity. CSD maps are useful for forming hypotheses about neural sources within the superficial cortex [10].

The fMRI technique has become another alternative to investigate brain ERPs since it can detect the hemodynamic of the brain. However, there are at least three shortcoming with this brain imaging modality: first, the temporal resolution is low; second, the activated areas based on hemodynamic techniques do not necessarily correspond to the neural activity identified by ERP measures; and, third, the fMRI is not sensitive to the type of stimulus (e.g. target, standard, or novel). It is considered that the state of the subject changes due to differences in the density of different stimulus types across blocks of trials. Target P300 refers to the P300 component elicited by events about which the subject has been instructed and to which the subject is required to generate some kind of response. A novel stimulus indicates a sole or irregular stimulus.

The ERP parameters such as amplitude and latency are the indicators of the function of the brain neurochemical systems and can potentially be used as predictors of the response of an individual to psychopharmacotrapy [11]. ERPs are also related to the circumscribed cognitive process. For example, there are interesting correlations between late-evoked positivities and memory, N400 and semantic processes, or the latencies of ERPs and the timing of cognitive processes. Therefore, the ERP parameters can be used as indicators of cognitive processes and dysfunctions not accessible to behavioural testing.

The fine-grained temporal resolution of ERPs has been traditionally limited. In addition, overlapping components within ERPs, which represent specific stages of information processing, are difficult to distinguish [12,13]. An example is the composite P300 wave, a positive ERP component, which occurs with a latency of about 300 ms after novel stimuli, or task-relevant stimuli, which requires an effortful response on the part of the individual under test [12–16].

The elicited ERPs are comprised of two main components: the mismatch negativity (MMN) and the novelty P300. Novelty P300 refers to the P300 component elicited by events about which the subject has not been instructed prior to the experiment. The MMN is the earliest ERP activity (which occurs within the first ten milliseconds after the stimulus) that indicates that the brain has detected a change in a background of brain homogeneous events. The MMN is thought to be generated in and around the primary auditory cortex [17]. The amplitude of the MMN is directly proportional, and its latency inversely related, to the degree of difference between standard and deviant stimuli. It is most clearly seen by subtraction of the ERP elicited by the standard stimulus from that elicited by the deviant stimulus during a passive oddball paradigm (OP), when both of those stimuli are unattended or ignored. Therefore, it is relatively automatic.

The P300 wave represents cognitive functions involved in orientation of attention, contextual updating, response modulation, and response resolution [12,14], and consists mainly of two overlapping subcomponents P3a and P3b [13,16,18]. P3a reflects an automatic orientation of attention to novel or salient stimuli independent of task relevance. Profrontal, frontal, and anterior temporal brain regions play the main role in generating P3a, giving it a frontocentral distribution [16]. In contrast, P3b has a greater centroparietal distribution due to its reliance on posterior temporal, parietal, and posterior cingulate cortex mechanisms [12,13]. P3a is also characterized by a shorter latency and more rapid habituation than P3b [18].

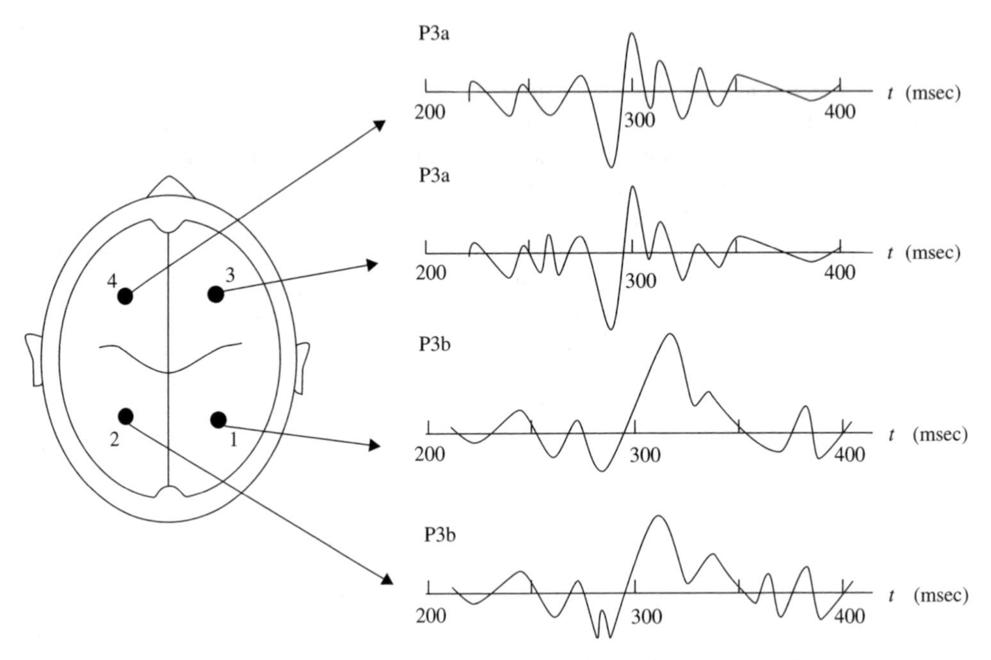

Figure 3.2 Typical P3a and P3b subcomponents of a P300 ERP signal and the measurement locations

A neural event is the frontal aspect of the novelty P300, i.e. P3a. For example, if the event is sufficiently deviant, the MMN is followed by the P3a. The eliciting events in this case are highly deviant environmental sounds such as a dog barking. Nonidentifiable sounds elicit larger P3a than identifiable sounds. Accordingly, a bigger MMN results in a dominant P3a. Figure 3.2 shows typical P3a and P3b subcomponents. P3b is also elicited by infrequent events, but, unlike P3a, it is task relevant or involves a decision to evoke this component.

The ERPs are the responses to different stimuli, i.e. novel or salient. It is important to distinguish between the ERPs when they are the response to novel or salient (i.e. what has already been experienced) stimuli, or when the degree of novelty changes.

The orienting response engendered by deviant or unexpected events consists of a characteristic ERP pattern, which is comprised sequentially of the mismatch negativity (MMN) and the novelty P300 or P3a. The novelty P3a has two spatial signatures, one frontal the other posterior, each with different cognitive or possibly neurological correlates.

The orienting response [19,20] is an involuntary shift of attention that appears to be a fundamental biological mechanism for survival. It is a rapid response to a new, unexpected, or unpredictable stimulus, which essentially functions as a what-is-it detector [21]. The plasticity of the orienting response has been demonstrated by showing that stimuli, which initially evoked the response, no longer did so with repeated presentation [20]. Habituation of the response is proposed to indicate that some type of memory for these prior events has been formed, which modifies the response to the repeated incidences.

Abnormalities in P300 are found in several psychiatric and neurological conditions [15]. However, the impact of the diseases on P3a and P3b may be different. Both audio and visual P300 (i.e. a P300 signal produced earlier due to an audio or a visual stimulation) are used. Audio and visual P300 appear to be differently affected by illnesses and respond differently to their treatment. This suggests differences in the underlying structures and neurotransmitter systems [13]. P300 has significant diagnostic and prognostic potential, especially when combined with other clinical symptoms and evidences.

In many applications such as human–computer interaction (HCI), muscular fatigue, visual fatigue, and mental fatigue are induced as a result of physical and mental activities. In order for the ERP signals and their subcomponents to be reliably used for clinical diagnosis, assessment of mental activities, fatigue during physical and mental activities, and for provision of the human–computer interface, very effective and reliable methods for their detection and parameter evaluation have to be developed. In the following section a number of established methods for the detection of ERP signals, especially P300 and its subcomponents, P3a and P3b, are explained.

3.1 Detection, Separation, Localization, and Classification of P300 Signals

Traditionally, EPs are synchronously averaged to enhance the evoked signal and suppress the background brain activity [22]. Step-wise discriminant analysis (SWDA) followed by peak picking and evaluation of the covariance was first introduced by Farwell and Dounchin [23]. Later, the discrete wavelet transform (DWT) was also added to the SWDA to localize efficiently the ERP components in both time and frequency [24].

Principal component analysis (PCA) has been employed to assess temporally overlapping EP components [25]. However, the resultant orthogonal representation does not necessarily coincide with the true component structure since the actual physiological components need not be orthogonal, i.e. the source signals may be correlated.

ICA was first applied to ERP analysis by Makeig *et al.* [26]. Infomax ICA [27] was used by Xu *et al.* [28] to detect the ERPs for the P300-based speller. In their approach, those ICs with relatively larger amplitudes in the latency range of P300 were kept, while the others were set to zero. Also, they exploited *a priori* knowledge about the spatial information of the ERPs and decided whether a component should be retained or wiped out. To manipulate the spatial information the ith row and jth column element in the inverse of the unmixing matrix, $\mathbf{W}^{-1}$, is denoted by w'_{ij}; therefore, $\boldsymbol{x}(n) = \mathbf{W}^{-1}\boldsymbol{u}(n)$. Then the jth column vector of W is denoted by $\boldsymbol{w}'_j$, which reflects the intensity distribution at each electrode for the jth IC $u_j(n)$ [26]. For convenience the spatial pattern $\mathbf{W}^{-1}$ is then transformed into an intensity order matrix $\mathbf{M} = \{m_{ij}\}$ with the same dimension. The value of the element m_{ij} in $\mathbf{M}$ is set to be the order number of the value w'_{ij} in the column vector $\boldsymbol{w}'_j$. For example, $m_{ij} = 1$ if w'_{ij} has the largest value, $m_{ij} = 2$ if it has the second largest value, and so on. Based on the spatial distribution of the brain activities, an electrode set $\mathbf{Q} = \{q_k\}$ of interest is selected in which q_k is the electrode number and is equal to the row index of the multichannel EEG signals $\boldsymbol{x}(n)$. For extraction of the P300 signals these electrodes are located around the vertex region (C_z, C_1, and C_2), since they are considered to have prominent P300. The spatial filtering of the ICs is simply performed as

$$\tilde{\boldsymbol{u}}_j(n) = \begin{cases} \boldsymbol{u}_j(n) & \text{if } \exists q_k \in \mathbf{Q} \text{ and } m_{q_k j} \leq T_r \\ 0 & \text{else} \end{cases} \tag{3.1}$$

where T_r is the threshold for the order numbers. T_r is introduced to retain the most prominent spatial information about the P300 signal. Therefore, $\tilde{\boldsymbol{u}}_j(n)$ holds most of the source information about P300; other irrelevant parts are set to zero. Finally, after the temporal and spatial manipulation of the ICs, the $\tilde{\boldsymbol{u}}_j(n)$, $j = 1, 2, \ldots, M$, where M is the number of ICs, are backprojected to the electrodes by using $\mathbf{W}^{-1}$ to obtain the scalp distribution of the P300 potential, i.e.

$$\tilde{\boldsymbol{x}}(n) = \mathbf{W}^{-1}\tilde{\boldsymbol{u}}(n) \tag{3.2}$$

where $\tilde{\boldsymbol{x}}(n)$ is the P300 enhanced EEG. The features of $\tilde{\boldsymbol{x}}(t)$ can then be measured for classification purposes [28].

3.1.1 Using ICA

ICA has also been used for the detection of P300 signals by Serby *et al.* [29]. Their work involved the application of a matched filter together with averaging and using a threshold technique for detecting the existence of the P300 signals. The block diagram in Figure 3.3 shows the method.

The IC corresponding to the P300 source is selected and segmented to form overlapping segments from 100 to 600 ms. Each segment is passed through a matched filter to give one feature that represents the maximum correlation between the segment and the average P300 template. However, the very obvious problem with this method is that the ICA system is very likely to be underdetermined (i.e. the number of sources is more than the number of sensors or observations) since only three mixtures are used. In this case the independent source signals are not separable.

3.1.2 Estimating Single Brain Potential Components by Modelling ERP Waveforms

The detection or estimation of evoked potentials (EPs) from only a single-trial EEG is very favourable since on-line processing of the signals can be performed. Unlike the averaging (multiple-trial) [30] scheme, in this approach the shape of the ERPs is first approximated and then used to recover the actual signals.

A decomposition technique that relies on the statistical nature of neural activity is one that efficiently separates the EEGs into their constituent components, including the ERPs. A neural activity may be delayed when passing through a number of synaptic nodes, each introducing a delay. Thus, the firing instants of many synchronized neurons may be assumed to be governed by Gaussian probability distributions. In a work by Lange *et al.*

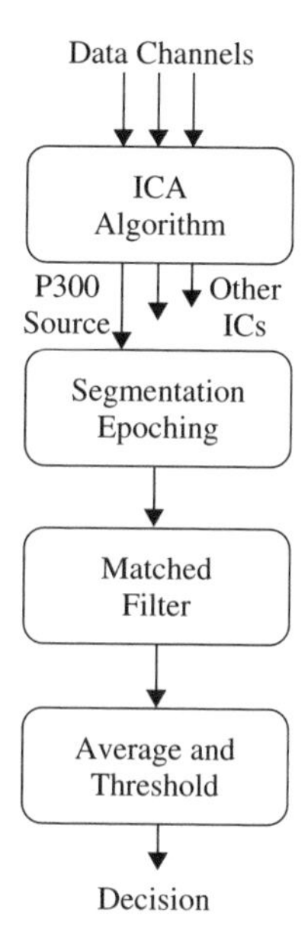

Figure 3.3 Block diagram of the ICA-based algorithm [29]. Three recorded data channels C_z, P_z, and F_z are the inputs to the ICA algorithm

[22] the evoked potential source waveform is assumed to consist of the superposition of p components u_i delayed by τ_i:

$$s(n) = \sum_{i=1}^{p} k_i u_i(n - \tau_i) \tag{3.3}$$

The model equation for constructing the measured data from a number of delayed templates can be given in the z-domain as

$$\mathbf{X}(z) = \sum_{i=1}^{p} \mathbf{B}_i(z)\mathbf{T}_i(z) + \mathbf{W}^{-1}(z)\mathbf{E}(z) \tag{3.4}$$

where $\mathbf{X}(z)$, $\mathbf{T}(z)$, and $\mathbf{E}(z)$ represent the observed process, the average EP, and a Gaussian white noise respectively. Assuming that the background EEG is statistically stationary, $\mathbf{W}(z)$ is identified from prestimulus data via AR modelling of the prestimulus interval, and used for poststimulus analysis [22]. The template and measured EEG are filtered through the identified $\mathbf{W}(z)$ to whiten the background EEG signal and thus a closed-form LS solution of the model is formulated. Therefore, the only parameters to be identify are the matrices $\mathbf{B}_i(z)$. This avoids any recursive identification of the model parameters. This can be represented using a regression-type equation as

$$\breve{x}(n) = \sum_{i=1}^{p} \sum_{j=-d}^{d} b_{i,j} \breve{T}_i(n - j) + e(n) \tag{3.5}$$

where d denotes the delay (latency), p is the number of templates, and $\breve{x}(n)$ and $\breve{T}(n)$ are whitened versions of $x(n)$ and $T(n)$ respectively. Let $\mathbf{A}^{\mathrm{T}}$ be the matrix of input templates and $\boldsymbol{b}^{\mathrm{T}}$ be the filter coefficient vector. Then

$$\breve{\boldsymbol{x}} = [\breve{x}(\mathrm{d}+1), \breve{x}(\mathrm{d}+2), \ldots, \breve{x}(N-d)]^{\mathrm{T}} \tag{3.6}$$

$$\mathbf{A} = \begin{bmatrix} \breve{T}_1(2d+1) & \breve{T}_1(2d+2) & \cdots & \breve{T}_1(N) \\ \breve{T}_1(2d) & \breve{T}_1(2d+1) & \cdots & \breve{T}_1(N-1) \\ \vdots & \vdots & & \vdots \\ \breve{T}_1(1) & \breve{T}_1(2) & \cdots & \breve{T}_1(N-2d) \\ \breve{T}_2(2d+1) & \breve{T}_2(2d+2) & & \breve{T}_2(N) \\ \vdots & \vdots & & \vdots \\ \breve{T}_2(1) & \breve{T}_2(2) & \cdots & \breve{T}_2(N-2d) \\ \vdots & & & \\ \breve{T}_p(2d+1) & \breve{T}_p(2d+2) & & \cdots \breve{T}_2(N) \\ \vdots & \vdots & & \vdots \\ \breve{T}_p(1) & \breve{T}_p(2) & \cdots & \breve{T}_p(N-2d) \end{bmatrix}^T \tag{3.7}$$

and

$$\boldsymbol{b} = [b_{1,-d}, b_{1,-d+1}, \ldots, b_{1,d}, b_{2,-d}, \ldots, b_{p,d}]^{\mathrm{T}} \tag{3.8}$$

The model can also be expressed in vector form as

$$\breve{\boldsymbol{x}} = \mathbf{A}\boldsymbol{b} + \boldsymbol{\varepsilon} \tag{3.9}$$

where $\boldsymbol{\varepsilon}$ is the vector of prediction errors:

$$\boldsymbol{\varepsilon} = [e(2d+1), e(2d+2), \ldots, e(N)]^{\mathrm{T}} \tag{3.10}$$

To solve for the model parameters the sum-squared error, defined as

$$\xi(\boldsymbol{b}) = \|\boldsymbol{\varepsilon}\|^2 \tag{3.11}$$

is minimized. Then, the optimal vector of parameters in the LS sense is obtained as

$$\hat{\boldsymbol{b}} = (\mathbf{A}^{\mathrm{T}}\mathbf{A})^{-1}\mathbf{A}^{\mathrm{T}}\breve{\boldsymbol{x}} \tag{3.12}$$

The model provides an easier way of analysing the ERP components from single-trial EEG signals and facilitates tracking of amplitude and latency shifts of such components. The experimental results show that the main ERP components can be extracted with high accuracy. The template signals may look like those in Figure 3.4.

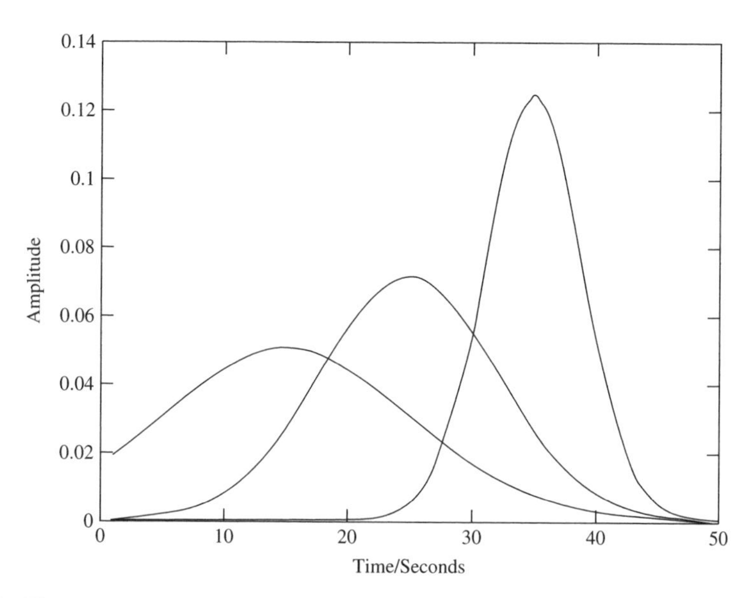

Figure 3.4 The ERP templates including a number of delayed Gaussian and exponential waveforms

A similar method was also developed to detect and track the P300 subcomponents, P3a and P3b, of the ERP signals [31]. This is described in the following section.

3.1.3 Source Tracking

The major drawback of BSS and dipole methods is that the number of sources needs to be known *a priori* to achieve good results. In a recent work [31] the P300 signals are modelled as spike-shaped Gaussian signals. The latencies and variances of these signals are, however, subject to change. The spikes serve as reference signals on to which the EEG data are projected. The existing spatiotemporal information is then used to find the closest representation of the reference in the data. The locations of all the extracted components within the brain are later computed using the least-squares (LS) method. Consider the $q \times T$ EEG signals $\mathbf{X}$ where n_e is the number of channels (electrodes) and T the number of time samples; then

$$\boldsymbol{x}(n) = \mathbf{H}\boldsymbol{s}(n) = \sum_{i=1}^{m} \boldsymbol{h}_i s_i(n) \tag{3.13}$$

where $\mathbf{H}$ is the $n_e \times m$ forward mixing matrix and $s_i(n)$ are the source signals. Initially, the sources are all considered as the ERP components that are directly relevant and time-locked to the stimulus and assumed to have a transient spiky shape. Then m filters $\{\boldsymbol{w}_i\}$ are needed (although the number of sources are not known beforehand) to satisfy

$$s_i(n) = \boldsymbol{w}_i^{\mathrm{T}} \boldsymbol{x}(n) \tag{3.14}$$

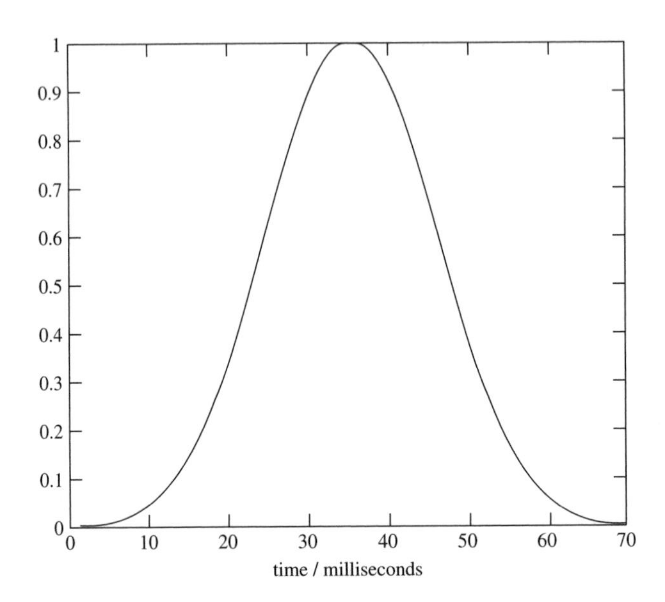

Figure 3.5 A spike model of the P300 sources. The amplitude and latency of this spike are subject to change

This can be achieved by the following minimization criterion:

$$\boldsymbol{w}_{\text{opt}} = \arg\min_{\boldsymbol{w}} \|\boldsymbol{s}_i - \boldsymbol{w}_i^{\text{T}}\mathbf{X}\|_2^2 \tag{3.15}$$

where $\mathbf{X} = [\boldsymbol{x}(1), \boldsymbol{x}(2), \ldots, \boldsymbol{x}(T)]$, which requires some knowledge about the shape of the sources. A Gaussian-type spike (such as that in Figure 3.5) defined as

$$s_i(n) = \mathrm{e}^{-(n-\tau_i)^2/\sigma_i^2} \tag{3.16}$$

where τ_i, $i = 1, 2, \ldots, T$, is the latency of the ith source and σ_i is its width, can be used as the model. The width is chosen as an approximation to the average width of the P3a and P3b subcomponents.

Then for each of the T filters ($T >> m$)

$$\boldsymbol{w}_{l\,\text{Opt}}^{\text{T}} = \arg\min_{\boldsymbol{w}_l} \|\boldsymbol{r}_l - \boldsymbol{w}_l^{\text{T}}\mathbf{X}\|_2^2, \qquad \boldsymbol{y}_l = \boldsymbol{w}_l^{\text{T}}\mathbf{X} \tag{3.17}$$

where $(.)^{\text{T}}$ denotes the transpose operation. Therefore, each $\boldsymbol{y}_l$ has a latency similar to that of a source. Figure 3.6 shows the results of the ERP detection algorithm. It can be seen that all possible peaks are detected first, but because of the limited number of ERP signals these outputs may be grouped into m clusters. To do that the following criteria for the latencies of the ith and $i-1$th estimated ERP components are examined according to the following rule:

for $l = 1$ to T,

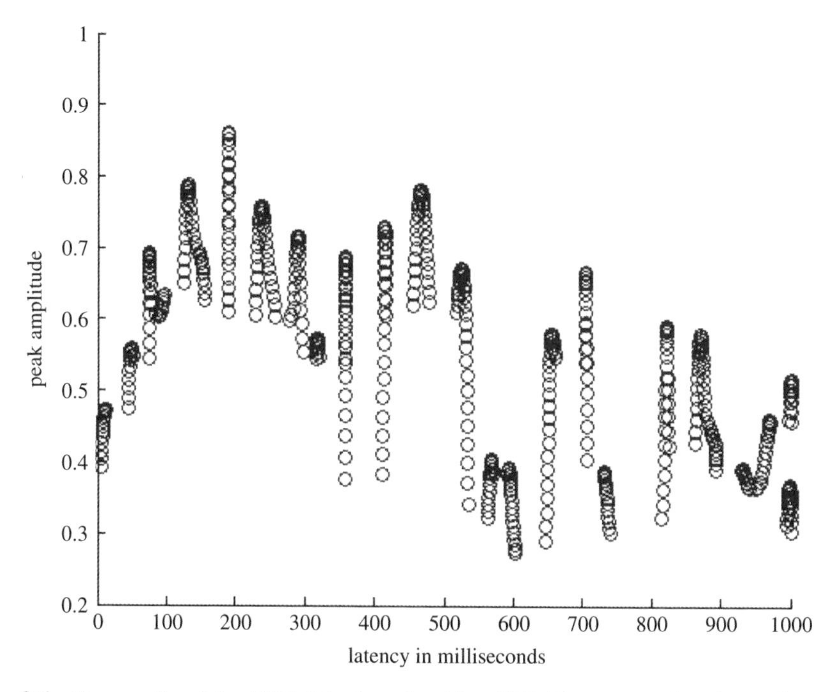

Figure 3.6 The results of the ERP detection algorithm [31]. The scatter plot shows the peaks at different latencies

if $l(i) - l(i-1) < \beta$, where β is a small empirical threshold, then $\boldsymbol{y}_i$ and $\boldsymbol{y}_{i-1}$ are assumed to belong to the same cluster;
if $l(i) - l(i-1) > \beta$ then $\boldsymbol{y}_i$ and $\boldsymbol{y}_{i-1}$ belong to different clusters.

Here $l(i)$ denotes the latency of the ith component. The signals within each cluster are then averaged to obtain the related ERP signal, $\boldsymbol{y}_c, c = 1, 2, \ldots, m$. Figure 3.7 presents the results after applying the above method to detect the subcomponents of a P300 signal from a single trial EEG for a healthy individual (control) and a schizophrenic patient. From these results it can be seen that the mean latency of P3a for the patient is less than that of the control. The difference, however, is less for P3b. On the other hand, the mean width of P3a is less for the control than for the patient, whereas the mean width of the P3b is more for the patient than that of the control. This demonstrates the effectiveness of the technique in classification of the healthy and schizophrenic individuals.

3.1.4 Localization of the ERP

In addition to the above tracking method, in order to investigate the location of the ERP sources a method based on least squares (LS) can be followed [31]. Using this method, the scalp maps (the column vectors of the forward matrix **H**) are estimated. Consider

$$\mathbf{R} = \mathbf{X}\mathbf{Y}^{\mathrm{T}} = \mathbf{H}\mathbf{S}\mathbf{Y}^{\mathrm{T}} \tag{3.18}$$

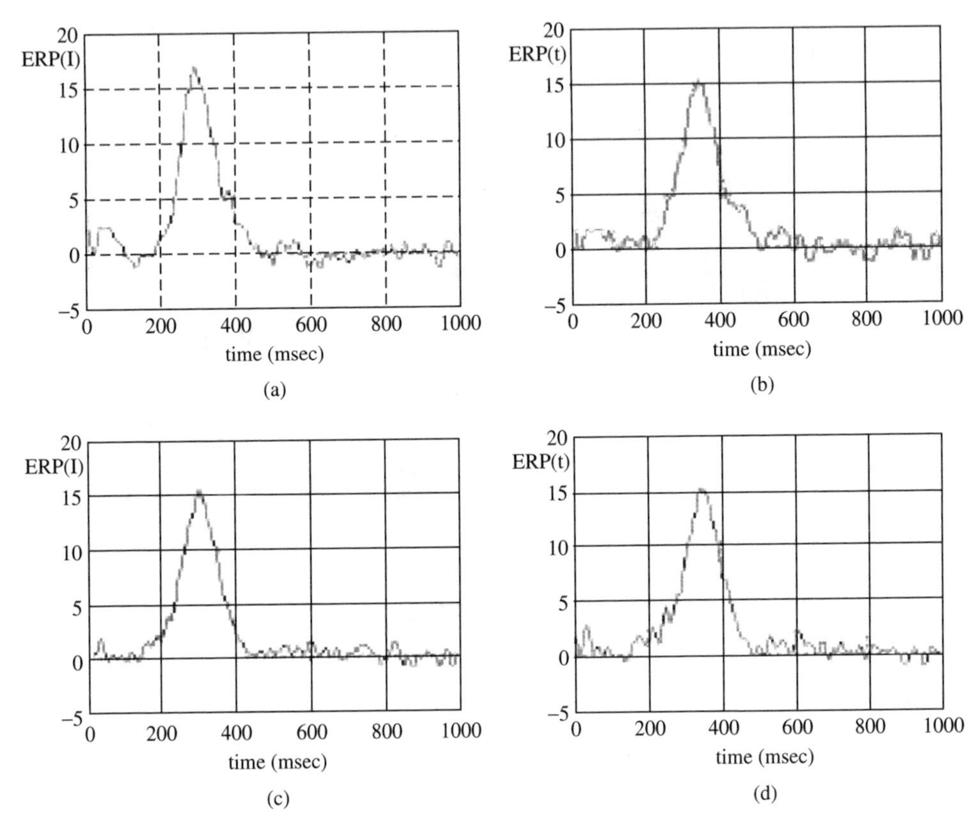

Figure 3.7 The average P3a and P3b for a patient (a) and (b) respectively, and for a healthy individual (c) and (d) respectively

where $\mathbf{Y}$ is a matrix with rows equal to $\boldsymbol{y}_c$ and $\mathbf{Y} = \mathbf{DS}$, where $\mathbf{D}$ is a diagonal scaling matrix:

$$\mathbf{D} = \begin{bmatrix} d_1 & 0 \ldots & 0 \\ 0 & d_2 \ldots & 0 \\ \vdots & & \ddots \vdots \\ 0 & 0 & d_m \end{bmatrix} \tag{3.19}$$

Postmultiplying $\mathbf{R}$ by $\mathbf{R}_y^{-1} = \mathbf{YY}^{-1}$,

$$\mathbf{RR}_y^{-1} = \mathbf{HSY}^{\mathrm{T}}(\mathbf{YY}^{\mathrm{T}})^{-1} = \mathbf{HD}^{-1}\mathbf{YY}^{\mathrm{T}}(\mathbf{YY}^{\mathrm{T}})^{-1} = \mathbf{HD}^{-1} \tag{3.20}$$

The order of the sources is arbitrary; therefore the permutation does not affect the overall process. Hence, the ith scaled scalp map corresponds to the scaled ith source. An LS method may exploit the information about the scalp maps to localize the ERP sources within the brain.

Assuming an isotropic propagation model of the head, the sources are attenuated with the third power of the distance [32], i.e. $d_j = 1/h_j^{1/3}$, where h_j is the jth element of a specific column of the **H** matrix. The source locations $\boldsymbol{q}$ are found as the solution to the following LS problem:

$$\min_{\boldsymbol{q},M} E(\boldsymbol{q}, M) = \min_{\boldsymbol{q},M} \sum_{j=1}^{n_e} \left(M\|\boldsymbol{q} - \boldsymbol{a}_j\|_2 - \boldsymbol{d}_j\right)^2 \tag{3.21}$$

where $\boldsymbol{a}_j$ are the positions of the electrodes, $\boldsymbol{d}_j$ are the scaled distances, and M (scalar) is the scale to be estimated together with $\boldsymbol{q}$. M and $\boldsymbol{q}$ are iteratively estimated according to

$$\boldsymbol{q}_{\rho+1} = \boldsymbol{q}_\rho - \mu_1 \nabla_{\boldsymbol{q}} E|_{\boldsymbol{q}=\boldsymbol{q}_\rho} \tag{3.22}$$

and

$$M_{\rho+1} = M_\rho - \mu_2 \nabla_M E|_{M=M_\rho} \tag{3.23}$$

where μ_1 and μ_2 are the learning rates and $\nabla_{\boldsymbol{q}}$ and ∇_M are respectively the gradients with respect to $\boldsymbol{q}$ and M, which are computed as [32]

$$\nabla_{\boldsymbol{q}} E(\boldsymbol{q}, M) = 2\sum_{j=1}^{n_e} (\boldsymbol{q} - \boldsymbol{a}_j)\left(M^2 - M\frac{\boldsymbol{d}_j}{\|\boldsymbol{q} - \boldsymbol{a}_j\|_2}\right) \tag{3.24}$$

$$\nabla_M E(\boldsymbol{q}, M) = 2\sum_{j=1}^{n_e} M\|\boldsymbol{q} - \boldsymbol{a}_j\|_2^2 - \|\boldsymbol{q} - \boldsymbol{a}_j\|_2 \boldsymbol{d}_j \tag{3.25}$$

The solutions to $\boldsymbol{q}$ and M are unique given $n_e \geq 3$ and $n_e > m$. Using the above localization algorithm, 10 sets of EEGs from five patients and five controls are each divided into 20 segments and examined. Figure 3.8 illustrates the locations of P3a and P3b sources for the patients. From this figure it is clear that the clusters representing the P3a and P3b locations are distinct. Figure 3.9, on the other hand, presents the sources for the control subjects. Unlike Figure 3.8, the P3a and P3b sources are randomly located within the brain.

In a recent work by Spyrou *et al.* [33] a special notch filter (or a beamformer) has been designed to maximize the similarity between the extracted ERP source and a generic spike-shaped template, and at the same time enforce a null at the location of the source. This method is based on a head and source model, which describes the propagation of the brain sources. The sources are modelled as magnetic dipoles and their propagation to the sensors is mathematically described by an appropriate forward model. In this model the EEG signal is considered as an $n_e \times T$ matrix, where n_e is the number of electrodes and T is the number of time samples for an EEG channel signal block:

$$\mathbf{X} = \mathbf{HMS} + \mathbf{N} = \sum_{i=1}^{m} \mathbf{H}_i \boldsymbol{m}_i \boldsymbol{s}_i + \mathbf{N} \tag{3.26}$$

where **H** is an $n_e \times 3m$ matrix describing the forward model of the m sources to the n_e electrodes. **H** is further decomposed into m matrices $\mathbf{H}_i$ as

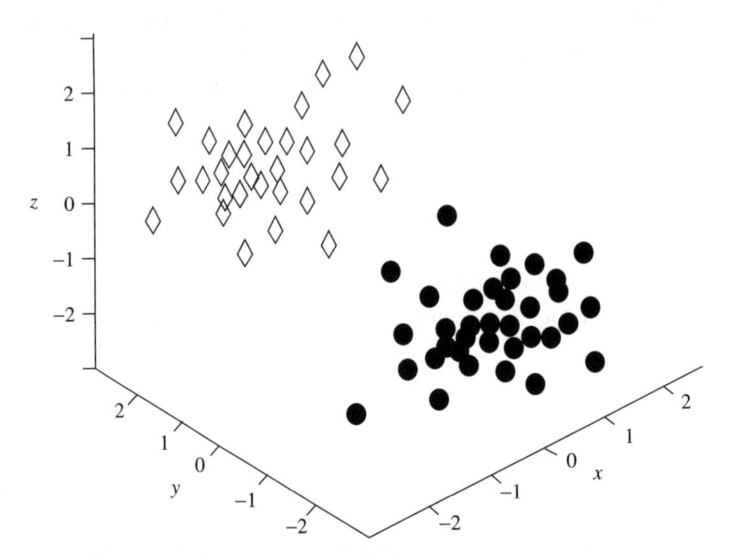

Figure 3.8 The locations of the P3a and P3b sources for five patients in a number of trials. The diamonds ◇ are the locations of the P3a sources and the circles • show the locations of P3b sources. The x axis denotes right to left, the y axis shows front to back, and the z axis denotes up and down. It is clear that the classes are distinct

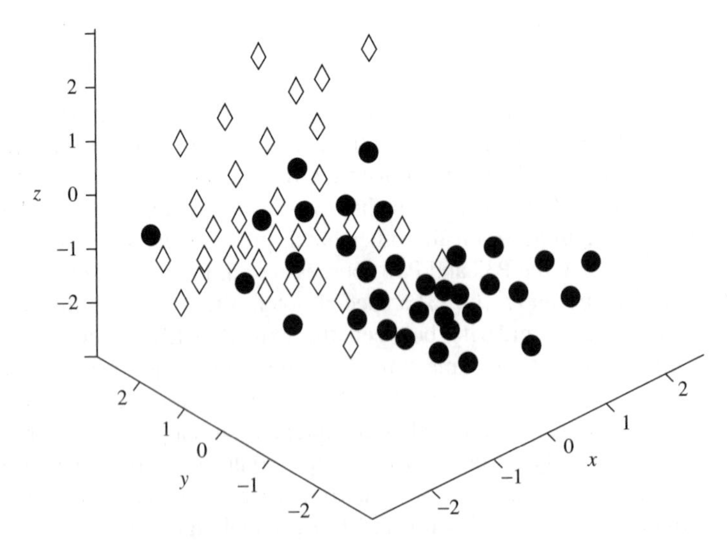

Figure 3.9 The locations of the P3a and P3b sources for five healthy individuals in a number of trials. The diamonds ◇ are the locations of the P3a sources and the circles • show the locations of P3b sources. The x axis denotes right to left, the y axis shows front to back, and the z axis denotes up and down

$$\mathbf{H} = [\mathbf{H}_1 \cdots \mathbf{H}_i \cdots \mathbf{H}_m] \tag{3.27}$$

where $\mathbf{H}_i$ are $n_e \times 3$ matrices where each column describes the potential at the electrodes due to the ith dipole for each of the three orthogonal orientations. For example, the first column of $\mathbf{H}_i$ describes the forward model of the x component of the ith dipole when y and z components are zero. Similarly, $\mathbf{M}$ is a $3m \times m$ matrix describing the orientation of the m dipoles and is decomposed as

$$\mathbf{M} = \begin{bmatrix} \boldsymbol{m}_1 & 0 & 0 & 0 & 0 \\ 0 & \cdots & 0 & 0 & 0 \\ 0 & 0 & \boldsymbol{m}_i & 0 & 0 \\ 0 & 0 & 0 & \cdots & 0 \\ 0 & 0 & 0 & 0 & \boldsymbol{m}_m \end{bmatrix} \tag{3.28}$$

where $\boldsymbol{m}_i$ is a 3×1 vector describing the orientation of the ith dipole; $\boldsymbol{s}_i$ is a $1 \times T$ vector where the time signal originated from the ith dipole, and $\mathbf{N}$ is the measurement noise and the modelling error. In addition, the ERP signal $\boldsymbol{r}_i$ for the ith source has been modelled similar to those in Equation (3.16) with variable width and latency.

A constrained optimization procedure is then followed in which the primary cost function is the Euclidean distance between the reference signal and the filtered EEG [33], i.e.

$$f_d(\boldsymbol{w}) = \|\boldsymbol{r}_i - \boldsymbol{w}^T\mathbf{X}\|_2^2 \tag{3.29}$$

The minimum point for this can be obtained by the classic LS minimization and is given by

$$\boldsymbol{w}_{opt} = (\mathbf{X}\mathbf{X}^T)^{-1}\mathbf{X}\mathbf{r}_i^T \tag{3.30}$$

This method designs an $n_e \times 1$ filter $\boldsymbol{w}_{opt}$, which gives an estimate of the reference signal.

However, this procedure does not include any spatial information unless the $\boldsymbol{w}_i\boldsymbol{s}$ for all the sources are taken into account. In this way, a matrix $\mathbf{W}$ can be constructed that is similar to the separating matrix in an ICA framework, which could be converted to the forward matrix $\mathbf{H}$. To proceed with this idea a constraint function is defined as

$$f_c(\boldsymbol{w}) = \boldsymbol{w}^T\mathbf{H}(p) = 0 \tag{3.31}$$

where $\mathbf{H}(p)$ is the forward matrix of a dipole at location p and a grid search is performed over a number of locations. The constrained optimization problem is then stated as [33]

$$\min f_d(\boldsymbol{w}) \text{ subject to } f_c(\boldsymbol{w}) = 0 \tag{3.32}$$

Without going through the details the filter (beamformer) $\boldsymbol{w}$ for extraction of a source at location j, as the desired source (ERP component), when there is neither noise nor correlation among the components, is described as

$$\boldsymbol{w}^T = \boldsymbol{r}_i \sum_{j \neq i} \boldsymbol{s}_i^T \boldsymbol{m}_i^T \mathbf{H}_i^T \mathbf{C}_x^{-1} \mathbf{H}_j (\mathbf{H}_j^T \mathbf{C}_x^{-1} \mathbf{H}_j)^{-1} \mathbf{H}_j^T \mathbf{C}_x^{-1} \tag{3.33}$$

where $\mathbf{C}_x = \mathbf{X}\mathbf{X}^{\mathrm{T}}$. In the case of correlation among the sources, the beamformer, $\mathbf{w}_{\mathrm{c}}$, for extraction of a source at location j includes another term as [31]

$$\boldsymbol{w}_{\mathrm{c}}^{\mathrm{T}} = \boldsymbol{w}^{\mathrm{T}} + \boldsymbol{r}_i\tilde{\mathbf{X}}^{\mathrm{T}}\mathbf{C}_x^{-1}\mathbf{H}(p)(\mathbf{H}^{\mathrm{T}}(p)\mathbf{C}_x^{-1}\mathbf{H}(p))^{-1}\mathbf{H}^{\mathrm{T}}(p)\mathbf{C}_x^{-1} \tag{3.34}$$

where $\tilde{\mathbf{X}}$ includes all the sources correlated with the desired source and $\boldsymbol{w}$ is given in equation (3.33). A similar expression can also be given when the noise is involved [33].

The algorithm finds the location of the desired brain ERP component (by manifesting a sharp null at the position of the desired source) with a high accuracy, even in the presence of interference and noise and where the correlation between the components is considerable.

3.1.5 *Time–Frequency Domain Analysis*

In cases where a signal contains frequency components emerging and vanishing in time-limited intervals, localization of the active signal components within the time–frequency (TF) domain is very useful [34]. The traditional method proposed for such an analysis is application of the short-time discrete Fourier transform (STFT) [35]. The STFT enables the time localization of a certain sinusoidal frequency but with an inherent limitation due to Heisenberg's uncertainty principle, which states that resolution in time and frequency cannot be both arbitrarily small because their product is lower bounded by $\Delta t \Delta\omega \geq 1/2$. Therefore, the wavelet transform (WT) has become very popular instead. An advantage of the WT over the STFT is that, for the STFT, the phase space is uniformly sampled, whereas in the wavelet transform the sampling in frequency is logarithmic which enables the analysis of higher frequencies in shorter windows and lower frequencies in longer windows in time [36].

As explained in Chapter 2 of this book, a multiresolution decomposition of signal $x(n)$ over I octaves is given as

$$x(n) = \sum_{i=1}^{\infty}\sum_{k\in Z} a_{i,k} g_i(n - 2^i k) + \sum_{k\in Z} b_{I,k} h_I(n - 2^I k) \tag{3.35}$$

where Z refers to integer values. The DWT computes the wavelet coefficients $a_{i,k}$ for $i = 1, \ldots, I$ and the scaling coefficients $b_{I,k}$ given by

$$a_{i,k} = \mathrm{DWT}\{x(n);\ 2^I, k2^I\} = \sum_{n=0}^{N} x(n) g_i^*(n - 2^i k) \tag{3.36}$$

and

$$b_{I,k} = \sum_{n=0}^{N} x(n) h_I^*(n - 2^I k) \tag{3.37}$$

The functions $g(.)$ and $h(.)$ perform as highpass and lowpass filters respectively. Among the wavelets discrete B-spline WTs have near-optimal TF localization [37] and also generally have antisymmetric properties, which make them more suitable for analysis of the

ERPs. The filters for an nth-order B-spline wavelet multiresolution decomposition are computed as

$$h(k) = \tfrac{1}{2}\left(b^{2n+1}\right)^{-1}(k)\uparrow_2 {}^*b^{2n+1}(k)^* p^n(k) \tag{3.38}$$

and

$$g(k+1) = \tfrac{1}{2}\left(b^{2n+1}\right)^{-1}(k)\uparrow_2 {}^*(-1)^k p^n(k) \tag{3.39}$$

where $\uparrow_2$ indicates upsampling by 2 and n is an integer. For the quadratic spline wavelet ($n = 2$) used by Ademoglu *et al.* to analyse pattern reversal VEPs [36], by substituting $n = 2$ in the above equations, the parameters can be derived mathematically as

$$[(b^5)^{-1}](k) = Z^{-1}\left(\frac{120}{z^2 + 26z + 66 + 26z^{-1} + z^{-2}}\right) \tag{3.40}$$

$$p^2(k) = \frac{1}{2^2} Z^{-1}\left(1 + 3z^{-1} + 3z^{-2} + z^{-3}\right) \tag{3.41}$$

where Z^{-1} denotes the inverse Z-transform. To find $[(b^5)^{-1}](k)$ the z-domain term can be factorized as

$$[(b^5)^{-1}](k) = Z^{-1}\left[\frac{\alpha_1\alpha_2}{(1-\alpha_1 z^{-1})(1-\alpha_1 z)(1-\alpha_2 z^{-1})(1-\alpha_2 z)}\right] \tag{3.42}$$

where $\alpha_1 = -0.04309$ and $\alpha_2 = -0.43057$. Therefore,

$$\left[(b^5)^{-1}\right](k) = \frac{\alpha_1\alpha_2}{(1-\alpha_1^2)(1-\alpha_2^2)(1-\alpha_1\alpha_2)(\alpha_1-\alpha_2)}\left[\alpha_1(1-\alpha_2^2)\alpha_1^{|k|} - \alpha_2(1-\alpha_1^2)\alpha_2^{|k|}\right] \tag{3.43}$$

On the other hand, $p(k)$ can also be computed easily by taking the inverse transform in Equation (3.41). Therefore the sample values of $h(n)$ and $g(n)$ can be obtained. Finally, construction of the WT is simply lowpass and highpass filtering followed by downsampling sequentially, as stated in Chapter 2. In such a multiresolution scheme the number of wavelet coefficients halves from one scale to the next, which requires longer time windows for lower frequencies and shorter time windows for higher frequencies.

In the above work the VEP waveform is recorded using a bipolar recording at positions C_z and O_z in the conventional 10–20 EEG system. The sampling rate is 1 kHz. Therefore, each scale covers the following frequency bands: 250–500 Hz, 125–250 Hz, 62.5–125 Hz, 31.3–62.5 Hz, 15.5–31.5 Hz, 7.8–15.6 Hz (including N70, P100, and N130), and 0–7.8 Hz (residual scale). Although this method is not to discriminate between normal and pathological subjects, the ERPs can be highlighted and the differences between normal and abnormal cases are observed. By using the above spline wavelet analysis, it is observed that the main effect of a latency shift of the N70–P100–N130 complex is reflected in the sign and magnitude of the second, third, and fourth wavelet coefficients within the 7.8–15.6 Hz band. Many other versions of the WT approach have been

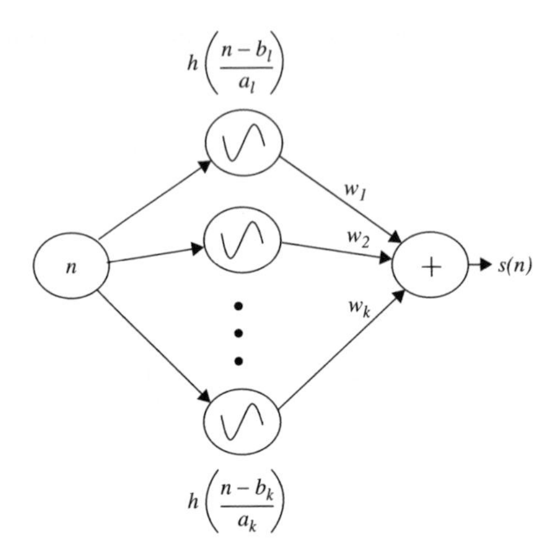

Figure 3.10 Construction of an ERP signal using a wavelet network (WN). The nodes in the hidden layer are represented by modified wavelet functions

introduced in the literature during the recent decade, such as in References [38], [39] and [40].

As an example, single-trial ERPs have been analysed using wavelet networks (WNs) and the developed algorithm applied to the study of attention deficit hyperactivity disorder (ADHD) [38]. This work is based on the fact that the ERP and the background EEG are correlated; i.e. the parameters that describe the signals before the stimulation do not hold for the post stimulation period [41]. In a wavelet network (WN) topology they presented an ERP, $\hat{s}(t)$, as a linear combination of K-modified wavelet functions $h(n)$:

$$\hat{s}(n) = \sum_{k=1}^{K} w_k h\left(\frac{n-b_k}{a_k}\right) \tag{3.44}$$

where b_k, a_k, and w_k, are the shift, scale, and weight parameters respectively and $s(n)$ are called wavelet nodes. Figure 3.10 shows the topology of a WN for signal representation. In the case of a Morlet wavelet,

$$\hat{s}(n) = \sum_{k=1}^{K} w_{1,k} \cos\left[\omega_k\left(\frac{n-b_k}{a_k}\right)\right] + w_{2,k} \sin\left[\omega_k\left(\frac{n-b_k}{a_k}\right)\right] \exp\left[-\frac{1}{2}\left(\frac{n-b_k}{a_k}\right)^2\right] \tag{3.45}$$

In this equation each of the frequency parameters ω_k and the corresponding scale parameter a_k define a node frequency $f_k = \omega_k/2\pi a_k$, and are optimized during the learning process.

In the above approach a WN is considered as a one hidden layer perceptron whose neuron activation function [42] is defined by wavelet functions. Therefore, the well-known backpropagation neural network algorithm [43] is used to simultaneously update the WN parameters (i.e.$w_{\cos,k}$, $w_{\sin,k}$, ω_k, a_k, and b_k of node k). The minimization is performed on the error between the node outputs and the bandpass filtered version of the residual error obtained using ARMA modelling. The bandpass filter and a tapering time-window are applied to ensure that the desired range is emphasized and the node sees its specific part of the TF plane [38]. It has been reported that for the study of ADHD this method enables the detection of the N100 ERP components in places where the traditional time averaging approach fails due to the position variation of this component.

In a recent work there has been an attempt to create a wavelet-like filter from simulation of the neural activities, which is then used in place of a specific wavelet to improve the performance of wavelet-based single-trial ERP detection. The main application of this has been reported to be in the area of BCI [44].

3.1.6 Adaptive Filtering Approach

The main problem in linear time invariant filtering of ERP signals is that the signal and noise frequency components highly overlap. Another reason for that is due to the fact that the ERP signals are transient and their frequency components may not fall into a certain range. Wiener filtering provides an optimum filtering in the mean-square sense, but it is not suitable for nonstationary signals such as ERPs.

With the assumption that the ERP signals are dynamically slowly varying processes, the future realizations are predictable from the past realizations. These changes can be studied using a state-space model. Kalman filtering and generic observation models have been used to denoise the ERP signals [45]. Considering a single-channel EEG observation vector at stimulus time m as $\boldsymbol{x}_n = [x_n(1), x_n(2), \ldots, x_n(N)]^{\mathrm{T}}$, the objective is to estimate a vector of parameters such as $\boldsymbol{\theta}_n$ of length k for which the map from $\boldsymbol{x}_n$ to $\boldsymbol{\theta}_n$ is (preferably) a linear map. An optimum estimator, $\hat{\boldsymbol{\theta}}_{n,\mathrm{opt}}$, that minimizes $E[||\hat{\boldsymbol{\theta}}_n - \boldsymbol{\theta}_n||^2]$ is $\hat{\boldsymbol{\theta}}_{n,\mathrm{opt}} = E[\boldsymbol{\theta}_n|\boldsymbol{x}_n]$ and can be shown to be

$$\hat{\boldsymbol{\theta}}_{n,\mathrm{opt}} = \boldsymbol{\mu}_{\boldsymbol{\theta}_n} + \mathrm{C}_{\boldsymbol{\theta}_n,\boldsymbol{x}_n}\mathrm{C}_{\boldsymbol{x}_n}^{-1}(\boldsymbol{x}_n - \boldsymbol{\mu}_{\boldsymbol{x}_n}) \tag{3.46}$$

where $\boldsymbol{\mu}_{\boldsymbol{\theta}_n}$ and $\boldsymbol{\mu}_{\boldsymbol{x}_n}$ are the means of $\boldsymbol{\theta}_n$ and $\boldsymbol{x}$ respectively. and $\mathrm{C}_{\boldsymbol{\theta}_n,\boldsymbol{x}_n}$ is the cross-covariance matrix of the observations and the parameters to be estimated and $\mathrm{C}_{\boldsymbol{x}_n}$ is the covariance matrix of the column vector $\boldsymbol{x}_n$. Such an estimator is independent of the relationship between $\boldsymbol{\theta}_n$ and $\boldsymbol{x}_n$. The covariance matrix of the estimated error $\boldsymbol{\varepsilon}_{\boldsymbol{\theta}} = \boldsymbol{\theta}_n - \hat{\boldsymbol{\theta}}_n$ can also be found as

$$\mathrm{C}_{\boldsymbol{\varepsilon}} = \mathrm{C}_{\boldsymbol{\theta}_n} - \mathrm{C}_{\boldsymbol{\theta}_n,\boldsymbol{x}_n}\mathrm{C}_{\boldsymbol{x}_n}^{-1}\mathrm{C}_{\boldsymbol{x}_n,\boldsymbol{\theta}_n} \tag{3.47}$$

In order to evaluate $\mathrm{C}_{\boldsymbol{\theta}_n,\boldsymbol{x}_n}$ some prior knowledge about the model is required. In this case the observations may be assumed to be of the form $\boldsymbol{x}_n = \boldsymbol{s}_n + \boldsymbol{v}_n$, where $\boldsymbol{s}_n$ and $\boldsymbol{v}_n$ are considered respectively as the response to the stimulus and the background EEG (irrelevant for the stimulus and ERP). Also, the ERP is modelled as

$$\boldsymbol{x}_n = \mathbf{H}_n\boldsymbol{\theta}_n + \boldsymbol{v}_n \tag{3.48}$$

where $\boldsymbol{H}_n$ is a deterministic $N \times k$ observation model matrix. The estimated ERP $\hat{\boldsymbol{s}}_n$ can then be obtained as

$$\hat{\boldsymbol{s}}_n = \mathbf{H}_n\hat{\boldsymbol{\theta}}_n \tag{3.49}$$

where the estimated parameters, $\hat{\boldsymbol{\theta}}_n$, of the observation model $\boldsymbol{H}_n$, using a linear mean square (MS) estimator and with the assumption of $\boldsymbol{\theta}_n$ and $\boldsymbol{v}_n$ being uncorrelated, is achieved as [46]

$$\hat{\boldsymbol{\theta}}_n = \left(\mathbf{H}_n^{\mathrm{T}} \mathrm{C}_{\boldsymbol{v}_n}^{-1} \mathbf{H}_n + \mathrm{C}_{\boldsymbol{\theta}_n}^{-1}\right)^{-1} \left(\mathbf{H}_n^{\mathrm{T}} \mathrm{C}_{\boldsymbol{v}_n}^{-1} \boldsymbol{x}_n + \mathrm{C}_{\boldsymbol{\theta}_n}^{-1} \boldsymbol{\mu}_{\boldsymbol{\theta}_n}\right) \tag{3.50}$$

This estimator is optimum if the joint distribution of $\boldsymbol{\theta}_n$ and $\boldsymbol{x}_n$ is Gaussian and the parameters and noise are uncorrelated.

However, in order to take into account the dynamics of the ERPs from trial to trial evolution of $\boldsymbol{\theta}_n$ has to be taken into account. Such evolution may be denoted as

$$\boldsymbol{\theta}_{n+1} = \mathbf{F}_n \boldsymbol{\theta}_n + \mathbf{G}_n \boldsymbol{w}_n \tag{3.51}$$

and some initial distribution for $\boldsymbol{\theta}_0$ assumed. Although the states are not observed directly the measurements are related to the observations through Equation (3.39). In this model it is assumed that $\mathbf{F}_n$, $\mathbf{H}_n$, and $\mathbf{G}_n$ are known matrices, $(\boldsymbol{\theta}_0, \boldsymbol{w}_n, \boldsymbol{v}_n)$ is a sequence of mutually uncorrelated random vectors with finite variance, $\mathrm{E}[\boldsymbol{w}_n] = \mathrm{E}[\boldsymbol{v}_n] = 0 \forall n$, and the covariance matrices $\mathrm{C}_{\boldsymbol{w}_n}$, $\mathrm{C}_{\boldsymbol{v}_n}$, and $\mathrm{C}_{\boldsymbol{w}_n, \boldsymbol{v}_n}$ are known.

With the above assumptions the Kalman filtering algorithm can be employed to estimate $\boldsymbol{\theta}$ as [47]

$$\hat{\boldsymbol{\theta}}_n = \left(\mathbf{H}_n^{\mathrm{T}} \mathrm{C}_{\boldsymbol{v}_n}^{-1} \mathbf{H}_n + \mathrm{C}_{\boldsymbol{\theta}_{n|n-1}}^{-1}\right)^{-1} \left(\mathbf{H}_n^{\mathrm{T}} \mathrm{C}_{\boldsymbol{v}_n}^{-1} \boldsymbol{x}_n + \mathrm{C}_{\boldsymbol{\theta}_{n|n-1}}^{-1} \hat{\boldsymbol{\theta}}_{n|n-1}\right) \tag{3.52}$$

where $\hat{\boldsymbol{\theta}}_{n|n-1} = E[\boldsymbol{\theta}_n | \boldsymbol{x}_{n-1}, \ldots, \boldsymbol{x}_1]$ is the prediction of $\boldsymbol{\theta}_n$ subject to $\hat{\boldsymbol{\theta}}_{n-1} = E[\boldsymbol{\theta}_{n-1} | \boldsymbol{x}_{n-1}, \ldots, \boldsymbol{x}_1]$, which is the optimum MS estimate at $n-1$. Such an estimator is the best sequential estimator if the Gaussian assumption is valid and is the best linear estimator disregarding the distribution. The overall algorithm for Kalman filtering may be summarized as follows:

$$\mathrm{C}_{\hat{\boldsymbol{\theta}}_0} = \mathrm{C}_{\boldsymbol{\theta}_0} \tag{3.53}$$

$$\hat{\boldsymbol{\theta}}_0 = E[\boldsymbol{\theta}_0] \tag{3.54}$$

$$\hat{\boldsymbol{\theta}}_{n|n-1} = \mathbf{F}_{n-1} \hat{\boldsymbol{\theta}}_{n-1} \tag{3.55}$$

$$\mathrm{C}_{\hat{\boldsymbol{\theta}}_{n|n-1}} = \mathbf{F}_{n-1} \mathrm{C}_{\hat{\boldsymbol{\theta}}_{n-1}} \mathbf{F}_{n-1}^{\mathrm{T}} + \mathbf{G}_{n-1} \mathrm{C}_{\mathbf{w}_{n-1}} \mathbf{G}_{n-1}^{\mathrm{T}} \tag{3.56}$$

$$\mathbf{K}_t = \mathbf{C}_{\hat{\boldsymbol{\theta}}_{n|n-1}} \mathbf{H}_n^{\mathrm{T}} \left(\mathbf{H}_n \mathrm{C}_{\hat{\boldsymbol{\theta}}_{n|n-1}} \mathbf{H}_n^{\mathrm{T}} + \mathrm{C}_{\boldsymbol{v}_n}\right)^{-1} \tag{3.57}$$

$$\mathrm{C}_{\hat{\boldsymbol{\theta}}_n} = (\mathbf{I} - \mathbf{K}_n \mathbf{H}_n) \mathrm{C}_{\hat{\boldsymbol{\theta}}_{n|n-1}} \tag{3.58}$$

$$\hat{\boldsymbol{\theta}}_n = \boldsymbol{\theta}_{n|n-1} + \mathbf{K}_n \left(\boldsymbol{x}_n - \mathbf{H}_n \hat{\boldsymbol{\theta}}_{n|n-1}\right) \tag{3.59}$$

In the work by Georgiadis *et al.* [45], however, a simpler observation model has been considered. In this model the state-space equations have the form

$$\boldsymbol{\theta}_{n+1} = \boldsymbol{\theta}_n + \boldsymbol{w}_n \tag{3.60}$$

$$\boldsymbol{x}_n = \mathbf{H}_n\boldsymbol{\theta}_n + \boldsymbol{v}_n \tag{3.61}$$

in which the state vector models a random walk process [48]. Now, if it is assumed that the conditional covariance matrix of the parameter estimation error is $\mathbf{P}_t = \mathrm{C}_{\boldsymbol{\theta}_t} + \mathrm{C}_{\boldsymbol{w}_t}$ the Kalman filter equations will be simplified to

$$\mathbf{K}_n = \mathbf{P}_{n-1}\mathbf{H}_n^{\mathrm{T}}\left(\mathbf{H}_n\mathbf{P}_{n-1}\mathbf{H}_n^{\mathrm{T}} + \mathrm{C}_{\boldsymbol{v}_n}\right)^{-1} \tag{3.62}$$

$$\mathbf{P}_n = (\mathrm{I} - \mathbf{K}_n\mathbf{H}_n)\,\mathbf{P}_{n-1} + \mathrm{C}_{\mathbf{w}_n} \tag{3.63}$$

$$\hat{\boldsymbol{\theta}}_n = \hat{\boldsymbol{\theta}}_{n-1} + \mathbf{K}_n\left(\mathbf{x}_n - \mathbf{H}_n\hat{\boldsymbol{\theta}}_{n-1}\right) \tag{3.64}$$

where $\mathbf{P}_n$ and $\mathbf{K}_n$ are called the recursive covariance matrix estimate and Kalman gain matrix respectively. At this stage it is interesting to compare these parameters for the four well-known recursive estimation algorithms, namely recursive least squares (RLS), least mean square (LMS), normalized least mean square (NLMS), and the Kalman estimators for the above simple model. This is depicted in Table 3.1 [45].

In practice $\mathrm{C}_{\mathbf{w}_n}$ is considered as a diagonal matrix such as $\mathrm{C}_{\mathbf{w}_n} = 0.1\mathbf{I}$. Also, it is logical to consider that $\mathbf{C}_{\boldsymbol{v}_n} = \mathbf{I}$ and, by assuming the background EEG to have Gaussian distribution, it is possible to use the Kalman filter to estimate the ERP signals dynamically. As an example, it is possible to consider $\mathbf{H}_n = \mathbf{I}$, so that the ERPs are recursively estimated as

$$\hat{s}_n = \hat{\boldsymbol{\theta}}_n = \hat{\boldsymbol{\theta}}_{n-1} + \mathbf{K}_n(\boldsymbol{x}_n - \hat{\boldsymbol{\theta}}_{n-1}) = \mathbf{K}_n\boldsymbol{x}_n + (\mathbf{I} - \mathbf{K}_n)\hat{s}_{n-1} \tag{3.65}$$

Such assumptions make the application of the algorithms in Table 3.1 easier for this application.

In a more advanced algorithm $\mathbf{H}_n$ may be adapted to the spatial topology of the electrodes. This means that the column vectors of $\mathbf{H}_n$ are weighted based on the expected locations of the ERP generators within the brain. Figure 3.11 compares the estimated ERPs for different stimuli for the LMS and recursive mean-square estimator (Kalman filter). From this figure it is clear that the estimates using the Kalman filter appear to be more robust and show very similar patterns for different stimulation instants.

Table 3.1 $\mathbf{K}_n$ and $\mathbf{P}_n$ for different recursive algorithms

RLS	$\mathrm{K}_n = \mathbf{P}_{n-1}\mathbf{H}_n^{\mathrm{T}}\left(\mathbf{H}_n\mathbf{P}_{n-1}\mathbf{H}_n^{\mathrm{T}} + \lambda_n\right)^{-1}$ $\mathbf{P}_n = \lambda_n^{-1}(\mathbf{I} - \mathbf{K}_n\mathbf{H}_n)\mathbf{P}_{n-1}$
LMS	$\mathbf{K}_n = \mu\mathbf{H}_n^{\mathrm{T}}$ $\mathbf{P}_n = \mu\mathbf{I}$
NLMS	$\mathbf{K}_n = \mu\mathbf{H}_n^{\mathrm{T}}(\mu\mathbf{H}_n\mathbf{H}_n^{\mathrm{T}} + 1)^{-1}$ $\mathbf{P}_n = \mu\mathbf{I}$
Kalman filter	$\mathbf{K}_n = \mathbf{P}_{n-1}\mathbf{H}_n^{\mathrm{T}}\left(\mathbf{H}_n\mathbf{P}_{n-1}\mathbf{H}_n^{\mathrm{T}} + \mathrm{C}_{\boldsymbol{v}_n}\right)^{-1}$ $\mathbf{P}_n = (\mathbf{I} - \mathbf{K}_n\mathbf{H}_n)\,\mathbf{P}_{n-1} + \mathrm{C}_{\mathbf{w}_n}$

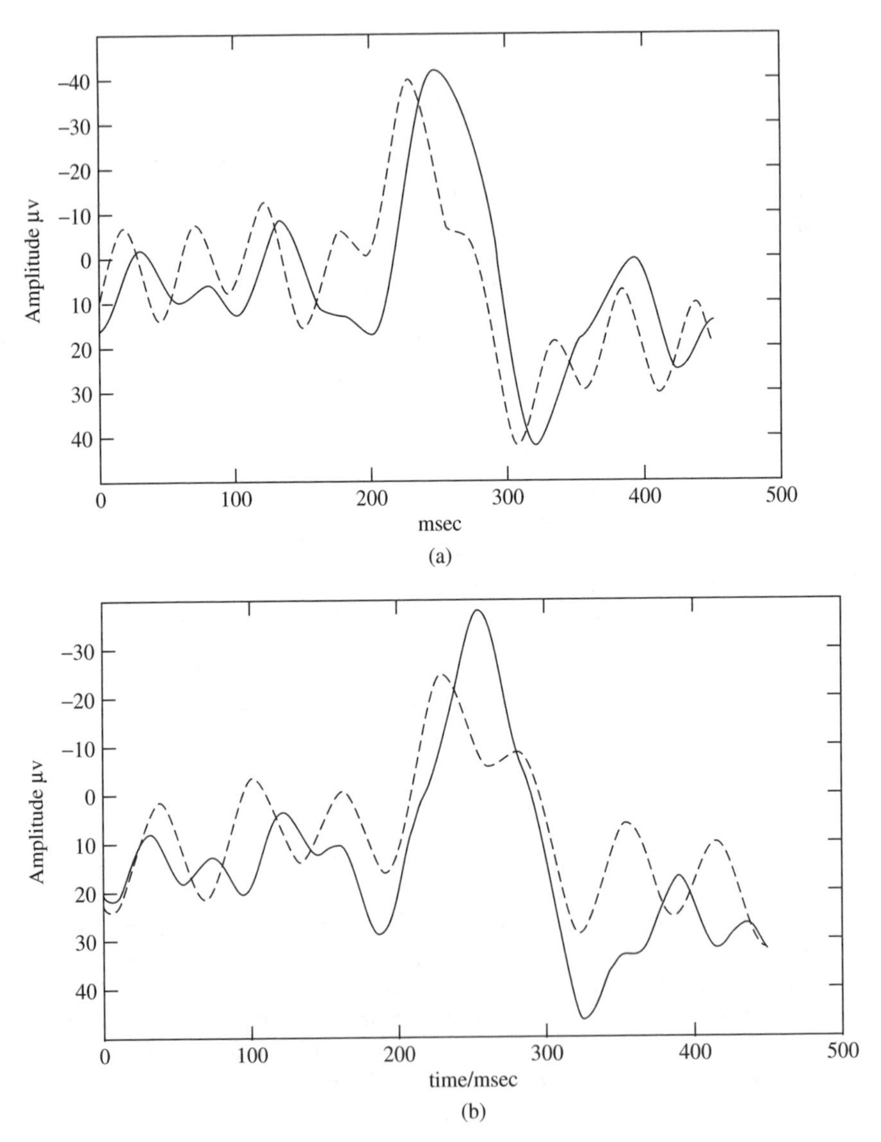

Figure 3.11 Dynamic variations of ERP signals: (a) first stimulus and (b) twentieth stimulus for the ERPs detected using LMS (dotted line) and Kalman filtering (thick line) approaches

3.1.7 Prony's Approach for Detection of P300 Signals

Development of Prony's method in signal modelling is explained in Chapter 2. Based on this approach, the source signals are modelled as a series expansion of damped sinusoidal basis functions. Prony's method is often used for estimation of pole positions when the noise is white. Also, in Reference [49] the basis functions are estimated using the least-squares Prony method, derived for coloured noise. This is mainly because after the set of basis functions are estimated using Prony's method, the optimal weights are calculated for each single trial ERP by minimizing the squared error.

The amplitude and latency of the P300 signal give information about the level of attention. In many psychiatric diseases this level changes. Therefore, a reliable algorithm for more accurate estimation of these parameters is very useful.

In the above attempt, the ERP signal is initially divided into the background EEG and ERP before and after the stimulus time instant respectively. The ERP is also divided into two segments, the early brain response, which is a low-level high-frequency signal, and the late response, which is a high-level low-frequency signal, i.e.

$$x(n) = \begin{cases} x_1(n) & -L \leq n < 0 \\ s_e(n - n_0) + x_2(n) & 0 \leq n \leq L_1 \\ s_l(n - n_0 - L_1) + x_3(n) & L_1 \leq n < N + L_1 \end{cases} \tag{3.66}$$

where $x_1(n)$, $x_2(n)$, and $x_3(n)$ are the background EEGs before the stimulus, during the early brain response respectively, and during the late brain response, and $s_e(n)$ and $s_l(n)$ are the early and late brain responses to a stimuli respectively. Therefore, to analyse the signals they are windowed (using a window $g(n)$) within $L_1 \leq n < N + L_1$ during the time the P300 signal exists. This gives

$$x_g(n) = x(n)g(n) \approx \begin{cases} s_l(n - L_1) + x_3(n) & L_1 \leq n < N + L_1 \\ 0 & \text{otherwise} \end{cases} \tag{3.67}$$

It can be considered that $s(n) = s_l(n - L_1)$ for $L_1 \leq n < N + L_1$, for simplicity. Based on Prony's algorithm the ERP signals are represented as the following overdetermined systems:

$$s(n) = \sum_{i=1}^{M} a_i \rho_i^n \mathrm{e}^{\mathrm{j}\omega_i n} = \sum_{i=1}^{M} a_i z_i^n \tag{3.68}$$

where $z_i = \rho_i \mathrm{e}^{\mathrm{j}\omega_i}$ and $M < N$. Using vector notation,

$$\boldsymbol{x}_g = \boldsymbol{Z}\boldsymbol{a} + \boldsymbol{x}_3 \tag{3.69}$$

where $\boldsymbol{x}_g = [x_g(L_1) \ldots x_g(N + L_1 - 1)]^{\mathrm{T}}, \boldsymbol{x}_3 = [x_3(L_1) \ldots x_3(N + L_1 - 1)]^{\mathrm{T}}, \boldsymbol{a} = [a_1 \ldots a_M]^{\mathrm{T}}$, and

$$\mathbf{Z} = \begin{bmatrix} \rho_1 \mathrm{e}^{\mathrm{j}\omega_1} & \ldots & \rho_M \mathrm{e}^{\mathrm{j}\omega_M} \\ \vdots & & \\ \rho_1 \mathrm{e}^{\mathrm{j}\omega_1 N} & \ldots & \rho_M \mathrm{e}^{\mathrm{j}\omega_M N} \end{bmatrix} \tag{3.70}$$

Therefore, in order to estimate the waveform $\boldsymbol{s} = \boldsymbol{Z}\boldsymbol{a}$ both $\mathbf{Z}$ and $\boldsymbol{a}$ have to be estimated. The criterion to be minimized is the weighted sum squared error criterion given by

$$\min_{\boldsymbol{Z},\boldsymbol{a}} \left[(\boldsymbol{x}_g - \boldsymbol{Z}\boldsymbol{a})^{\mathrm{H}} \mathbf{R}_{\boldsymbol{x}_3}^{-1} (\boldsymbol{x}_g - \boldsymbol{Z}\boldsymbol{a}) \right] \tag{3.71}$$

where $\mathbf{R}_{x_3}$ is the covariance matrix of the EEG (as a disturbing signal) after stimulation. For a fixed $\mathbf{Z}$ the optimum parameter values are obtained as

$$\boldsymbol{a}_{\text{opt}} = \left(\mathbf{Z}^{\text{H}}\mathbf{R}_{x_3}^{-1}\mathbf{Z}\right)^{-1}\mathbf{Z}^{\text{H}}\mathbf{R}_{x_3}^{-1}\boldsymbol{x}_g \tag{3.72}$$

Then the corresponding estimate of the single-trial ERP is achieved as

$$\boldsymbol{s} = \boldsymbol{Z}\boldsymbol{a}_{\text{opt}} \tag{3.73}$$

Prony's method is employed to minimize Equation (3.71) for coloured noise. In order to do that a matrix $\mathbf{F}$ orthogonal to $\mathbf{Z}$ is considered, i.e. $\mathbf{F}^{\text{H}}\mathbf{Z} = 0$ [50]. The ith column of $\mathbf{F}$ is

$$col\{\mathbf{F}\}_i = [\underbrace{0\ldots0}_{i-1}\ f_0\ldots f_M\ \underbrace{0\ldots0}_{N-M-i}]^{\text{T}}$$

The roots of the polynomial $f_0 + f_1 z^{-1} + \ldots + f_M z^{-M}$ are the elements z_i in $\mathbf{Z}$ [50]. The minimization in Equation (3.71) is then converted to

$$\min_{\mathbf{F}}\left[\mathbf{x}_g^{\text{H}}\mathbf{F}(\mathbf{F}^{\text{H}}\mathbf{R}_{x_3}\mathbf{F})^{-1}\mathbf{F}^{\text{H}}\boldsymbol{x}_g\right] \tag{3.74}$$

In practice a linear model is constructed from the above nonlinear system using [50,51]

$$\mathbf{F}\left(\mathbf{F}^{\text{H}}\mathbf{R}_{x_3}\mathbf{F}\right)^{-1}\mathbf{F}^{\text{H}} \approx \tilde{\mathbf{F}}\mathbf{R}_{x_3}^{-1}\tilde{\mathbf{F}}^{\text{H}} \tag{3.75}$$

where the ith column of $\tilde{\mathbf{F}}$ is

$$col\{\tilde{\mathbf{F}}\}_i = [\underbrace{0\ldots0}_{i-1}\ f_0\ldots f_p\ \underbrace{0\ldots0}_{N-p-i}]^{\text{T}}$$

where $p >> M$ and therefore it includes the roots of $\mathbf{F}$ plus $p - M$ spurious poles. Then, the minimization in Equation (3.74) changes to

$$\min_{\tilde{\mathbf{F}}}\left(\boldsymbol{e}^{\text{H}}\mathbf{R}_{x_3}^{-1}\boldsymbol{e}\right) = \min_{\tilde{\mathbf{F}}}\{\boldsymbol{x}_g^{\text{H}}\tilde{\mathbf{F}}\mathbf{R}_{x_3}^{-1}\tilde{\mathbf{F}}^{\text{H}}\boldsymbol{x}_g\} \tag{3.76}$$

where $\boldsymbol{e} = \tilde{\mathbf{F}}^{\text{H}}\boldsymbol{x}_g = \mathbf{X}\boldsymbol{f}$. In this equation

$$\mathbf{X} = \begin{bmatrix} x_g(p) & x_g(p-1) & \ldots & x_g(0) \\ \vdots & \vdots & & \vdots \\ x_g(N-1) & x_g(N-2) & \ldots & x_g(N-p-1) \end{bmatrix} = [\tilde{\boldsymbol{x}}\quad \tilde{\mathbf{X}}] \tag{3.77}$$

where $\tilde{\boldsymbol{x}}$ is the first column of $\mathbf{X}$ and includes the spurious poles. The original Prony method for solving Equation (3.76) is found by replacing $\mathbf{R}_{x_3} = \mathbf{I}$ with $f_0 = 1$ or $f_1 = 1$. However, the background EEG cannot be white noise; therefore, in this approach a coloured noise, $\mathbf{R}_{x_3} \neq \mathbf{I}$, is considered. Hence the column vector $\boldsymbol{f}$ including the poles or the resonance frequencies of the basis functions in Equation (3.68) and is estimated as

$$\boldsymbol{f} = [1; -[\tilde{\mathbf{X}}^{\mathrm{H}}\mathbf{R}_{x_3}^{-1}\tilde{\mathbf{X}}]_M^{-1}[\tilde{\mathbf{X}}^{\mathrm{H}}\mathbf{G}^{-\mathrm{H}}]_M\mathbf{G}^{-1}\tilde{\boldsymbol{x}}] \tag{3.78}$$

$\mathbf{G}\mathbf{G}^{\mathrm{H}} = \mathbf{R}_{x_3}$ and the notation $[.]_M$ denotes a rank M approximation matrix. The rank reduction is performed using a singular-value decomposition (SVD). Although several realizations of the ERP signals may have similar frequency components, $\boldsymbol{a}$ may vary from trial to trial. Therefore, instead of averaging the data to find $\boldsymbol{a}$, the estimated covariance matrix is averaged over D realizations of the data. An analysis based on a forward–backward Prony solution is finally employed to separate the spurious poles from the actual poles. The parameter $\boldsymbol{a}$ can then be computed using Equation (3.72).

Generally, the results achieved using Prony-based approaches for a single-trial ERP extraction are rather reliable and consistent with the physiological and clinical expectations. The above algorithm has been examined for a number of cases including prestimulus, after stimulation and during poststimulus. Clear peaks corresponding to mainly P300 sources are evident in the extracted signals.

In another attempt similar to the above approach it is considered that the frequency, amplitude, and phase characteristics of the ERP signals vary with time; therefore a piecewise Prony method (PPM) has been proposed [52]. The reasons for using PPM are that first this method enables nonmonotonically growing or decaying components with nonzero onset time to be modelled. Therefore, by using PPM it is assumed that the sinusoidal components have growing or decaying envelopes, and abrupt changes in amplitude and phase. The PPM uses variable-length windows (previously suggested by [52] and [53]) and variable sampling frequencies (performed also by Kulp [54] for adjusting sampling rate to the frequencies to be modelled) to overcome the limitations of the original Prony method. Also, the window size is determined based on the signal characteristics. Signals with large bandwidths are modelled in several steps, focusing on smaller frequency bands per step, as proposed in Reference [55]. Such varying-length windows try to include the conventional Prony components. Therefore, there is more consistency in detecting the ERP features (frequency, amplitude, and phase). Finally, it is reasonable to assume that the signal components obtained from adjacent windows with identical frequency are part of the same component and can be combined into one component.

3.1.8 Adaptive Time–Frequency Methods

A combination of an adaptive signal estimation technique and time–frequency signal representation can enhance the performance of the ERP and EP detections. This normally refers to modelling the signals with variable-parameter systems and application of adaptive estimators to estimate suitable parameters.

Using the short-time discrete Fourier transform (STFT) the information about the duration of the activities is not exploited and therefore it cannot describe structures much shorter or much longer than the window length. The wavelet transform (WT) can overcome this limitation by allowing for variable window lengths, but there is still a reciprocal relation between the central frequency of the wavelet and its window length. Therefore, the WT does not precisely estimate the low-frequency components with short-time duration or narrow-band high-frequency components.

In a very recent work the adaptive chirplet transform (ACT) has been utilized to characterize the time-dependent behaviour of the VEPs from their transient to steady state section [56]. Generally, this approach uses the matching pursuit (MP) algorithm to estimate the chirplets and a maximum likelihood (ML) algorithm to refine the results and enable estimation of the signal with a low signal-to-noise ratio (SNR). Using this method it is possible to visualize the early moments of a VEP response.

The ACT attempts to decompose the signals into Gaussian chirplet basis functions with four adjustable parameters of time spread, chirp rate, time centre and frequency centre. Moreover, it is shown that only three chirplets can be used to represent a VEP response.

In this approach the transient VEP, which appears first following the onset of the visual stimulus [57], is analysed together with the steady-state VEP [58,59]. Identification of the steady-state VEP has had many clinical applications and can help diagnosing sensory dysfunction [60,61]. The model using only steady-state VEPs is, however, incomplete without considering the transient pattern. A true steady-state VEP is also difficult to achieve in the cases where the mental situation of the patient is not stable.

Chirplets are windowed rapidly swept sinusoidal signals also called chirps [62]. The bases for a Gaussian chirplet transform (CT) are derived from a simple Gaussian function $\pi^{-1/4}\exp(-t^2/2)$ through four operations–scaling, chirping, time shifting, and frequency shifting–which (as for wavelets) produce a family of wave packets with four adjustable parameters [62]. Such a continuous time chirplet may be represented as

$$g(t) = \pi^{-1/4}\Delta_t^{-1/2} e^{-(1/2)[t-t_c/\Delta_t]^2} e^{j[c(t-t_c)+\omega_c](t-t_c)} \tag{3.79}$$

where t_c is the time centre, ω_c is the frequency centre, $\Delta_t > 0$ is the effective time spread, and c is the chirp rate, which refers to the speed of changing the frequency. The CT of a signal $f(t)$ is then defined as

$$a(t_c, \omega_c, c, \Delta_t) = \int_{-\infty}^{\infty} f(t)g^*(t)\,dt \tag{3.80}$$

Based on this approach, the signal $f(t)$ is reconstructed as a linear combination of Gaussian chirplets as

$$f(t) = \sum_{n=1}^{p} a(t_c, \omega_c, c, \Delta_t)g(t) + e(t) \tag{3.81}$$

where p denotes the approximation order and $e(t)$ is the residual signal. Figure 3.12 illustrates the chirplets extracted from an EEG-type waveform. To estimate $g(t)$ at each round (iteration) six parameters have to be estimated (consider the a complex). An optimal estimation of these parameters is generally impossible. However, there have been some suboptimal solutions such as in References [63] to [66]. In the approach by Cui and Wong [56] a coarse estimation of the chirplets is obtained using the MP algorithm. Then the maximum likelihood estimation (MLE) is performed iteratively to refine the results.

In order to perform this the algorithm is initialized by setting $e(n) = f(n)$. A dictionary is constructed using a set of predefined chirplets to cover the entire time–frequency plane [63]. The MP algorithm projects the residual $e(n)$ to each chirplet in the dictionary and

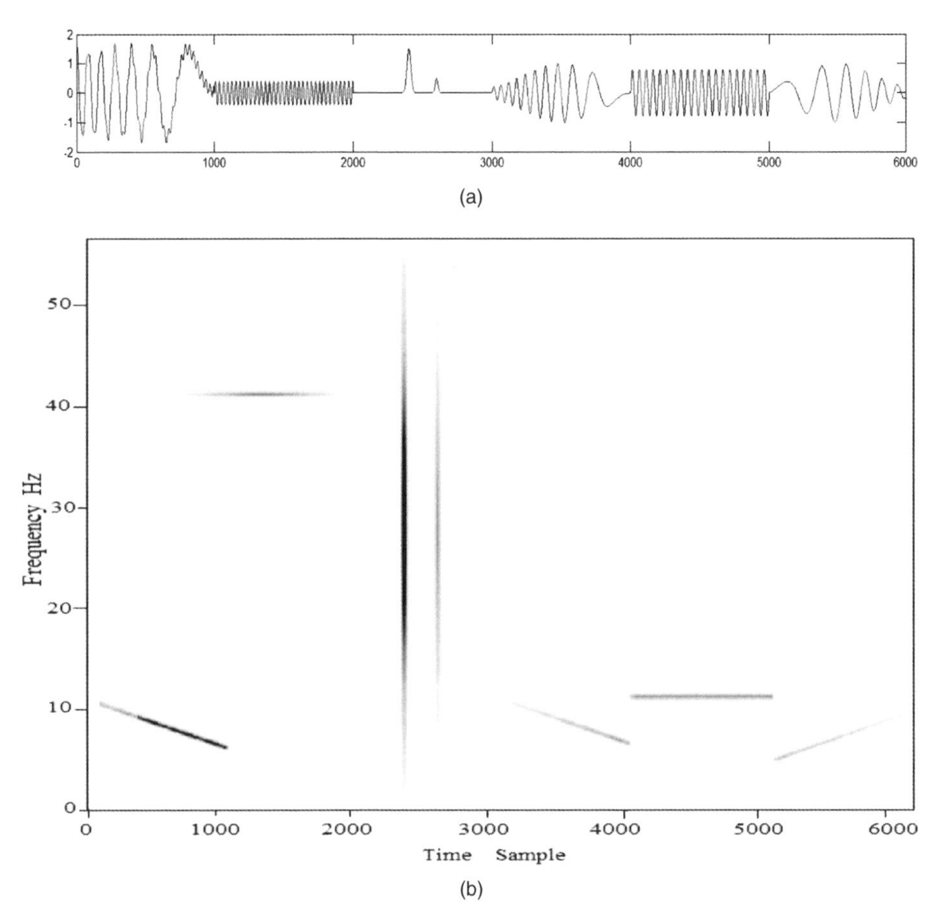

Figure 3.12 The chirplets extracted from a simulated EEG-type waveform [53], where (a) represents the combined waveforms and (b) shows the TF representation of the corresponding chirplets

the optimal chirplet is decided based on the projection amplitude. In the next stage, a Newton–Raphson method is used to refine the results and achieve the optimum match.

Under a low SNR situation a postprocessing step may be needed to follow the MLE concept [67] followed by the expectation maximization (EM) method proposed in Reference [56].

3.2 Brain Activity Assessment Using ERP

The P300 components have been found to play a significant role in identifying the depth of cognitive information processing. For example, it has been reported that the P300 amplitude elicited by the mental task loading decreases with the increase in the perceptual/cognitive difficulty of the task [68]. Assessment of mental fatigue using P300 signals by measuring the amplitude and latency of the signals and also the alpha-band power have shown that [69] the P300 amplitude tends to decrease immediately after the experimental task whereas the latency decreases at this time. The amplitude decrease is an indicator of decreasing the attention level of the subject. The increase in the latency might be indicative of the prolonged temporal processing due to the difficulty of cognitive information processing [68,70,71]. Therefore, mental and physical fatigues can be due to this problem. In this work, it is also pointed out that one aspect of the mentally and physically fatigued could be because of decreased activity of the central nervous system (CNS), which appears in both the temporal prolongation of cognitive information processing and the decrease in attention level. It has also been demonstrated that the alpha-band power decreases immediately after the task. This indicates that the activity of the CNS is decelerated with the accumulation of mental and physical fatigue. However, the appearance of fatigue is reflected more strongly in the amplitude and latency of the P300 signal rather than the alpha-band power.

3.3 Application of P300 to BCI

The electrical cortical activities used in BCI may be divided into the five following categories:

(a) beta (β) and mu (μ) rhythms. These activities range respectively within 8–12 Hz and 12–30 Hz frequent bands. These signals are associated with those cortical areas most directly connected to the motor output of the brain and can willingly be modulated with an imaginary movement [72].
(b) P300 evoked potential (EP). It is a late appearing component of an auditory, visual, or somatosensory ERP, as explained earlier.
(c) Visual N100 and P200. The ERPs with short latency that represent the exogenous response of the brain to a rapid visual stimulus. These potentials are used as clues, indicating the direction of the gaze of the user [73,74].
(d) Steady-state visual evoked potentials (SSVEP). These signals are natural responses to visual stimulations at specific frequencies. When the retina is excited by a visual stimulus ranging from 3.5 to 75 Hz, the brain generates an electrical activity at the same (or multiples of) frequency of the visual stimulus. They are used for understanding which stimulus the subject is looking at in the case of stimuli with different flashing frequencies [75,76].
(e) Slow cortical potentials (SCP). They are slow potential variations generated in the cortex after 0.5–10 seconds of presenting the stimulus. Negative SCPs are generated by movement whereas positive SCPs are associated with reduced cortical activation. The adequately trained users can control these potentials and use them to control the movement of a cursor on the computer screen [77].

To enable application of these waves, especially P300 for a P300-based BCI, the data have to be initially processed to reduce noise and enforce P300-related information. A pattern recognition algorithm has to be developed later in order to check the presence of the P300 wave in the recorded ERP epochs and label them. Also, a feedback mechanism has to be established to send the user a visible signal on the monitor correlated with the recorded epoch. Finally, the parameters of the pattern recognition algorithm have to be made adaptable to the user characteristics.

The P300 speller, as the first P300-BCI, described by Farwell and Dounchin, adapted the OP as the operating principle of BCI [23]. In this paradigm the participant is presented with a Bernoulli sequence of events, each belonging to one of two categories. The participant is assigned a task that cannot be performed without a correct classification of the events, each belonging to one of two categories. Using this speller, a 6×6 matrix is, for example, displayed to the subject. The system is operated by briefly intensifying each row and column of the matrix and the attended row and column elicits a P300 response. In a later work [78] it was found that some people who suffer from amyotrophic lateral sclerosis (ALS) can respond better to the system with a smaller size matrix (less than 6×6).

In addition, measurement and evaluation of movement-related events, such as event-related desynchronization (ERD) from the EEGs, can improve the diagnosis of functional deficits in patients with cerebrovascular disorders and Parkinson's disease (PD). There is a high correlation between morphological (such as CT) and functional (such as EEG) findings in cerebrovascular disorders. For example, the ERD reduces over the affected hemisphere.

The premovement ERD in PD is less lateralized over the contralateral sensorimotor area and starts later than in control subjects. Also, postmovement beta ERS is of a smaller magnitude and delayed in PD as compared to controls. It has been shown [79] that based on only two parameters, namely the central ERD within 6–10 Hz and postmovement event-related synchronization (ERS) within 16–20 Hz, it is possible to discriminate PD patients with a Hoehn and Yahr scale of 1–3 from age-matched controls by a linear discriminant analysis. Also, following the above procedure in P300 detection and classification, two user-adaptive BCI systems based on SSVEP and P300 have been proposed by Beverina *et al.* [80].

The P300 component has also been detected (separated) using ICA for the BCI purpose in another attempt [28]. It has also been empirically confirmed that the visual spatial attention modulation of the SSVEP can be used as a control mechanism in a real-time independent BCI (i.e. when there is no dependency on peripheral muscles or nerves) [81].

P300 signals have also been used in the Wandsworth BCI development [82]. In this application a similar method to that proposed by Farwell and Donchin [23] is used for highlighting and detection of the P300 signals. It has been shown that this well established system can be used by severely disabled people in their homes with minimal ongoing technical oversight.

In another attempt, P300 signals have been used for the design of a speller (text-input application). Twenty-five channels around C3, C4, Cz, CPz, and FCz electrodes are manually selected to obtain the best result. In addition P7 and P8 electrodes are used. The number of channels is later reduced to 20 using the principal component analysis (PCA) and selecting the largest eigenvalues. During this process the data are also whitened and the effect of the eye-blink is removed. Support vector machines (SVMs) are then used to classify the principal components for each character, using a Gaussian kernel [83].

3.4 Summary and Conclusions

ERP signals indicate the types and states of many brain abnormalities and mental disorders. These signals are characterized by their spatial, temporal, and spectrum locations. They are also often considered as independent sources within the brain. The ERP signals can be characterized by their amplitudes, latencies, source locations, and frequency contents. Automatic extraction and classification of these signals, however, requires sufficient knowledge and expertise in the development of mathematical and signal processing algorithms. Although to date there has not been any robust and well-established method to detect and characterize these signals, some recent methods have been described here and their potentials supported by evaluation on real EEG signals.

References

[1] Walter, W. G., 'Contingent negative variation: an electrical sign of sensorimotor association and expectancy in the human brain', *Nature*, **203**, 1964, 380–384.

[2] Sutton, S., Braren, M., Zoubin, J., and John, E. R., 'Evoked potential correlates of stimulus uncertainty', *Science*, **150**, 1965, 1187–1188.

[3] Johnson, Jr, R., 'Event-related brain potentials', in *Progressive Supranuclear Palsy: Clinical and Research Approaches*, Eds I. Litvan and Y. Agid Oxford University Press, New York, 1992, 122–154.

[4] Visser, S. L., Stam, F. C., Van Tilburg, W., Jonker, C., and De Rijke, W., 'Visual evoked response in senile and presenile dementia', *Electroencephalogr. Clin. Neurophysiol.*, **40**(4), 1976, 385–392.

[5] Cosi, V., Vitelli, E., Gozzoli, E., Corona, A., Cerroni, M., and Callieco, R., 'Visual evoked potentials in aging of the brain', *Adv. Neurol.*, **32**, 1982, 109–115.

[6] Coben, L. A., Danziger, W. L., and Hughes, C. P., 'Visual evoked potentials in mild senile dementia of Alzheimer type', *Electroencephalogr. Clin. Neurophysiol.*, **52**, 1981, 100.

[7] Visser, S. L., Van Tilburg, W., Hoojir, C., Op den Velde, W., Blom, J. L., and De Rijke, W., 'Visual evoked potentials (VEPs) in senile dementia (Alzheimer type) and in nonorganic behavioural disorders in the elderly: comparison with EEG parameters', *Electroencephalogr. Clin. Neurophysiol.*, **60**(2), 1985, 115–121.

[8] Nunez, P. L., *Electric Fields of the Brain*, Oxford University Press, New York, 1981.

[9] Picton, T. W., Lins, D. O., and Scherg, M., 'The recording and analysis of event-related potentials', in *Hand Book of Neurophysiology*, Vol. 10, Eds F. Boller and J. Grafman, Amsterdam, Elsevier, 1995, 3–73.

[10] Perrin, P., Pernier, J., Bertrand, O., and Echallier, J. F., 'Spherical splines for scalp potential and current density mapping', *Electroencephalogr. Clin. Neurophysiol.*, **72**, 1989, 184–187.

[11] Hegerl, U., 'Event-related potentials in psychiatry', Chapter 31, in *Electroencephalography, Basic Principles, Clinical Applications, and Related Fields*, Eds E. Niedermayer and F. Lopes da Silva, 4th edn., Eds Lippen cott, Williams and Wilkins, Philadelphia, Pennsylvania, 1999, 621–636.

[12] Diez, J., Spencer, K., and Donchin, E., 'Localization of the event-related potential novelty response as defined by principal component analysis', *Brain Res. Cognition*, **17** (3), 2003, 637–650.

[13] Frodl-Bauch, T., Bottlender, R., and Hegerl, U., 'Neurochemical substrates and neuro-anatomical generators of the event-related P300', *Neuropsychobiology*, **40**, 1999, 86–94.

[14] Kok, A., Ramautar, J., De Ruiter, M., Band, G., and Ridderinkhof, K., 'ERP components associated with successful and unsuccessful stopping in a stop-signal task', *Psychophysiology*, **41**(1), 2004, 9–20.

[15] Polich, J., 'Clinical application of the P300 event-related brain potential', *Phys. Med. Rehab. Clin. N. Am.*, **15**(1), 2004, 133–161.

[16] Friedman, D., Cycowics, Y., and Gaeta, H., 'The novelty P3: an event-related brain potential (ERP) sign of the brains evaluation of novelty', *Neurosci. Biobehav. Rev.*, **25**(4), 2001, 355–373.

[17] Kropotov, J. D., Alho, K., Näätänen, R., Ponomarev, V. A., Kropotova, O. V., Anichkov, A. D. and Nechaev, V. B., 'Human auditory-cortex mechanisms of preattentive sound discrimination', *Neurosci. Lett.*, **280**, 2000, 87–90.

[18] Comerchero, M., and Polich, J., 'P3a and P3b from typical auditory and visual stimuli', *Clin. Neurophysiol.*, **110**(1), 1999, 24–30.

[19] Luria, A. R., *The Working Brain*, Basic Books, New York, 1973.

[20] Sokolov, E. N., *Perception and the Conditioned Reflex*, Pergamon Press, Oxford, 1963.

[21] Friedman, D., Cycowicz, Y. M., and Gaeta, H., 'The novelty P3: an event-related brain potential (ERP) sign of the brain's evaluation of novelty', *Neuroscience and Biobehavioral Reviews*, Volume 25, Number 4, Elsevier, June 2001, pp. 355–373.

[22] Lange, D. H., Pratt, H., and Inbar, G. F., 'Modeling and estimation of single evoked brain potential components', *IEEE Trans. Biomed. Engng.*, **44**(9), 1997, 791–799.

[23] Farwell, L. A., and Dounchin, E., 'Talking off the top of your heard: toward a mental prosthesis utilizing event-related brain potentials', *Electroencephalogr. Clin. Neurophysiol.*, **70**, 1998, 510–523.

[24] Donchin, E., Spencer, K. M., and Wijesingle, R., 'The mental prosthesis: assessing the speed of a P300-based brain–computer interface', *IEEE Trans. Rehabil. Engng.*, **8**, 2000, 174–179.

[25] McGillem, C. D. and Aunon, J. I., 'Measurement of signal components in single visually evoked brain potentials', *IEEE Trans. Biomed. Engng.*, **24**, May 1977, 232–241.

[26] Makeig, S., Jung, T. P., Bell, A. J., and Sejnowsky, T. J., 'Blind separation of auditory event-related brain responses into independent components', *Proc. Natl Acad. Sci.*, **94**, 1997, 10979–10984.

[27] Bell, A. J., and Sejnowsky, T. J., 'An information maximization approach to blind separation and blind deconvolution', *Neural Computation*, **7**, 1995, 1129–1159.

[28] Xu, N., Gao, X., Hong, B., Miao, X., Gao, S., and Yang, F., 'BCI competition 2003-data set IIb: enhancing P300 wave detection using ICA-based subspace projections for BCI applications', *Trans. Biomed. Engng.*, **51**(6), 2004, 1067–1072.

[29] Serby, H., E. Yom-Tov, and Inbar, G. F., 'An improved P300-based brain–computer interface', *IEEE Trans. Neural Systems and Rehabil. Engng.*, **13**(1), March 2005, 89–98.

[30] Aunon, J. I., McGillem, C. D., and Childers, D. G., 'Signal processing in evoked potential research: averaging and modelling', *CRC Crit. Rev. Bioengng*, **5**, 1981, 323–367.

[31] Spyrou, L., Sanei, S., and C. Cheong Took, 'Estimation and location tracking of the P300 subcomponents from single-trial EEG', in Proceedings of the IEEE International Conference on *Acoustics, Speech, and Signal Processing (ICASSP)*, Vol. II, 2007, 1149–1152.

[32] Sarvas, J., 'Basic mathematical and electromagnetic concepts of the biomagnetic inverse problem', *Phys. Med. Biol.*, **32**(1), 1987, 11–22.

[33] Spyrou, L., and Sanei, S., 'Localization of event related potentials incorporating spatial notch filters', *IEEE Trans. Biomed. Engng.*, January 2007 (to be published subject to revision).

[34] Marple Jr., S. L., *Digital Spectral Analysis with Applications*, Prentice-Hall, Englewood Cliffs, New Jersey, 1987.

[35] Bertrand, O., Bohorquez, J., and Pernier, J., 'Time–frequency digital filtering based on an invertible wavelet transform: an application to evoked potentials', *IEEE Trans. Biomed. Engng.*, **41**(1), 1994, 77–88.

[36] Ademoglu, A., Micheli-Tzanakou, E., and Istefanopulos, Y., 'Analysis of pattern reversal visual evoked potentials (PRVEPs) by Spline wavelets', *IEEE Trans. Biomed. Engng.*, **44**(9), 1997, 881–890.

[37] Unser, M., Aldroubi, A., and Eden, M., 'On the asymptotic convergence of B-spline wavelets to Gabor functions', *IEEE Trans. Inform. Theory*, **38**(2), 1992, 864–872.

[38] Heinrich, H., Dickhaus, H., Rothenberger, A., Heinrich, V., and Moll, G. H., 'Single sweep analysis of event-related potentials by wavelet networks–methodological basis and clinical application', *IEEE Trans. Biomed. Engng.*, **46**(7), 1999, 867–878.

[39] Quiroga, R. Q., and Garcia, H., 'Single-trial event-related potentials with wavelet denoising', *Clin. Neurophysiol.*, **114**, 2003, 376–390.

[40] Bartnik, E. A., Blinowska, K., and Durka, P. J., 'Single evoked potential reconstruction by means of wavelet transform', *Biolog. Cybern.*, **67**(2), 1992, 175–181.

[41] Basar, E., 'EEG dynamics and evoked potentials in sensory and cognitive processing by brain', in *Dynamics of Sensory and Cognitive Processing by Brain*, Ed. E. Basar, Springer, 1988.

[42] Zhang, Q., and Benveniste, A., 'Wavelet networks', *IEEE Trans. Neural Networks*, **3**, 1992, 889–898.

[43] Rumelhart, D., Hinton, G. E., and Williams, R. J., 'Learning internal representations by error propagation', in *Parallel Distributed Processing*, Eds D. Rumelhart and J. L. McClelland, Vol.1, MIT Press, Cambridge, Massachusetts, 1986, 318–362.

[44] Glassman, E. L., 'A wavelet-like filter based on neuron action potentials for analysis of human scalp electroencephalographs', *IEEE Trans. Biomed. Engng.*, **52**(11), 2005, 1851–1862.

[45] Georgiadis, S. D., Ranta-aho, P. O., Tarvainen, M. P., and Karjalainen, P. A., 'Single-trial dynamical estimation of event-related potentials: a Kalman filter-based approach.' *IEEE Trans. Biomed. Engng.*, **52**(8), 2005, 1397–1406.

[46] Sorenson, H. W., *Parameter Estimation: Principles and Problems*, Marcel Dekker, New York, 1980.

[47] Melsa, J., and Cohn, D., *Decision and Estimation Theory*, McGraw-Hill, New York, 1978.

[48] Yates, R. D. and Goodman, D. J., *Probability and Stochastic Processes*, 2nd edn., John Wiley & Sons, Ltd, Chichester, 2005.

[49] Hansson, M., Gansler, T., and Salomonsson, C., 'Estimation of single event-related potentials utilizing the Prony method', *IEEE Trans. Biomed. Engng.*, **43**(10), 1996, 973–978.

[50] Scharf, L. L., *Statistical Signal Processing*, Addison-Wesley, Reading, Massachusetts, 1991.

[51] Garoosi, V., and Jansen, B. H., 'Development and evaluation of the piecewise Prony method for evoked potential analysis', *IEEE Trans. Biomed. Engng.*, **47**(12), 2000, 1549–1554.

[52] Barone, P., Massaro, E., and Polichetti, A., 'The segmented Prony method for the analysis of nonstationary time series', *Astronomy and Astrophys.*, **209**, 1989, 435–444.

[53] Meyer, J. U., Burkhard, P. M., Secomb, T. W., and Intaglietta, M., 'The Prony spectral line estimation (PSLE) method for the analysis of vascular oscillations', *IEEE Trans. Biomed. Engng.*, **36**, 1989, 968–971.

[54] Kulp, R. W., 'An optimum sampling procedure for use with the Prony method', *IEEE Trans Electromagn. Compatibility* **EMC-23**(2), 1981, 67–71.

[55] Steedly, W. M., Ying, C. J., and Moses, R. L., 'A modified TLS–Prony method using data decimation', *IEEE Trans. Signal Process.*, **42**, 1994, 2292–2303.

[56] Cui, J., and Wong, W., 'The adaptive chirplet transform and visual evoked potentials', *IEEE Trans. Biomed. Engng.*, **53**(7), July 2006, 1378–1384.

[57] Regan, D., *Human Brain Electrophysiology: Evoked Potentials and Evoked Magnetic Fields in Science and Medicine*, Elsevier, New York, 1989.

[58] Middendorf, M., McMillan, G., Galhoum, G., and Jones, K. S., 'Brain computer interfaces based on steady-state visual-evoked responses', *IEEE Trans. Rehabil. Engng.*, **8**(2), 2000, 211–214.

[59] Cheng, M., Gao, X. R., Gao, S. G., and Xu, D. F., 'Design and implementation of a brain–computer interface with high transfer rates', *IEEE Trans. Biomed. Engng.*, **49**(10), 2002, 1181–1186.

[60] Holliday, A. M., *Evoked Potentials in Clinical Testing*, 2nd edn, Churchill Livingston, Edinburgh, 1992.

[61] Heckenlively, J. R., and Arden, J. B., *Principles and Practice of Clinical Electrophysiology of Vision*, Mosby Year Book, St Louis, Missouri, 1991.

[62] Mann, S., and Haykin, S., 'The chirplet transform–physical considerations', *IEEE Trans. Signal Process.*, **43**(11), November 1995, 2745–2761.

[63] Bultan, A., 'A four-parameter atomic decomposition of chirplets', *IEEE Trans. Signal Process.*, **47**(3), March 1999, 731–745.

[64] Mallat, S. G. and Zhang, Z., 'Matching-pursuit with time-frequency dictionaries', *IEEE Trans. Signal Process.*, **41**(12), December 1993, 3397–3415.

[65] Gribonval, R., 'Fast matching pursuit with a multiscale dictionary of Gaussian chirps', *IEEE Trans. Signal Process.*, **49**(5), May 2001, 994–1001.

[66] Qian, S., Chen, D. P., and Yin, Q. Y., 'Adaptive chirplet based signal approximation', in Proceeding of the IEEE International Conference on *Acoustics, Speech, and Signal Processing*, 1998, Vols **1–6**, 1781–1784.

[67] O'Neil, J. C., and Flandrin, P., 'Chirp hunting', in Proceedings of the IEEE-SP Internationl Symposium on Time–Frequency and Time–Scale Analysis, 1998, 425–428.

[68] Ullsperger, P., Metz, A. M., and Gille, H. G., 'The P300 component of the event-related brain potential and mental effort', *Ergonomics*, **31**, 1998, 1127–1137.

[69] Uetake, A., and Murata, A., 'Assessment of mental fatigue during VDT task using event-related potential (P300)', in Proceeding of the IEEE International Workshop on *Root and Human Interactive Communications*, Osaka, Japan, 27–29 September 2000, 235–240.

[70] Ulsperger, P., Neumann, U., Gille, H. G., and Pictschan, M., 'P300 component of the ERP as an index of processing difficulty', in *Human Memory and Cognitive Capabilities*, Eds F. Flix and H. Hagendorf, Amsterdam, 1986, 723–773.

[71] Neumann, U., Ulsperger, P., and Erdman, U., 'Effects of graduated processing difficulty on P300 component of the event-related potential', *Z. Psychology*, **194**, 1986, 25–37.

[72] McFarland, D. J., Miner, L. A., Vaughan, T. M., and Wolpaw, J. R., 'Mu and beta rhythm topographies during motor imagery and actual movements', *Brain Topography*, Volume 12, Number 3, Springer, 2000, 177–186.

[73] Vidal, J. J., 'Real-time detection of brain events in EEG', *IEEE Proc.*, **65**, 1977, 633–664, Special Issue on Biological Signal Processing and Analysis.

[74] Sutter, E. E., 'The brain response interface communication through visually induced electrical brain response', *J. Microcomp. Applic*, **15**, 1992, 31–45.

[75] Morgan, S. T., Hansen, J. C., and Hillyard, S. A., 'Selective attention to the stimulus location modulates the steady state visual evoked potential', *Neurobiology*, **93**, 1996, 4770–4774.

[76] Muller, M. M., and Hillyard, S. A., 'Effect of spatial selective attention on the steady-state visual evoked potential in the 20-28 Hz range', *Cognitive Brain Res.*, **6**, 1997, 249–261.

[77] Birbaumer, N., Hinterberger, T., Kubler, A., and Neumann, N., 'The thought-translation device (ttd): neurobehavioral mechanisms and clinical outcome', *IEEE Trans. Neural System Rehabil. Engng.*, **11**, 2003, 120–123.

[78] Sellers, E. W., Kubler, A., and Donchin, E., 'Brain–computer interface research at the University of South Florida Cognitive Psychology Laboratory: the P300 speller', *IEEE Trans. Neural Systems and Rehabil. Engng.*, 2006, 221–224 (to appear).

[79] Diez, J., Pfurtscheller, G., Reisecker, F., Ortmayr, B., Zalaudek, K., 'Event-related desynchronization and synchronization in idiopathic Parkinson's disease', *Electroencephalogr. and Clin. Neurophysiol.*, **103**(1), 1997, 155–155.

[80] Beverina, F., Palmas, G., Silvoni, S., Piccione, F., and Giove, S., 'User adaptive BCIs: SSVEP and P300 based interfaces', *J. Psychol.*, **1**(4), 2003, 331–354.

[81] Kelly, S. P., Lalor, E. C., Finucane, C., McDarby, G., and Reilly, R. B., 'Visual spatial attention control in an independent brain–computer interface', *IEEE Trans. Biomed. Engng.*, **52**(9), Sept. 2005, 1588–1596.

[82] Vaughan, T. M., McFarland, D. J., Schalk, G., Sarnacki, W. A., Krusienski, D. J., Sellers, E. W., and Wolpaw, J. R., 'The Wandsworth BCI Research and Development Program: at home with BCI', *IEEE Trans. Neural Systems and Rehabil. Engng.*, **14**(2), June 2006, 229–233.

[83] Thulasidas, M., Cuntai, G., and Wu, J., 'Robust classification of EEG signal for brain–computer interface', *IEEE Trans. Neural Systems and Rehabil. Engng.*, **14**(1), 2006, 24–29.

4

Seizure Signal Analysis

An original impression and the fundamental concepts of epilepsy were refined and developed in ancient Indian medicine during the Vedic period of 4500–1500BC. In the Ayurvedic literature of Charaka Samhita (around 400BC, and the oldest existing description of the complete Ayurvedic medical system), epilepsy is described as ‘apasmara’, which means ‘loss of consciousness’. The word ‘epilepsy’ is derived from the Greek word *epilambanein*, which means ‘to seize or attack’. It is now known, however, that seizures are the result of sudden, usually brief, excessive electrical discharges in a group of brain cells (neurons) and that different parts of the brain can be the site of such discharges. The clinical manifestations of seizures therefore vary and depend on where in the brain the disturbance first starts and how far it spreads. Transient symptoms can occur, such as loss of awareness or consciousness and disturbances of movement, sensation (including vision, hearing, and taste), mood, or mental function.

The literature of Charaka Samhita contains abundant references to all aspects of epilepsy including symptomatology, etiology, diagnosis, and treatment. Another ancient and detailed account of epilepsy is on a Babylonian tablet in the British Museum in London. This is a chapter from a Babylonian textbook of medicine comprising 40 tablets dating as far back as 2000BC. The tablet accurately records many of the different seizure types recognized today. In contrast to the Ayurvedic medicine of Charaka Samhita, however, it emphasizes the supernatural nature of epilepsy, with each seizure type associated with the name of a spirit or god – usually evil. Treatment was, therefore, largely a spiritual matter. The Babylonian view was the forerunner of the Greek concept of ‘the sacred disease’, as described in the famous treatise by Hippocrates (dated to the 5th century BC). The term ‘seleniazetai’ was also often used to describe people with epilepsy because they were thought to be affected by the phases of the moon or by the moon god (Selene), and hence the notion of ‘moonstruck’ or ‘lunatic’ (the Latinized version) arose. Hippocrates, however, believed that epilepsy was not sacred, but a disorder of the brain. He recommended physical treatments and stated that if the disease became chronic, it was incurable. However, the perception that epilepsy was a brain disorder did not begin to take root until the 18th and 19th centuries AD. The intervening 2000 years were dominated by more supernatural views. In Europe, for example, St. Valentine has been the patron saint of people with epilepsy since mediaeval times. During this time people with epilepsy were

EEG Signal Processing S. Sanei and J. Chambers

viewed with fear, suspicion, and misunderstanding and were subjected to enormous social stigma. People with epilepsy were treated as outcasts and punished. Some, however, succeeded and became famous the world over. Among these people were Julius Caesar, Czar Peter the Great of Russia, Pope Pius IX, the writer Fedor Dostoevsky, and the poet Lord Byron.

During the 19th century, as neurology emerged as a new discipline distinct from psychiatry, the concept of epilepsy as a brain disorder became more widely accepted, especially in Europe and the United States of America (USA). This helped to reduce the stigma associated with the disorder. Bromide, introduced in 1857 as the world's first effective antiepileptic drug, became widely used in Europe and the USA during the second half of the last century. The first hospital centre for the 'paralysed and epileptic' was established in London in 1857. At the same time a more humanitarian approach to the social problems of epilepsy resulted in the establishment of epilepsy 'colonies' for care and employment.

The new understanding of epilepsy (pathophysiology) was also established in the 19th century with the work of neurologist Hughlings Jackson in 1873, who proposed that seizures were the result of sudden brief electrochemical discharges in the brain. He also suggested that the character of the seizures depended on the location and function of the site of the discharges. Soon afterwards the electrical excitability of the brain in animals and man was discovered by David Ferrier in London and Gustav Theodor Fritsch and Eduard Hitzig in Germany.

Hans Berger, a psychiatrist, developed the human EEG. Its important application from the 1930s onwards was in the field of epilepsy. The EEG revealed the presence of electrical discharges in the brain. It also showed different patterns of brainwave discharges associated with different seizure types. The EEG also helped to locate the site of seizure discharges and expanded the possibilities of neurosurgical treatments, which became much more widely available from the 1950s onwards in London, Montreal, and Paris.

During the first half of the 20th century the main drugs for treatment were phenobarbitone (first used in 1912) and phenytoin (first used in 1938). Since the 1960s, there has been an accelerating process of drug discovery, based in part on a much greater understanding of the electrochemical activities of the brain, especially the excitatory and inhibitory neurotransmitters. In developed countries in recent years, several new drugs have come into the market and seizures can now be controlled in 70–80 % of newly diagnosed children and adults.

Neuroimaging techniques such as fMRI and position emission tomography (PET) boost the success in diagnosis of epilepsy. Such technology has revealed many of the more subtle brain lesions responsible for epilepsy. Several brain lesions such as trauma, congenital, developmental, infection, vascular, and tumour might lead to epilepsy in some people. Of the 50 million people in the world with epilepsy, some 35 million have no access to appropriate treatment. This is either because services are nonexistent or because epilepsy is not viewed as a medical problem or a treatable brain disorder. As a major campaign for the treatment of epilepsy, the International League Against Epilepsy (ILAE) and the International Bureau for Epilepsy (IBE) joined forces with the World Health Organization in 1997 to establish the Global Campaign Against Epilepsy to address these issues.

Epilepsy is a sudden and recurrent brain malfunction and is a disease that reflects an excessive and hypersynchronous activity of the neurons within the brain. It is probably the most prevalent brain disorder among adults and children, second only to stroke. Over

50 million people worldwide are diagnosed with epilepsy, whose hallmark is recurrent seizures [1]. The prevalence of epileptic seizures changes from one geographic area to another [2].

The seizures occur at random to impair the normal function of the brain. Epilepsy can be treated in many cases and the most important treatment today is pharmacological. The patient takes anticonvulsant drugs on a daily basis, trying to achieve a steady-state concentration in the blood, which are chosen to provide the most effective seizure control. Surgical intervention is an alternative for carefully selected cases that are refractory to medical therapy. However, in almost 25 % of the total number of patients diagnosed with epilepsy, seizures cannot be controlled by any available therapy. Furthermore, side effects from both pharmacological and surgical treatments have been reported.

An epileptic seizure can be characterized by paroxysmal occurrence of synchronous oscillations. Such seizures can be classified into two main categories depending on the extent of involvement of various brain regions: focal (or partial) and generalized. Generalized seizures involve most areas of the brain whereas focal seizures originate from a circumscribed region of the brain, often called epileptic foci [3]. Figure 4.1 shows two segments of EEG signals involving generalized and focal seizures respectively.

Successful surgical treatment of focal epilepsies requires exact localization of the epileptic focus and its delineation from functionally relevant areas [4]. The physiological aspects of seizure generation and the treatment and monitoring of a seizure, including presurgical examinations, have been well established [3] and medical literature provided.

EEG, MEG, and recently fMRI are the major neuroimaging modalities used for seizure detection. The blood-oxygenation-level-dependent (BOLD) regions in fMRI of the head clearly show the epileptic foci. However, the number of fMRI machines is limited in each area, they are costly, and a full body scan is time consuming. Therefore, using fMRI for all patients at all times is not feasible. MEG, on the other hand, is noisy and since the patient under care has to be steady during the recording, it is hard to achieve clear data for moderate and severe cases using current MEG machines.

Therefore, EEG remains the most useful and cost effective modality for the study of epilepsy. Although for generalized seizure the duration of seizure can be easily detected using a naked eye, for most focal epilepsies such intervals are difficult to recognize.

From the pathological point of view, there are clear classification schemes for seizures. ‘Partial’ is used to describe isolated phenomena that reflect focal cortical activity, either evident clinically or by EEG. The term ‘simple’ indicates that consciousness is not impaired. For example, a seizure visible as a momentarily twitching upper extremity, which subsides, would be termed a simple partial seizure with motor activity. Partial seizures may have motor, somatosensory, psychic, or autonomic symptoms [5].

The term ‘complex’ defines an alteration of consciousness associated with the seizure. ‘Generalization’ is a term used to denote the spread from a focal area of the cortex, which could be evident clinically by EEG, and involves all areas of the cortex with resulting generalized motor convulsion. It is known that in adults the most common seizure type is that of initial activation of one area of the cortex with subsequent spread to all areas of the cortex; frequently this occurs too quickly to be appreciated by bedside observation.

The other major grouping of seizure types is for generalized seizures, which may be termed convulsive or nonconvulsive. For this type, all areas of the cortex are activated at once. This, for example, is seen with absence seizures and myoclonic seizures [6].

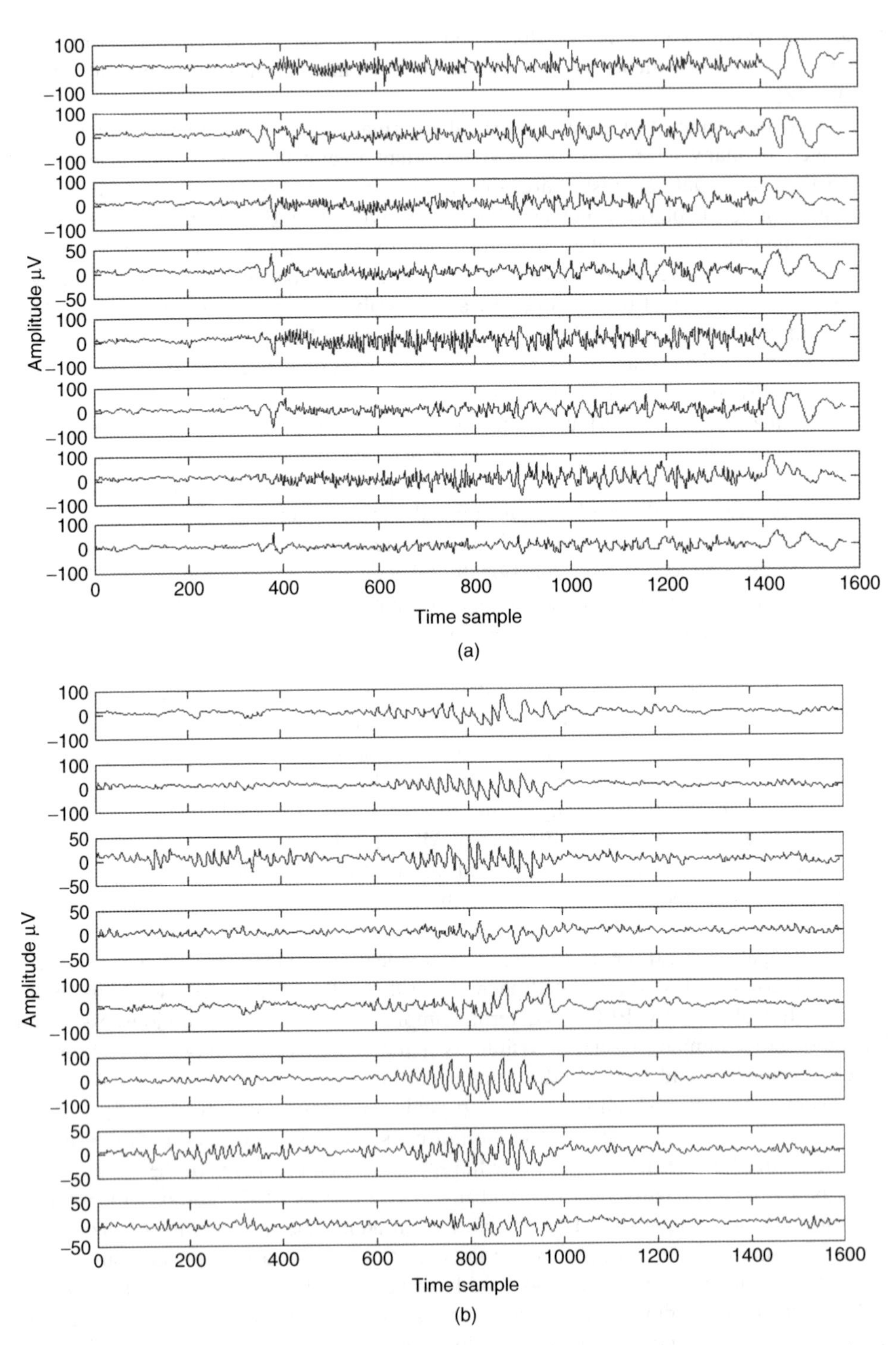

Figure 4.1 Two segments of EEG signals each from a patient suffering (a) generalized seizure and (b) focal seizure onset

Tonic-clonic (grand mal) seizures are more common and because of that other types of seizures may escape detection. They were diagnosed for the first time in 1827 by Bravais [7]. The recurrent behaviour of the spike waveforms in the EEG signals triggers investigation for a possible seizure disorder. Seizures of frontal or temporal cortical origin with nonclassical motor movements are fairly commonly encountered. The patient may show some seemingly organized motor activity without the usually in-phase jerking movements more typical of generalized seizures. Also complicating the problem is that clouding or alteration of consciousness may occur without complete loss of consciousness.

The term 'aura' is used to represent ongoing seizure activity limited to a focal area of the cortex. In this case the abnormal electrical activity associated with the seizure does not spread or generalize to the entire cerebral cortex but remains localized and persists in one abnormal focus.

One seizure type may evolve into another seizure type. For example, a simple motor seizure may evolve into a complex partial seizure with altered consciousness; the terminology for this would be 'partial complex status epilepticus' [6].

Absence seizures (also known as petit mal) are a primarily generalized seizure type involving all cortical areas at once; this is typically a seizure disorder of childhood with a characteristic EEG pattern [6]. At times, absence seizures may persist with minimal motor movements and altered consciousness for hours or days. Absence epileptic seizure and complex partial epileptic seizure are often grouped under the term 'nonconvulsive status epilepticus' and are referred to at times as twilight or fugue states.

The term 'subtle status epilepticus' is more correctly used to indicate patients that have evolved from generalized convulsive epileptic seizure or are in a comatose state with epileptiform activity.

The character of an epileptic seizure is determined based on the region of the brain involved and the underlying basic epileptic condition, which are mostly age-determined. As a conclusion, clinical classification of epileptic seizures is summarized as:

1. Partial (focal) seizures. These seizures arise from an electric discharge of one or more localized areas of the brain regardless of whether the seizure is secondarily generalized. Depending on their type, they may or may not impair consciousness. Whether seizures are partial or focal, they begin in a localized area of the brain, but then may spread to the whole brain causing a generalized seizure. They are divided in to:
   (a) simple partial;
   (b) complex partial:
      - simple partial onset followed by alteration of consciousness;
      - partial evolving to secondarily generalized.
2. Generalized seizures (convulsive and nonconvulsive). The electrical discharge which leads to these seizures involves the whole brain and may cause loss of consciousness and/or muscle contractions or stiffness. They include what used to be known as 'grand mal' convulsion and also the brief 'petit mal' absence of consciousness. These seizures are further divided into:
   (a) absence (typical and atypical);
   (b) clonic, tonic, or tonic–clonic (grand mal);

   (c) myoclonic;
   (d) atonic (astatic).
3. Epileptic seizures with unknown waveform patterns.
4. Seizures precipitated by external triggering events [6].

Moreover, epileptic seizures may be chronic or acute. However, so far no automatic classification of these seizures based on the EEG waveforms has been reported.

Most traditional epilepsy analysis methods, based on the EEG, are focused on the detection and classification of epileptic seizures, among which the best method of analysis is still the visual inspection of the EEG by a highly skilled electroencephalographer or neurophysiologist. However, with the advent of new signal processing methodologies several computerized techniques have been proposed to detect and localize epileptic seizures. In addition, based on the mathematical theory of nonlinear dynamics [8], there has been an increased interest in the analysis of the EEG for the prediction of epileptic seizures.

Seizure detection and classification using signal processing methods has been an important issue of research for the last two decades [9–12]. Researchers have tried to highlight different signal characteristics within various domains and classify the signal segments based on the measured features. Adult seizure is more robust than neonate (newborn) seizure. Therefore, its detection, labelling, and classification is not very difficult. On the other hand, neonate seizure is more chaotic and although some methods have been suggested for the detection of such events the problem still remains open. Therefore, various automated spike detection approaches have been developed [13].

Recently, predictability of seizure from long EEG recordings has attracted many researchers. It has been shown that epileptic sources gradually tend to be less chaotic from a few minutes before the seizure onset. This finding is clinically very important since the patients do not need to be under anticonvulsant administration permanently, but from just a few minutes before seizure. In the following sections the major research topics in the areas of epileptic seizure detection and prediction are discussed.

4.1 Seizure Detection

4.1.1 Adult Seizure Detection

In clinics, for patients with medically intractable partial epilepsies, time-consuming video-EEG monitoring of spontaneous seizures is often necessary [11]. Visual analysis of interictal EEG is, however, time intensive. Application of invasive methods for monitoring the seizure signals and identification of an epileptic zone is hazardous and involves risk for the patient.

Before designing any automated seizure detection system the characteristics of the EEG signals before, during, and after the seizure have to be determined and evaluated. Several features have been identified to describe the features better. These may represent the static behaviour of the signals within a short time interval, such as signal energy, or the dynamic behaviour of the signals, such as chaoticity and the change in frequency during the seizure onset.

Automatic seizure detection quantification and recognition has been an area of interest and research within clinical, physics, and engineering communities since the 1970s

[14–27]. In some early works in spike detection [16,21–23] a number of parameters such as relative amplitude, sharpness, and duration of EEG waves were measured from the EEG signals and evaluated. The method is sensitive to various artefacts. In these attempts different states such as active wakefulness or desynchronized EEG were defined, in which typical nonepileptic transients were supposed to occur [22,23]. A multistage system to detect the epileptiform activities from the EEGs was developed by Dingle *et al.* [18]. They combined a mimetic approach with a rule-based expert system, and thereby considered and exploited both the spatial and temporal systems. In another approach [19] multichannel EEGs were used and a correlation-based algorithm was attempted to reduce the muscle artefacts. Following this method, approximately 67 % of the spikes can be detected. By incorporating both multichannel temporal and spatial information, and including the electrocardiogram, electromyogram, and electrooculogram information into a rule-based system [20], a higher detection rate was achieved. A two-stage automatic system was developed by Davey *et al.* [17]. In the first stage a feature extractor and in the second stage a classifier were introduced. A 70 % sensitivity was claimed for this system.

Artificial neural networks (ANNs) have been used for seizure detection by many researchers [26,28]. The Kohonen self-organizing feature map ANN [29,30] was used for spike detection by Kurth *et al.* [28]. In this work, for each patient three different-sized neural networks (NNs) have been examined. The training vector included a number of signals with typical spikes, a number of eye-blinking artefact signals, some signals of muscle artefacts, and also background EEG signals. The major problem with these methods is that the epileptic seizure signals do not follow similar patterns. Presenting all types of seizure patterns to the ANN, on the other hand, reduces the sensitivity of the overall detection system. Therefore, a clever feature detection followed by a robust classifier often provides an improved result.

Among recent works, time–frequency (TF) approaches effectively use the fact that the seizure sources are localized in the time–frequency domain. Most of these methods are mainly for detection of neural spikes [31] of different types. Different TF methods following different classification strategies have been proposed by many researchers [32,33] in this area. The methods are especially useful since the EEG signals are statistically nonstationary. The discrete wavelet transform (DWT) obtains a better TF representation than the TF based on the short-term Fourier transform due to its multiscale (multilevel) characteristics; i.e. it can model the signal according to its coarseness. The DWT analyses the signal over different frequency bands, with different resolutions, by decomposing the signal into a coarse approximation and detail information. In a recent approach by Subasi [33], a DWT-based TF method followed by an ANN has been suggested. The ANN classifies the energy of various resolution (detail) levels. Using this technique, it is possible to detect more than 80 % of adult epileptic seizures. Other TF distributions such as the pseudo-Wigner–Ville can also be used for the same purpose [34].

In an established work by Osorio *et al.* [35,36] a digital seizure detection algorithm has been proposed and implemented. The system is capable of accurate real-time detection, quantitative analysis, and very short-term prediction of the clinical onset of seizure. This system computes a measure, namely ‘foreground’, of the median signal energy in the frequencies between 8 and 42 Hz in a short window of specific length (e.g. two seconds). The foreground is calculated through the following steps: (1) decomposing the signals into epileptiform (containing epileptic seizures) and nonepileptiform (without any seizure) components using a 22-coefficient wavelet filter (DAUB4, level 3), which separates the frequency subbands from 8 to 40 Hz; (2) the epileptiform component is squared, and 3) the squared components are median filtered. On the other hand, a 'background' reference signal is obtained as an estimate of the median energy of a longer time (approximately 30 minutes) of the signal. A large ratio between the foreground and background will then show the event of seizure [37]. An analogue system was later developed to improve the technical drawbacks of the above system such as speed and noise [37].

There are many features that can be detected/measured from the EEGs for detection of epileptic seizures. Often seizures increase the average energy of the signals during the onset. For a windowed segment of the signal this can be measured as

$$E(n) = \frac{1}{L} \sum_{p=m-L/2}^{n-1+L/2} x^2(p) \tag{4.1}$$

where L is the window length and the time index n is the window centre. The seizure signals have a major cyclic component and therefore generally exhibit a dominant peak in the frequency domain. The frequency of this peak, however, decays with time during the onset of seizure. Therefore the slope of decay is a significant factor in the detection of seizure onset. Considering $\mathrm{X}(f, n)$, the estimated spectrum of the windowed signal $x(n)$ centred at n, the peak at time n will be

$$f_d(n) = \arg\max_f(|X(f, n)|) \tag{4.2}$$

The spectrum is commonly estimated using autoregressive modelling [38]. From the spectrum the peak frequency is measured. The slope of the decay in the peak frequency can then be measured and used as a feature. The cyclic nature of the EEG signals can also be measured and used as an indication of seizure. This can be best identified by incorporating certain higher order statistics of the data. One such indicator is related to the second- and fourth order statistics of the measurements as follows [39]:

$$I = |C_2^0(0)|^{-4} \sum_{\alpha \neq 0} |P^\alpha|^2 \tag{4.3}$$

where $P^\alpha = C_4^\alpha(0, 0, 0)$ represents the Fourier coefficients of the fourth-order cyclic cumulant at zero lag and can be estimated as follows:

$$\hat{C}_4^\alpha(0, 0, 0) = \frac{1}{N} \sum_{n=0}^{N-1} x_c^4(n) \mathrm{e}^{-\mathrm{j}2\pi n\alpha/N} - 3 \sum_{\beta=0}^{\alpha} C_2^{\alpha-\beta}(0) C_2^\beta(0) \tag{4.4}$$

where $x_c(n)$ is the zeroed mean version of $x(n)$, and an estimation of $C_2^\alpha(0)$ is calculated as

$$\hat{C}_2^\alpha(0,) = \frac{1}{N} \sum_{n=0}^{N-1} x_c(n) \mathrm{e}^{-\mathrm{j}2\pi n\alpha/N} \tag{4.5}$$

This indicator is also measured with respect to the time index n since it is calculated for each signal window centred at n. I in Equation (4.3) measures, for a frame centred at n, the spread of the energy. Over the range of frequencies before seizure onset, the EEG is chaotic and no frequency appears to control its trace. During seizure, the EEG becomes more ordered (rhythmic) and therefore the spectrum has a large peak.

The above features have been measured and classified using a support vector machine (SVM) classifier [38]. It has been illustrated that for both tonic–clonic and complex partial seizures the classification rate can be as high as 100 %.

In a robust detection of seizure, however, all statistical measures from the EEGs together with all other symptoms, such as blood morphology, body movement, pain, changes in metabolism, heart rate variability, and respiration before, during, and after seizure, have to be quantified and effectively taken into account. This is a challenge for future signal processing/data fusion based approaches.

In another attempt for seizure detection a cascade of classifiers based on artificial neural networks (ANNs) has been used [40]. The classification is performed in three stages. In the first stage the following six features feed two perceptrons to classify peaks into *definite epileptiform, definite nonepileptiform*, and *unknown* waveforms. The features are selected based on the expected shape of an epileptic spike, such as the one depicted in Figure 4.2. These features are:

(a) first half-wave amplitude (FHWA),
(b) second half-wave amplitude (SHWA),
(c) first half-wave duration (FHWD),
(d) second half-wave duration (SHWD),
(e) first half-wave slope (FHWS $\approx$ FHWA/FHWD), and
(f) second half-wave slope (SHWS $\approx$ SHWA/SHWD).

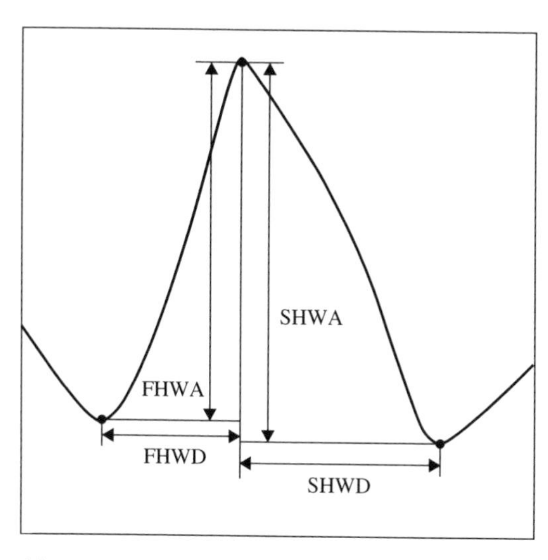

Figure 4.2 An epileptiform spike with features used in the classification of epileptic and nonepileptic seizures

Since three outputs are needed after the first stage (also called the preclassifier) two single-layer perceptrons are used in parallel. One perceptron is trained to give +1 for *definite nonepileptiform* and −1 otherwise. The second network produces +1 for *definite epileptiform* and −1 otherwise. A segment that produces −1 at the output of both networks is assigned to the *unknown* group.

The update equation to find $\boldsymbol{w}$, the vector of the weights of the perceptron ANNs, at the kth iteration can be simply presented as

$$\boldsymbol{w}_k = \boldsymbol{w}_{k-1} + \mu(\boldsymbol{d} - \boldsymbol{y})\boldsymbol{x} \tag{4.6}$$

where $\boldsymbol{x}$ is the input feature vector, $\boldsymbol{d}$ is the expected feature vector, $\boldsymbol{y}$ is the output, calculated as $\boldsymbol{y} = \text{sign}(\boldsymbol{w}^T\boldsymbol{x} - T_r)$, where T_r is an empirical threshold, and μ is the learning rate set empirically or adaptively.

In the second stage the *unknown* waveforms (spikes) are classified using a radial-basis function (RBF) neural network. An RBF has been empirically shown to have a better performance than other ANNs for this stage of classification. The inputs are the segments of actual waveforms. The output is selected by hard-limiting the output of the last layer (after normalization).

Finally, in the third stage a multidimensional SVM with a nonlinear (RBF) kernel is used to process the data from its multichannel input. An accuracy of up to 100 % has been reported for detection of generalized and most focal seizures [40].

Epilepsy is often characterized by the sudden occurrence of synchronous activity within relatively large brain neuronal networks that disturb the normal working of the brain. Therefore, some measures of dynamical change have also been used for seizure detection. These measures significantly change in the transition between the preictal and ictal states or even in the transition between the interictal and ictal states. In the latter case the transition can occur either as a continuous change in phase, such as in some cases of mesial temporal lobe epilepsy (MTLE), or as a sudden leap, e.g. in most cases of absence seizures. In the approaches based on chaos measurement by estimation of attractor dimension (as discussed later in this chapter), for the first case the attractor of the system gradually deforms from an interictal to an ictal attractor. In the second case, where a sharp critical transition takes place, it can be assumed that the system has at least two simultaneous interictal and ictal attractors all the time. In a study by Lopes da Silva *et al.* [41] three states (routes) have been characterized as illustrative models of epileptic seizures:

1. An abrupt transition of the bifurcation type, caused by a random perturbation. An interesting example of this is the epileptic syndrome characterized by paroxysmal spike-and-wave discharges in the EEG and nonconvolusive absence type of seizures.
2. A route where a deformation of the attractor is caused by an external perturbation (photosensitive epilepsy).
3. A deformation of the attractor leading to a gradual evolution into the ictal state, e.g. temporal lobe epilepsy (TLE).

The authors concluded that under these routes it is possible to understand those circumstances where the transition from the ongoing (interictal) activity mode to the ictal (seizure) mode may, or may not, be anticipated. Also, any of the three routes is possible, depending on the type of underlying epilepsy. Seizures may be generally unpredictable, as most often in absence-type seizures of idiopathic (primary) generalized epilepsy, or predictable, preceded by a gradual change in dynamics, detectable some time before the manifestation of seizure, as in TLE.

Chaotic behaviour of the EEGs is discussed in Section 4.3 and is mainly exploited for the prediction of seizure, as a new research direction.

4.1.2 Detection of Neonate Seizure

Seizure occurs in approximately 0.5 % of newborn (the first four weeks of life) babies. It represents a distinctive indicator of abnormality in the central nervous system (CNS). There are many causes for this abnormality, with the majority due to biochemical imbalances within the CNS, intracranial haemorrhage and infection, developmental (structural) defects, and passive drug addiction and withdrawal [42]. Analysis of neonatal EEG is a very difficult issue in the biomedical signal processing context. Unlike in adults, the presence of spikes may not be an indication of seizure. The clinical signs in the newborn are not always as obvious as those for an adult, where seizure is often accompanied by uncontrollable, repetitive, or jerky movements of the body, or the tonic flexion of muscles. The less obvious symptoms in the newborn, i.e. subtle seizures, may include sustained eye opening with ocular fixation, repetitive blinking or fluttering of eyelids, drooling, sucking, or other slight facial expressions or body movements. Therefore, detection of epileptic seizures for the newborn is far more complex than for adults and so far only a few approaches have been attempted. These approaches are briefly discussed in this section.

Although there is a significant spectral range for the newborn seizure signal [43] in most cases the seizure frequency band lies within the delta and theta bands (1–7 Hz). However, TF approaches are more popular due to the statistical nonstationarity of the data, but they can also be inadequate since the spikes may have less amplitude than the average amplitude of a normal EEG signal.

In most cases the signals are preprocessed before application of any seizure detection algorithm. Eye blinking and body movements are the major sources of artefacts. Conventional adaptive filtering methods (with or without a reference signal) may be implemented to remove interference [44,45]. This may be followed by the calculation of a TF representation of seizure [46] or used in a variable-time model of the epileptiform signal.

A model-based approach was proposed to model a seizure segment and the model parameters were estimated [47]. The steps are as follows:

1. The periodogram of the observed vector of an EEG signal, $\boldsymbol{x} = [x(1), x(2), \ldots, x(N)]^{\mathrm{T}}$, is given as

$$I_{xx}(k) = \frac{1}{2\pi N}\left|\sum_{n=1}^{N} \mathrm{e}^{-\mathrm{j}\lambda_k n} x(n)\right|^2 \tag{4.7}$$

where $\lambda_k = 2\pi k/N$.

2. A discrete approximation to the log-likelihood function for estimation of a parameter vector of the model, $\boldsymbol{\theta}$, is computed as

$$L_N(\boldsymbol{x}, \boldsymbol{\theta}) = -\sum_{k=0}^{\lfloor (N-1)/2 \rfloor} \left[\log(2\pi)^2 S_{xx}(\lambda_k, \boldsymbol{\theta}) + \frac{I_{xx}(\lambda_k)}{S_{xx}(\lambda_k, \boldsymbol{\theta})} \right] \tag{4.8}$$

where $S_{xx}(\lambda_k, \boldsymbol{\theta})$ is considered as the spectral density of a Gaussian vector process $\boldsymbol{x}$ with parameters $\boldsymbol{\theta}$, $\lambda_k = 2\pi k/N$, and $\lfloor \cdot \rfloor$ denotes the floor operator. L_N needs to be maximized with respect to $\boldsymbol{\theta}$. As $N \to \infty$ this approximation approaches that of Whittle's [48].
The parameter estimate of Whittle is given by

$$\hat{\boldsymbol{\theta}} = \arg\max_{\boldsymbol{\theta}}(L_N(\boldsymbol{x}, \boldsymbol{\theta});\ \boldsymbol{\theta} \in \boldsymbol{\Theta}) \tag{4.9}$$

The parameter space $\boldsymbol{\Theta}$ may include any property of either the background EEG or the seizure, such as nonnegative values for postsynaptic pulse shaping parameters. This is the major drawback of the method since these parameters are highly dependent on the model of both seizure spikes and the background EEG. A self-generating model [49] followed by a quadratic discriminant function was used to distinguish between the spectrum density of background EEG segments ($S_k^{\text{Background}}$) and seizure segments (S_k^{Seizure}); k represents the segment number. This model has already been explained in Chapter 2.

3. Based on the model above, the power of the background EEG and the seizure segment is calculated [47] and their ratio is tested against an empirical threshold level, say $\boldsymbol{\Gamma}$; i.e.

$$\boldsymbol{\Gamma} = \frac{P_{\text{Seizure}}}{P_{\text{Background}}} \tag{4.10}$$

where

$$P_{\text{Background}} = \sum_{k=0}^{N-1} S_k^{\text{Background}} \quad \text{and} \quad P_{\text{Seizure}} = \sum_{k=0}^{N-1} S_k^{\text{Seizure}}$$

It is claimed that for some well-adjusted model parameters a false alarm percentage as low as 20 % can be achieved [47].

A TF approach for neonate seizure detection has been suggested [50] in which a template in the TF domain is defined as

$$Z_{\text{ref}}(\tau, \upsilon) = \sum_{i=1}^{L} \exp\left[\frac{-(\upsilon - \alpha_i \tau)^2}{2\sigma^2} \right] \tag{4.11}$$

where the time scales α_i and variance σ^2 can change respectively, with the position and width of the template. Z_{ref}, with variable position and variance in the TF domain, resembles a seizure waveform and is used as a template. This template is convolved with the TF domain EEG, $X(t, f)$, in both the time and frequency domains, i.e.

$$\eta_{\alpha,\sigma^2}(n, f) = X(n, f) ** Z_{\text{ref}}(n, f) \tag{4.12}$$

where ** represents the two-dimensional convolution with respect to discrete time and frequency, and $X(n, k)$ for a real discrete signal x is defined as

$$X(n, k) = \sum_{q=n/2}^{N-n/2} x(q + n/2)x(q - n/2)e^{-j2\pi kq} \tag{4.13}$$

where N is the signal length in terms of samples. Assuming the variance is fixed, the α_i may be estimated in order to maximize a test criterion to minimize the difference between Z_{ref} and the desired waveform. This criterion is tested against a threshold level to decide whether there is a seizure. The method has been applied to real neonate EEGs and a false detection rate (FDR) as low as 15 % in some cases has been reported [50].

The neonate EEG signals unfortunately do not manifest any distinct TF pattern at the onset of seizure. Moreover, they include many seizure type spikes due to immature mental activities of newborns and their rapid changes in brain metabolism. Therefore, although the above methods work well for some synthetic data, due to the nature of neonate seizures, as described before, they oftentimes fail to detect effectively all types of neonate seizures.

In a recent work by Karayiannis *et al.* a cascaded rule-based neural network algorithm for detection of epileptic seizure segments in neonatal EEG has been developed [51]. In this method it is assumed that the neonate seizures are manifested as subtle but somehow stereotype repetitive waveforms that evolve in amplitude and frequency before eventually decaying. Three different morphologies of seizure patterns for pseudosinusoidal, complex morphology, and rhythmic runs of spike-like waves have been considered. These patterns are illustrated in Figure 4.3 [51].

The automated detection of neonate seizure is then carried out in three stages. Each EEG channel is treated separately, and spatial and interchannel information is not exploited. In the first stage the spectrum amplitude is used to separate bursts of rhythmic activities. In the second stage the artefacts, such as the results of patting, sucking, respiratory function, and EKG (or ECG), are mitigated, and in the last stage a clustering operation is performed to distinguish between epileptic seizures, nonepileptic seizures, and the normal EEG affected by different artefacts. As a result of this stage, isolated and inconsistent candidate seizure segments are eliminated, and the final seizure segments are recognized. The performances of conventional feedforward neural networks (FFNN) [52] as well as quantum neural networks (QNN) [53] have been compared for classification of some frequency-domain features in all of the above stages. These features are denoted as first dominant frequency, second dominant frequency, width of dominant frequency, percentage of power contributed to the first dominant frequency, percentage of power contributed to the second dominant frequency, peak ratio, and stability ratio (a time-domain parameter that measures the amplitude stability of the EEG segment). It has also been shown that there is no significant difference in using an FFNN or QNN and the results will be approximately the same [54]. The overall algorithm is very straightforward to implement and both its sensitivity and specificity have been shown to be above 80 %.

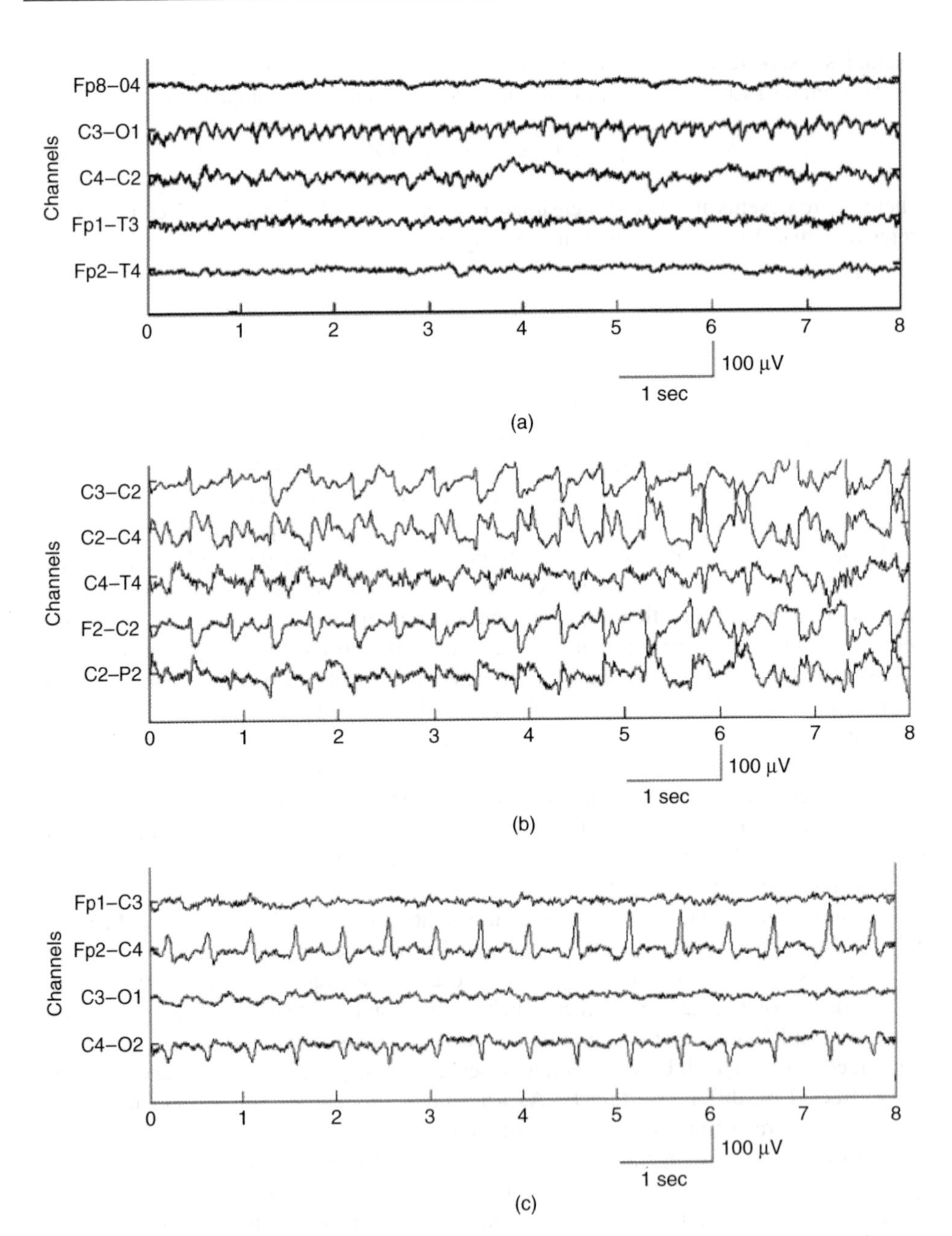

Figure 4.3 The main three neonate seizure patterns: (a) low amplitude depressed brain type discharge around the channels C3-O1 and FP1-T3, (b) repetitive complex slow waves with superimposed higher frequencies in all the channels, and (c) repetitive or periodic runs of sharp transients in channels FP2-C4 and C4-O2

4.2 Chaotic Behaviour of EEG Sources

Nonlinear analysis techniques provide insights into many processes that cannot be directly formulated or exactly modelled using state machines. This requires time series analysis of long sequences. A state space reconstruction of the chaotic data may be performed based on embedding methods [54,55]. Using the original time series and its time-delayed copies, i.e. $\boldsymbol{x}(n) = [x(n), x(n+T), \ldots, x(n+(d_E-1)T)]$, an appropriate state space can be reconstructed. Here $\boldsymbol{x}(n)$ is the original one-dimensional data, T is the time delay, and d_E is the embedding dimension. The time delay T is calculated from the first minimum of the average mutual information (AMI) function [56]. The minimum point of the AMI function provides adjacent delay coordinates with a minimum of redundancy. The embedding dimension (d_E) can be computed from a global false nearest neighbours (GFNN) analysis [57], which compares the distances between neighbouring trajectories at successively higher dimensions. The false neighbours occur when trajectories that overlap in dimension d_i are distinguished in dimension d_{i+1}. As i increases, the total percentage of false neighbours declines and d_E is chosen where this percentage approaches zero.

The statistical stationarity of the EEG patterns may be investigated and established by evaluating recurrence plots generated by calculation of the Euclidean distances between all pairs of points $\boldsymbol{x}(i)$ and $\boldsymbol{x}(j)$ in the embedded state space, and then such points are plotted in the (i, j) plane where $\delta_{i,j}$ is less than a specific radius ρ:

$$\delta_{ij} = \|\boldsymbol{x}(i) - \boldsymbol{x}(j)\|_2 < \rho \tag{4.14}$$

where $\|\cdot\|_2$ denotes the Euclidean distance. Since i and j are time instants, the recurrence plots convey natural and subtle information about temporal correlations in the original time series [58]. Nonstationarities in the time series are manifested as gross homogeneities in the recurrent plot. The value of ρ for each time series is normally taken as a small percentage of the total dataset size.

The scale-invariant self-similarity, as one of the hallmarks of low dimensional deterministic chaos [59], results in a linear variation of the logarithm of the correlation sum, $\log[C(r, N)]$, with respect to $\log(r)$ as $r \to 0$. In places where such similar dynamics exist, the *correlation dimension*, D, is defined as

$$D = \lim_{N\to\infty} \lim_{r\to 0} \frac{\log[C(r, N)]}{\log(r)} \tag{4.15}$$

in which $C(r, N)$ is calculated as

$$C(r, N) = \frac{1}{(N-n)(N-n-1)} \sum_{i=1}^{N} \sum_{j=1}^{N} H(r - \|\boldsymbol{x}(i) - \boldsymbol{x}(j)\|_2) \tag{4.16}$$

where r represents the volume of points being considered and $H(.)$ is a Heaviside step function. Computation of $C(r, N)$ is susceptible to noise and nonstationarities. It is also dominated by the finite length of the dataset. The linearity of the relationship between $\log[C(r, N)]$ and $\log(r)$ can be examined from the local slopes of $\log[C(r, N)]$ versus $\log(r)$.

Generally, traditional methods such as the Kolmogorov entropy, the correlation dimension, or Lyapunov exponents [5] can be used to quantify the dynamical changes of the brain. Finite-time Lyapunov exponents quantify the average exponential rate of divergence of neighbouring trajectories in state space, and thus provide a direct measure of the sensitivity of the system to infinitesimal perturbations.

A method based on the calculation of the largest Lyapunov exponent has often been used for evaluation of chaos in intracranial EEG signals. In a p-dimensional system there are p different Lyapunov exponents, λ_i. They measure the exponential rate of convergence or divergence of the different directions in the phase space. If one of the exponents is positive, the system is chaotic. Thus, two close initial conditions will diverge exponentially in the direction defined by that positive exponent. Since these exponents are ordered, $\lambda_1 \geq \lambda_2 \ldots \geq \lambda_d$, to study the chaotic behaviour of a system it is sufficient to study the changes in the largest Lyapunov exponent, λ_1. Therefore, emphasis is placed on the changes in the value of λ_1 as the epileptic brain moves from one state to another.

The maximum Lyapunov exponent (MLE) (λ_1) for a dynamical system can be defined from Reference [60] as

$$d(n) = d_0 \mathrm{e}^{\lambda_1 n} \tag{4.17}$$

where $d(n)$ is the mean divergence between neighbouring trajectories in state space at time n and d_0 is the initial separation between neighbouring points. Finite-time exponents (λ^*) are distinguished from true Lyapunov exponents λ_1, which are strictly defined only in the dual limit as $n \to \infty$ and $d_0 \to 0$ in Equation (4.17). For the finite length observation λ^* is the average of the Lyapunov exponents.

A practical procedure for the estimation of λ_1 from a time series was proposed by Wolf *et al.* [61]. This procedure gives a global estimate of λ_1 for stationary data. Since the EEG data are nonstationary [62], the algorithm to estimate λ_1 from the EEG should be capable of automatically identifying and appropriately weighting the transients of the EEG signals. Therefore, a modification of Wolf's algorithm, proposed in Reference [63], which mainly modifies the searching procedure to account for the nonstationarity of the EEG data, may be used. This estimate is called the short-term largest Lyapunov exponent (STL_{max}) and the changes of the brain dynamics can be studied by the time evolution of the STL_{max} values at different electrode sites. Estimation of the STL_{max} for time sequences using Wolf's algorithm has already been explained in Chapter 2.

4.3 Predictability of Seizure from the EEGs

Although most seizures are not life threatening, they are an unpredictable source of annoyance and embarrassment. They occur when a massive group of neurons in the cerebral cortex begins to discharge in a very organized way, leading to a temporary synchronized electrical activity that disrupts the normal activity of the brain. Sometimes, such disruption manifests itself in a brief impairment of consciousness, but it can also produce a more or less complex series of abnormal sensory and motor manifestations.

The brain is assumed to be a dynamical system, since epileptic neuronal networks are essentially complex nonlinear structures and their interactions are thus expected to exhibit nonlinear behaviour. These methods have substantiated the hypothesis that quantification of the brain's dynamical changes from the EEG might enable the prediction of epileptic seizures, while traditional methods of analysis have failed to recognize specific changes prior to seizure.

Iasemidis *et al.* [64] were the first group to apply nonlinear dynamics to clinical epilepsy. The main concept in their studies is that a seizure represents a transition of the epileptic brain from chaotic to a more ordered state, and therefore the spatiotemporal dynamical properties of the epileptic brain are different for different clinical states. Further studies of the same group, based on the temporal evolution of the short-term largest Lyapunov exponent (LLE) (a modification of the LLE to account for the nonstationarity of the EEG) for patients with temporal lobe epilepsy (TLE) [63], suggested that the EEG activity becomes progressively less chaotic as the seizure approaches. Therefore, the idea that seizures were abrupt transitions in and out of an abnormal state was substituted by the idea that the brain follows a dynamical transition to seizure for at least some kinds of epilepsy. Since these pioneering studies, nonlinear methods derived from the theory of dynamical systems have been employed to quantify the changes in the brain dynamics before the onset of seizures, providing evidence to the hypothesis of a *route* to seizure. Lehnertz and Elger [65] focused their studies on the decrease of complexity in neuronal networks prior to seizure. They used the information provided by changes in the *neuronal complexity loss* that summarizes the complex information content of the correlation dimension profiles in just a single number. Lerner [66] observed that changes in the correlation integral could be used to track accurately the onset of seizure for a patient with TLE. However, Osorio *et al.* [67] demonstrated that these changes in the correlation integral could be perfectly explained by changes in the amplitude and frequency of the EEG signals. Van Quyen *et al.* [68] found a decrease in the dynamical similarity during the period prior to seizure and that this behaviour became more and more pronounced as the onset of seizure approached. Moser *et al.* [69] employed four different nonlinear quantities within the framework of the Lyapunov theory and found strongly significant preictal changes. Litt *et al.* [70] demonstrated that the energy of the EEG signals increases as seizure approaches. In their later works, they provided evidence of seizure predictability based on the selection of different linear and nonlinear features of the EEG [71]. Iasemidis and coworkers [72,73], by using the spatiotemporal evolution of the short-term largest Lyapunov exponent, demonstrated that minutes or even hours before seizure, multiple regions of the cerebral cortex progressively approach a similar degree of chaoticity of their dynamical states. They called it *dynamical entrainment* and hypothesized that several critical sites have to be locked with the epileptogenic focus over a common period of time in order for a seizure to take place. Based on this hypothesis they presented an adaptive seizure prediction algorithm that analyses continuous EEG recordings for the prediction of temporal lobe epilepsy when only the occurrence of the first seizure is known [74].

Most of these studies for the prediction of epilepsy are based on intracranial EEG recordings. Two main challenges face the previous methods in their application to scalp EEG data: 1) the scalp signals are more subject to environmental noise and artefacts than the intracranial EEG and (2) the meaningful signals are attenuated and mixed in their propagation through soft tissue and bone. Traditional nonlinear methods (TNMs), such as the Kolmogorov entropy or the Lyapunov exponents, may be affected by the above two difficulties and therefore they may not distinguish between slightly different chaotic regimes of the scalp EEG [75]. One approach to circumvent these difficulties is based on the definition of different nonlinear measures that yield better performance over the TNM for the scalp EEG. This is the approach followed by Hively and Protopopescu [76]. They proposed a method based on the *phase-space dissimilarity measures* (PSDMs) for forewarning of epileptic events from scalp EEG. The approach of Iasemidis *et al.* of dynamical entrainment has also been shown to work well on scalp unfiltered EEG data for seizure predictability [77–79].

In principle, a nonlinear system can lie in a high-dimensional or infinite-dimensional phase space. Nonetheless, when the system comes into a steady state, portions of the phase space are revisited over time and the system lies in a subset of the phase space with a finite and generally small dimension, called an *attractor*. When this attractor has sensitive dependence to initial conditions (it is chaotic), it is termed a *strange* attractor and its geometrical complexity is reflected by its dimension, D_a. In practice, the system's equations are not available and there are only discrete measurements of a single observable, $u(n)$, representing the system. If the system comes into such a steady state, a p-dimensional phase space can be reconstructed by generating p different scalar signals, $x_i(n)$, from the original observable, $u(n)$, and embedding them into a p-dimensional vector:

$$\boldsymbol{x}(n) = [x_1(n), x_2(n), \ldots, x_p(n)]^T \tag{4.18}$$

According to Takens [80], if p is chosen large enough, a good phase portrait of the attractor can be generally obtained and therefore good estimates of the nonlinear quantities can be obtained. In particular, Takens' theorem states that the embedding dimension, p, should be at least equal to $2D_a + 1$. The easiest and probably the best way to obtain the embedding vector $\boldsymbol{x}(n)$ from $u(n)$ is by the method of delays. According to this method, p different time delays, $n_0 = 0, n_1 = \tau, n_2 = 2\tau, \ldots, n_p = (p-1)\tau$, are selected and the p different scalar signals are obtained as $x_i(n) = x(n + n_i)$ for $i = 0, \ldots, p-1$. If τ is chosen carefully, a good phase portrait of the attractor is obtained and therefore so too are good estimates of the parameters of nonlinear behaviour.

Since the brain is a nonstationary system, it is never in a steady state in the strictly dynamical sense. However, it can be considered as a dynamical system that constantly moves from one stable steady state to another. Therefore, local estimates of nonlinear measures should be possible and the changes of these quantities should be representative of the dynamical changes of the brain.

Previous studies have demonstrated a more ordered state of the epileptic brain during seizure than before or after it. The correlation dimension has been used to estimate the dimension, d, of the ictal state [64]. The values obtained range between 2 and 3, demonstrate and the existence of a low-dimensional attractor. Therefore, an embedding dimension of seven should be enough to obtain a good image of this attractor and a good space portrait of the ictal state. As concluded in Chapter 2, increasing the value of p more than what is strictly necessary increases the effect of noise and thus higher values of p are not recommended.

In a new approach [81] it has been shown that the TNM can be applied to the scalp EEGs indirectly. This requires the underlying sources of the brain to be correctly separated from the observed electrode signals without any *a priori* information about the source signals or the way the signals are combined. The effects of noise and other internal and external artefacts are also highly mitigated by following the same strategy. It is expected that signals similar to the intracranial recordings will be obtained to which TNM can be applied. To do so, the signal segments are initially separated into their constituent sources using blind source separation (BSS) (assuming that the sources are independent), since for a practical prediction algorithm, the nonlinear dynamics have to be quantified over long-term EEG recordings. After using a block-based BSS algorithm the continuity has to be maintained for the entire recording. This problem turns out not to be easy due to the two inherent ambiguities of BSS: (1) the variances (energies) of the independent components are unknown and (2) due to the inherent permutation problem of the BSS algorithms the order of the independent components cannot be determined.

The first ambiguity states that the sources can be estimated up to a scalar factor. Therefore, when moving from one block to another, the amplitude of the sources will be generally different and the signals can be inverted. This ambiguity can be solved as explained below, so its effect can be avoided. The nonlinear dynamics are quantified by the largest Lyapunov exponent λ_1. The calculation of λ_1 estimates for each block is based on ratios of distances between points within the block. Consequently, as long as λ_1 is estimated for the sources obtained by applying BSS to each block of EEG data individually, there is no need to adjust the energy of the sources.

The second ambiguity, however, severely affects the algorithm. The order in which the estimated sources appear, as a result of the BSS algorithm, changes from block to block. Therefore, a procedure is needed to reorder the signals to align the same signal from one block to another and maintain the continuity for the entire recording. The next section explains the approach followed in the present algorithm for this purpose.

An overlap window approach is followed to maintain the continuity of the estimated sources, solving both indeterminacies simultaneously. Instead of dividing the EEG recordings into sequential and discontinuous blocks, a sliding window of fixed length, L, is employed with an overlap of $L - N$ samples ($N < L$), and the BSS algorithm is applied to the block of data within that window. Therefore, it is assumed that $\boldsymbol{x}(n) = [x_1(n), x_2(n), \ldots, x_m(n)]^{\mathrm{T}}$ represents the entire scalp EEG recording, where m is the number of sensors. Two consecutive windows of data are selected as $\boldsymbol{x}_1(n) = \boldsymbol{x}(n_0 + n)$ and $\boldsymbol{x}_2(n) = \boldsymbol{x}(n_0 + N + n)$ for $t = 1, \ldots, L$, where $n_0 \geq 0$. Therefore,

$$\boldsymbol{x}_2(n) = \boldsymbol{x}_1(N + n) \quad \text{for } n = 1, \ldots, L - N \tag{4.19}$$

Once the BSS algorithm has been applied to $\boldsymbol{x}_1(n)$ and $\boldsymbol{x}_2(n)$, two windows of estimated sources $\hat{\boldsymbol{s}}_1(n) = [\hat{s}_1(n), \hat{s}_2(n), \ldots, \hat{s}_m(n)]^{\mathrm{T}}$ and $\hat{\boldsymbol{s}}_2(n) = [\hat{s}'_1(n), \hat{s}'_2(n), \ldots, \hat{s}'_m(n)]^{\mathrm{T}}$ will be obtained respectively, where m is the number of sources. These two windows overlap within a time interval, but due to the inherent ambiguities of BSS, $\hat{\boldsymbol{s}}_1(n)$ and $\hat{\boldsymbol{s}}_2(n)$ are not equal in this interval. Instead,

$$\hat{\boldsymbol{s}}_2(n) = \mathbf{P} \cdot \mathbf{D} \cdot \hat{\boldsymbol{s}}_1(n + N) \quad \text{for } n = 1, \ldots, L - N$$

where $\mathbf{P}$ is an $n \times n$ permutation matrix and $\mathbf{D} = \mathrm{diag}\{d_1, d_2, \ldots, d_n\}$ is the scaling matrix. Therefore, $\hat{\boldsymbol{s}}_2(t)$ is just a copy of $\hat{\boldsymbol{s}}_1(t)$ in the overlap block, with the rows (sources) permuted, and each of them is scaled by a real number d_i that accounts for the scaling ambiguity of BSS. A measure of similarity has also been used between the rows of $\hat{s}_1(t)$ and $\hat{s}_2(t)$ within the overlap region for this purpose. The cross-correlation between two zero mean wide sense stationary random signals $x(t)$ and $y(t)$ is defined as

$$r_{xy}(\tau) = E[x(n)y(n+\tau)] \tag{4.20}$$

where $E[.]$ denotes the expectation operation. This measure gives an idea of the similarity between $x(n)$ and $y(n)$, but its values are not bounded and depend on the amplitude of the signal. Therefore, it is preferable to use a normalization of r_{xy} given by the cross-correlation coefficient

$$\rho_{xy} = \frac{r_{xy}}{\sigma_x \sigma_y} \tag{4.21}$$

where σ_x and σ_y are the standard deviations of $x(n)$ and $y(n)$ respectively and the cross-correlation coefficient satisfies

$$-1 \le \rho_{xy} \le 1 \tag{4.22}$$

Furthermore, if $\rho_{xy} = 1$, then $y = ax$, with $a > 0$, and $x(t)$ and $y(t)$ are perfectly correlated; if $\rho_{xy} = -1$, then $y = -ax$ and $x(n)$ and $y(n)$ are perfectly anticorrelated. When the two signals have no information in common, $\rho_{xy} = 0$, they are said to be uncorrelated.

Theoretically, the BSS algorithm gives independent sources at the output. Since two independent signals are uncorrelated, if the cross-correlation coefficient ρ is calculated between one row $\hat{s}_i(n)$ of the overlap block of $\hat{\mathbf{S}}_1(n)$ and all the rows $\hat{s}'_j(t)$ of the overlap block of $\hat{\mathbf{S}}_2(n)$, all the values equal to zero should be obtained except one for which $|\rho_{ij}| = 1$. In other words, if the matrix $\mathbf{\Gamma} = \{\gamma_{ij}\}$ is defined with elements equal to the absolute value of ρ between the ith row of the overlap segment of $\hat{\mathbf{S}}_1(n)$ and the jth row of the overlap segment of $\hat{\mathbf{S}}_2(n)$, then $\mathbf{\Gamma} = \mathbf{P}^{\mathrm{T}}$ and the permutation problem can be solved.

Once the permutation problem has been solved, each of the signals $\hat{s}_i(n)$ corresponds to only one of the signals $\hat{s}'_j(n)$, but the latter signals are scaled and possibly inverted versions of the former signals, due to the first inherent ambiguity of BSS. The BSS algorithm sets the variances of the output sources to one and therefore $\hat{s}_i(n)$ and $\hat{s}'_j(n)$ both have equal variance. Since the signals only share an overlap of $L - N$ samples, the energy of the overlap segment of these signals will generally be different and therefore can be used to solve the amplitude ambiguity. In particular,

$$\hat{s}_i(n+N) = \mathrm{sign}(\rho_{ij})\frac{\sigma_i}{\sigma'_j}\hat{s}'_j(n) \quad \text{for } n = 1, \ldots, L-N \tag{4.23}$$

where ρ_{ij} is calculated for $\hat{s}_i(n)$ and $\hat{s}'_j(n)$ within the overlap segment, and σ_i and σ'_j are the standard deviations of $\hat{s}_i(n)$ and $\hat{s}'_j(n)$ respectively, within the overlap segment. This should solve the scaling ambiguity of the BSS algorithm.

In practice, the estimated sources are not completely uncorrelated and therefore $\mathbf{\Gamma} \neq \mathbf{P}^{\mathrm{T}}$. However, for each row it is expected that only one of the elements γ_{ij} will be obtained close to unity, corresponding to $j = j_0$, and the rest close to zero. Therefore, the algorithm can still be applied to maintain continuity of the signals. After the sources are estimated TNM can be applied to track the dynamics of the estimated source signals.

Using simultaneous scalp and intracranial EEG recordings the performance of the above system has been observed. The intracranial recordings were obtained from multicontact foramen ovale (FO) electrodes. Electrode bundles are introduced bilaterally through the FO under fluoroscopic guidance. The deepest electrodes within each bundle lie next to medial temporal structures, whereas the most superficial electrodes lie at or just below the FO [82]. As FO electrodes are introduced via anatomical holes, they provide a unique opportunity to record simultaneously from scalp and medial temporal structures without disrupting the conducting properties of the brain coverings by burr holes and wounds, which can otherwise make simultaneous scalp and intracranial recordings unrepresentative of the habitual EEG [83]. Simultaneously, scalp EEG recordings were obtained from standard silver cup electrodes applied according to the 'Maudsley' electrode placement system [84], which is a modification of the extended 10–20 system. The advantage of the Maudsley system with respect to the standard 10–20 system is that it provides a more extensive coverage of the lower part of the cerebral convexity, increasing the sensitivity for the recording from basal subtemporal structures.

For the scalp EEG the overlap window approach was used to maintain the continuity of the underlying sources, and once the continuity was maintained the resulting sources were divided into nonoverlapping segments of 2048 samples. The largest Lyapunov exponent, λ_1, was estimated for each of these segments. The intracranial signals were also divided into segments of the same size and λ_1 was estimated for each of these segments. In both cases, the parameters used for the estimation of λ_1 were those used by Iasemidis *et al.* for the estimation of STL_{max}, as explained in Reference [72]. After the λ_1 values are calculated for different segments, they are included in a time sequence and smoothed by time averaging.

Within the simultaneous intracranial and scalp EEG recording of 5 minutes and 38 seconds containing a focal seizure, the seizure is discernible in the intracranial electrodes (Figure 4.4) from around 308 seconds, and the ictal state lasts throughout the observation. Figure 4.4 shows a segment of the signals recorded by the scalp electrodes during the seizure. The signals are contaminated by noise and artefact signals such as eye blinking and electrocardiogram (ECG), and the seizure is not clearly discernible. Figure 4.5 shows the signals obtained after applying the BSS-based prediction algorithm to the same segment of scalp EEGs. The first and second estimated sources seem to record the seizure components while the noise and artefacts are separated into the other two sources. Figures 4.6(a) and (b) illustrate the smoothed λ_1 variations for two intracranial electrodes located in the focal area. The smoothed λ_1 is calculated by averaging the current value of λ_1 and the previous two values. These two electrodes show a clear drop in the value of λ_1 at the occurrence of seizure, starting prior to the onset. However, the intracranial EEG was contaminated by a high-frequency activity that causes fluctuations of λ_1 for the entire recording. Figures 4.6(c) to (f) illustrate the smoothed λ_1 evolution for four scalp electrodes once the baseline was removed. The value of λ_1 presents large fluctuations that can be due to the presence of noise and artefacts. Although the values seem to be lower as the seizure approaches, there is not a clear trend before seizure in any of the electrodes.

Figure 4.7 shows the results obtained for two of the estimated sources after the application of the proposed BSS algorithm. The algorithm efficiently separates the underlying

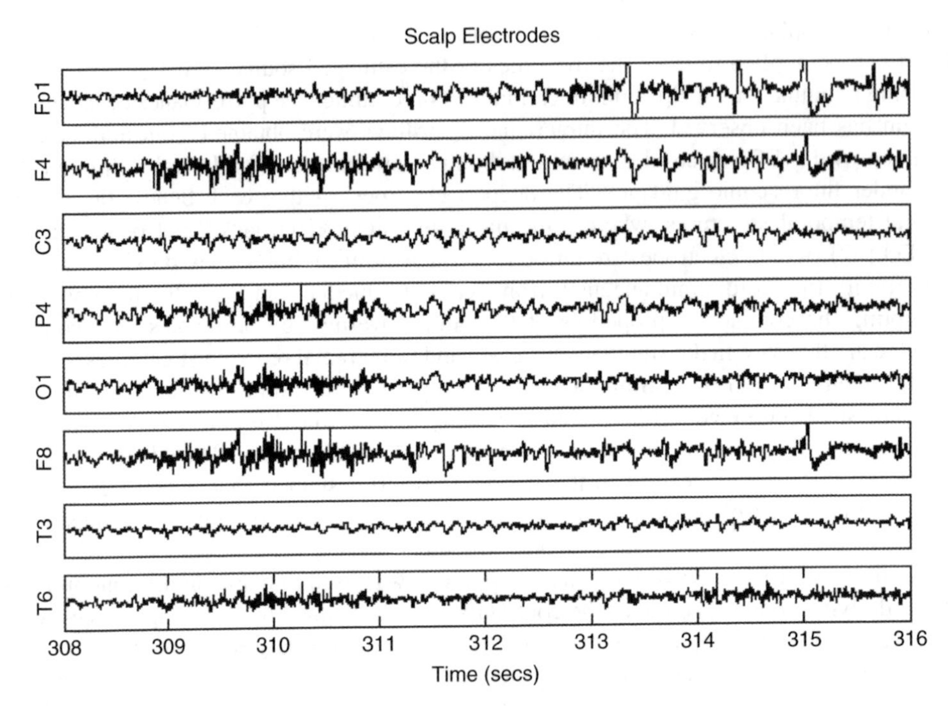

Figure 4.4 Eight seconds of EEG signals from 8 out of 16 scalp electrodes during a seizure (DC removed)

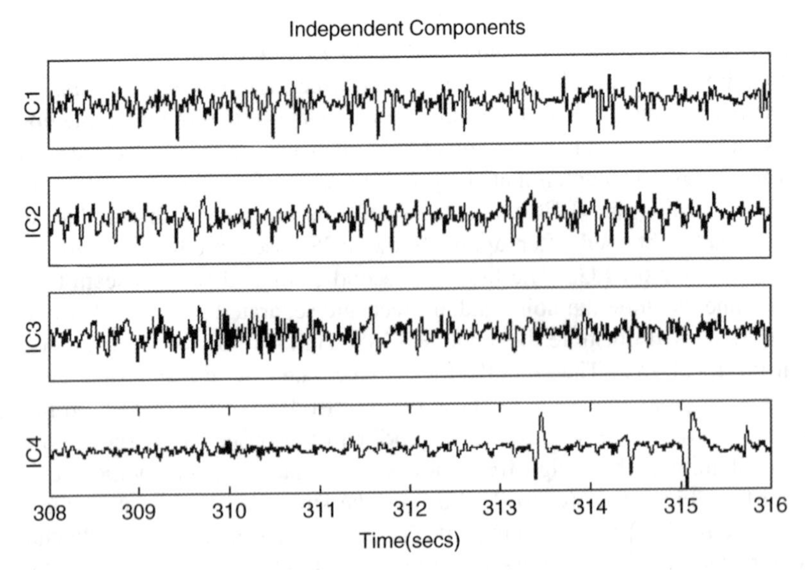

Figure 4.5 The four independent components obtained by applying BSS to the scalp electrode signals shown in Figure 4.4

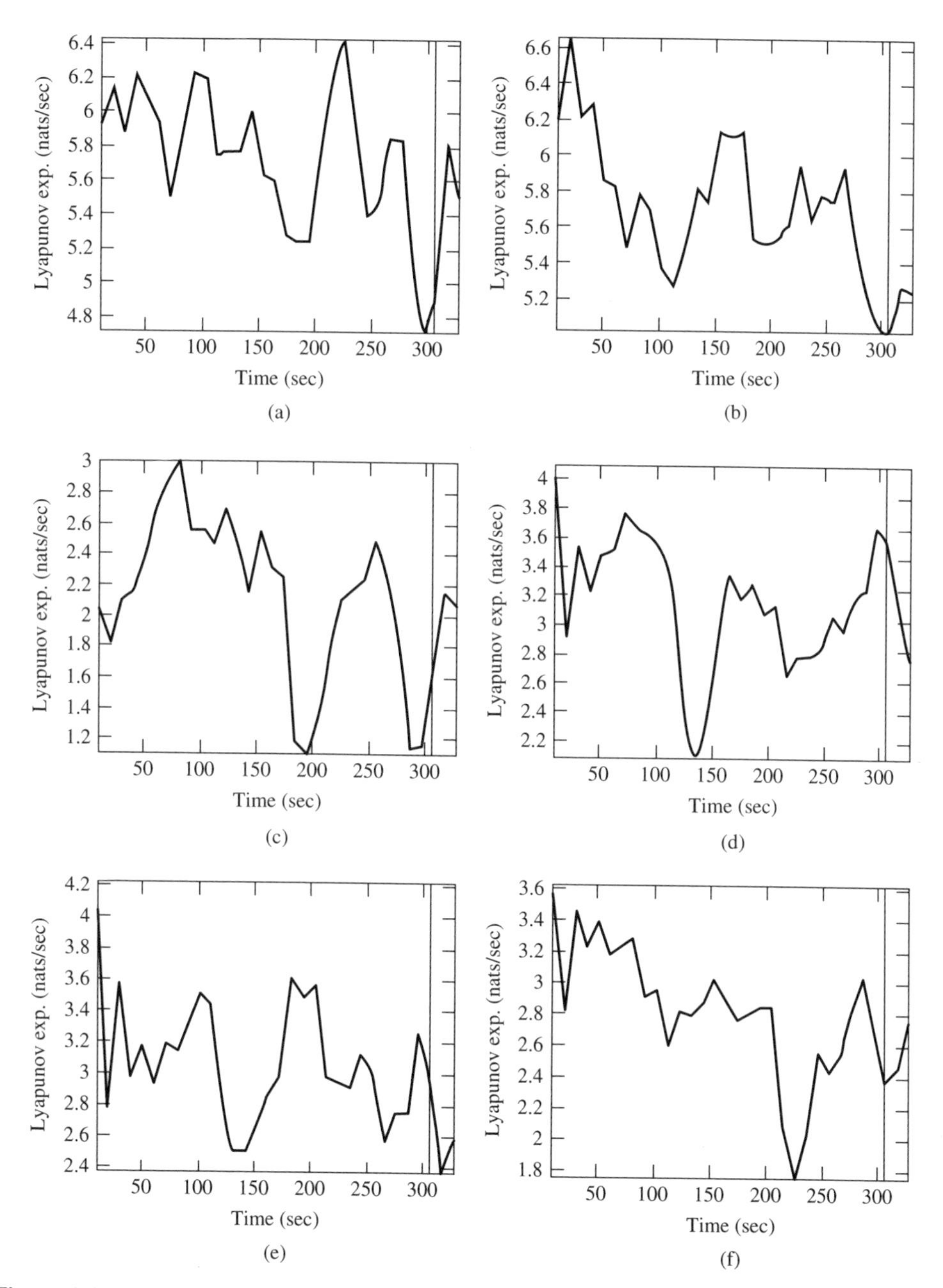

Figure 4.6 The smoothed λ_1 evolution over time for two intracranial electrodes located in the focal area: (a) for the LF4 electrode and (b) for the LF6 electrode; (c), (d), (e), and (f) show the smoothed λ_1 evolutions for four scalp electrodes. The length of the recording is 338 seconds and the seizure occurs at 306 seconds (marked by the vertical line)

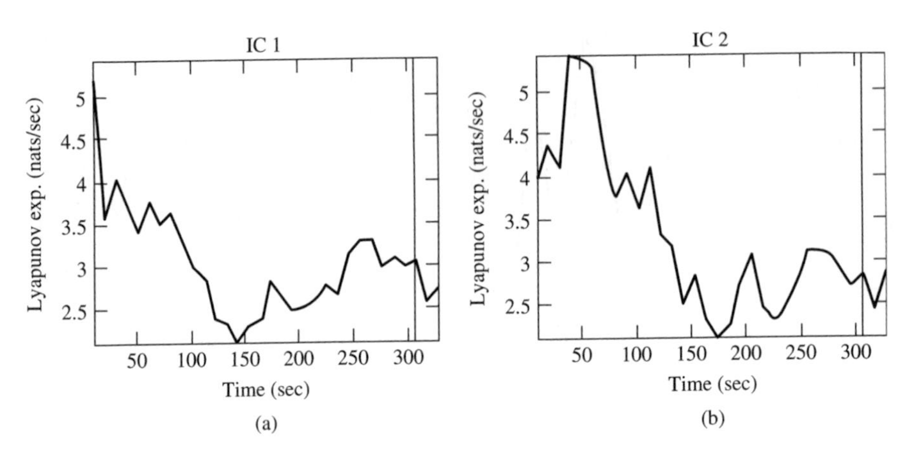

Figure 4.7 Smoothed λ_1 evolution over time for two independent components IC1 and IC2, for which ictal activity is prominent (see Figure 4.4)

sources from the eye blinking artefacts and noise. Both figures show how the BSS algorithm efficiently separates the epileptic components. The value of λ_1 reaches the minimum value more than one minute prior to seizure, remaining low until the end of the recording. This corresponds with the seizure lasting until the end of the recording.

Figures 4.8 to 4.10 illustrate the results obtained from a recording of duration 5 minutes and 34 seconds. In this particular case the epileptic component was not clearly visible by visual inspection of the intracranial electrode signals. The intracranial electrodes may not have recorded the electrical activity of the epileptic focus because of their location. Figure 4.8(a) shows a segment of the signals recorded by eight scalp electrodes during the seizure. Although the signals are contaminated by noise and artefacts, the seizure components are discernible in several electrodes. Figure 4.8(b) illustrates the signals obtained for the same segment of data after the BSS algorithm. In this case the seizure component seems to be separated from noise and artefacts in the third estimated source. Figure 4.9(a) displays the evolution of the smoothed λ_1 for four different intracranial electrodes. The values fluctuate during the recording but there is a gradual drop in λ_1, starting at the beginning of the recording. A large drop in the value of λ_1 is observed for the four electrodes around 250 seconds and reaches a minimum value around 275 seconds. However, the onset of seizure occurs around 225 seconds and therefore none of the intracranial electrodes is able to predict the seizure. Figure 4.9(b) shows the variation of smoothed λ_1 for four scalp electrodes. Likewise, for the intracranial electrodes, λ_1 values have large fluctuations but present a gradual drop towards seizure. Similarly, the drop to the lowest value of λ_1 starts after 250 seconds and therefore the signals from these electrodes are not used for seizure prediction.

Figure 4.10 illustrates the changes in the smoothed λ_1 for the third estimated source obtained after the application of BSS. λ_1 starts decreasing approximately two minutes before the onset of seizure. The minimum of λ_1 is obtained around the same time as that for the intracranial and scalp electrodes. However, a local minimum is clear at the onset of seizure and the values are clearly lower during the seizure than at the beginning of the recording. The BSS algorithm seems to separate the epileptic component in one of the

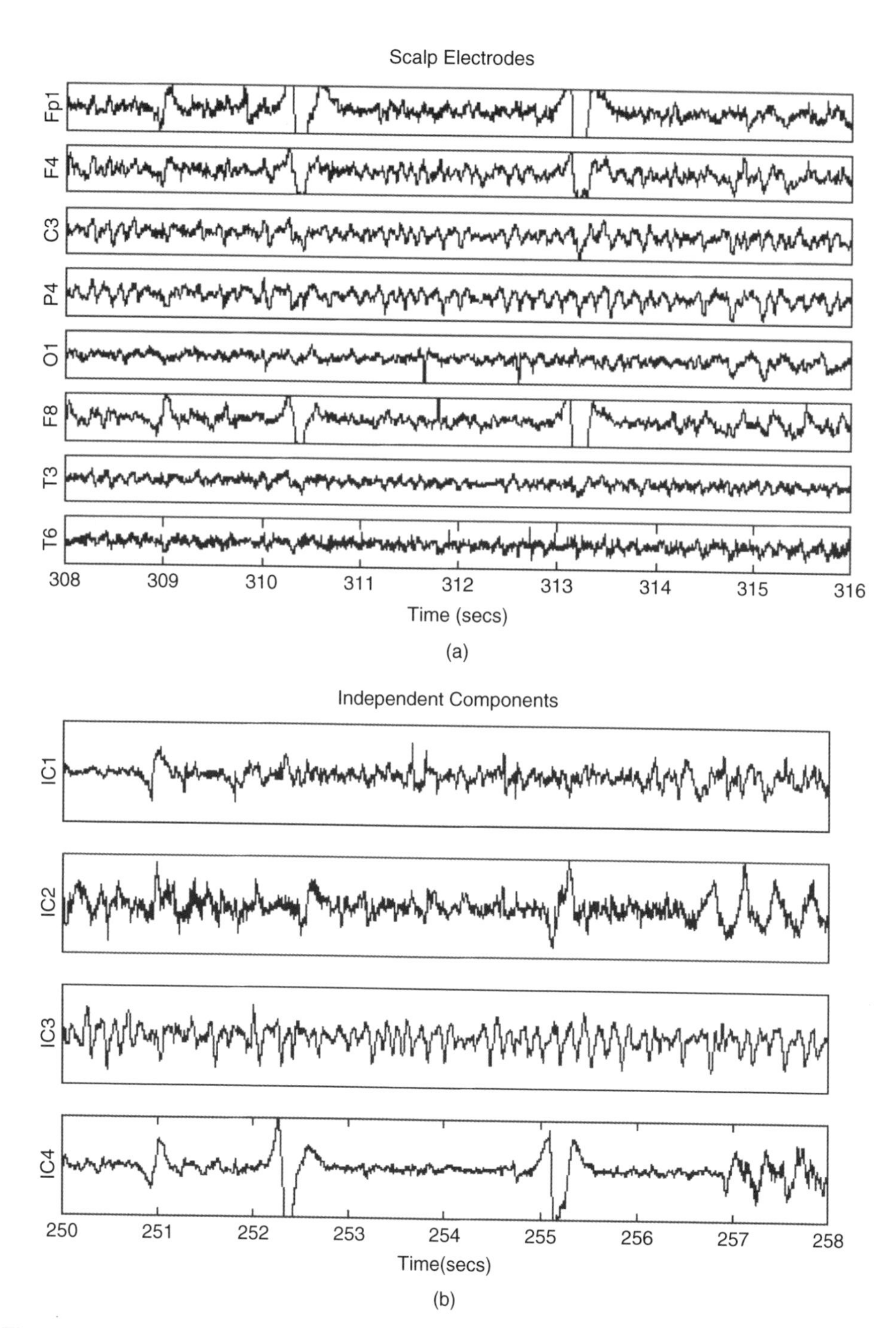

Figure 4.8 (a) A segment of 8 seconds of EEG signals (with zero mean) for 8 out of 16 scalp electrodes during the seizure and (b) the four independent components obtained by applying BSS to the scalp electrode signals shown in (a). The epileptic activity seems to be discernible in IC3 and its λ_1 evolution is shown in Figure 4.10

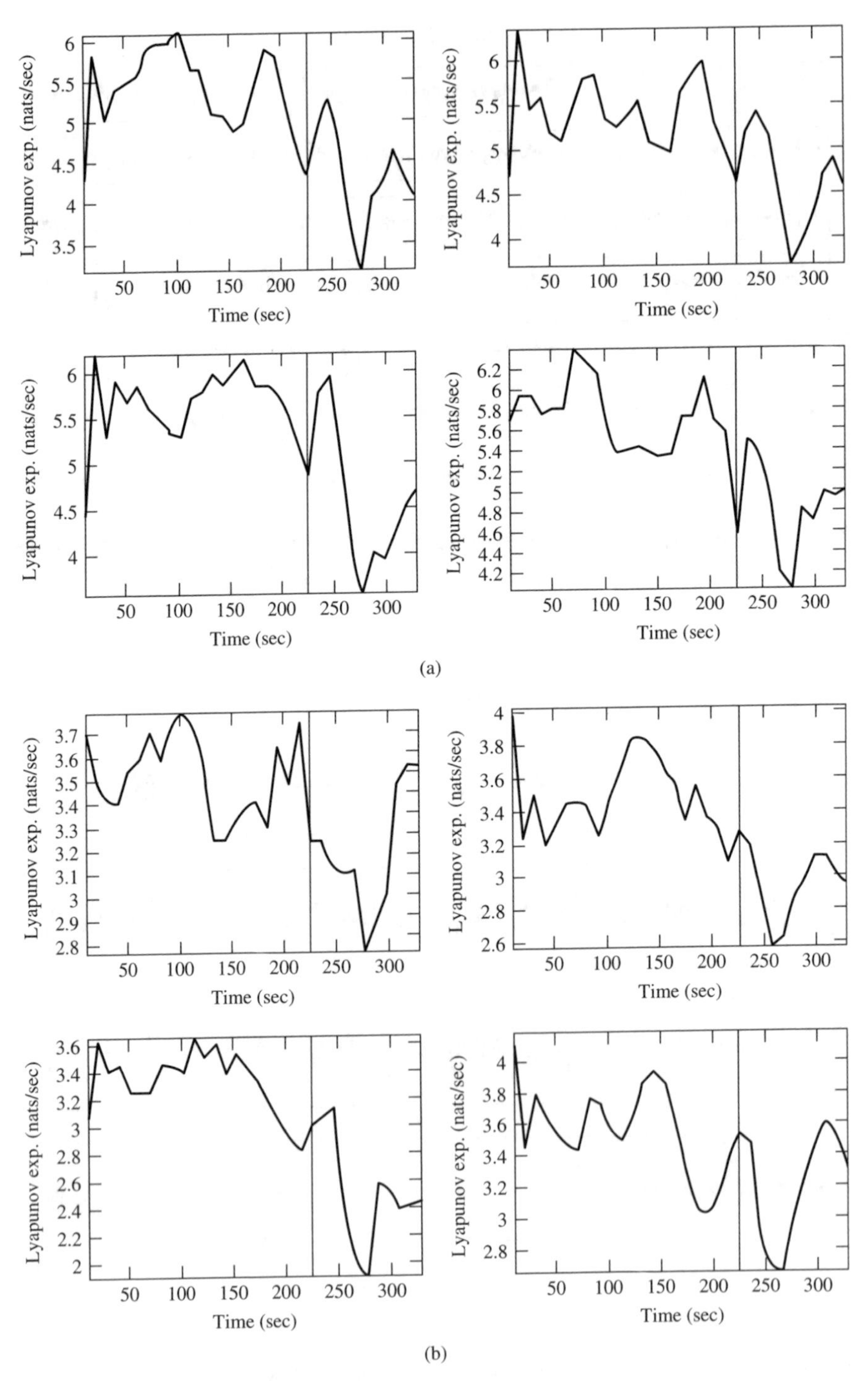

Figure 4.9 (a) Intracranial EEG analysis: 3-point smoothed λ_1 evolution from focal seizure. In this case the electrical activity of the epileptic focus seemed not to be directly recorded by the intracranial electrodes. (b) Scalp EEG analysis: the smoothed λ_1 evolution for four scalp electrodes. The length of the recording is 334 seconds. The seizure occurs at 225 seconds (marked by the vertical line)

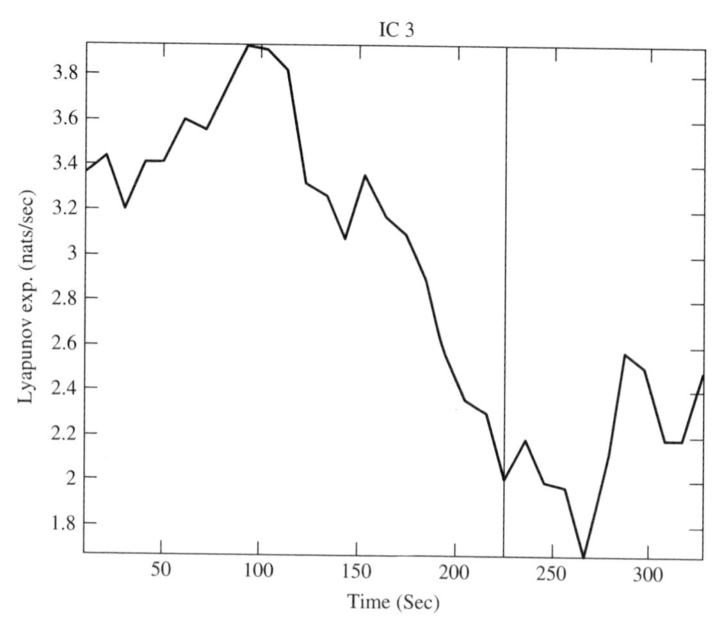

Figure 4.10 Smoothed λ_1 evolution for a focal seizure estimated from the independent component IC3 (i.e. the one where ictal activity is discernible in Figure 4.8(b)). The electrical activity of the epileptic focus was not directly recorded by the intracranial electrodes. The length of the recording is 334 seconds and the seizure occurs at 225 seconds (marked by the vertical line)

estimated sources, allowing the prediction of seizure even when this is not possible from the intracranial electrodes.

Figures 4.11(a) and (b) show the results obtained for a third EEG recording lasting 5 minutes and 37 seconds. The electrodes recorded a generalized seizure. Figure 4.11(a) illustrates the results for four intracranial electrodes. The value of λ_1 does not show any clear decrease until 250 seconds when there is a sudden drop in λ_1 for all the electrodes. The minimum value is obtained several seconds later; however, the onset of seizure was clearly discernible from the intracranial electrodes at around 236 seconds. Therefore, as an important conclusion, the intracranial EEG is not able to predict the onset of seizure in such cases and they are only able to detect the seizure after its onset. There is a clear drop in the value of λ_1 but it does not occur soon enough to predict the seizure.

Figure 4.11(b) shows the results obtained after the application of BSS. The evolution of λ_1 is similar to the evolution for intracranial recordings. However, the drop in the value of λ_1 for the estimated source seems to start decreasing before it does for the intracranial electrodes. The minimum λ_1 for the estimated source occurs before such a minimum for λ_1 is achieved for the intracranial electrodes. This means that by preprocessing the estimated sources using the BSS-based method, the occurrence time of seizure can be estimated more accurately.

There is no doubt that local epileptic seizures are predictable from the EEGs. Scalp EEG recordings seem to contain enough information about the seizure; however, this information is mixed with the signals from the other sources within the brain and is

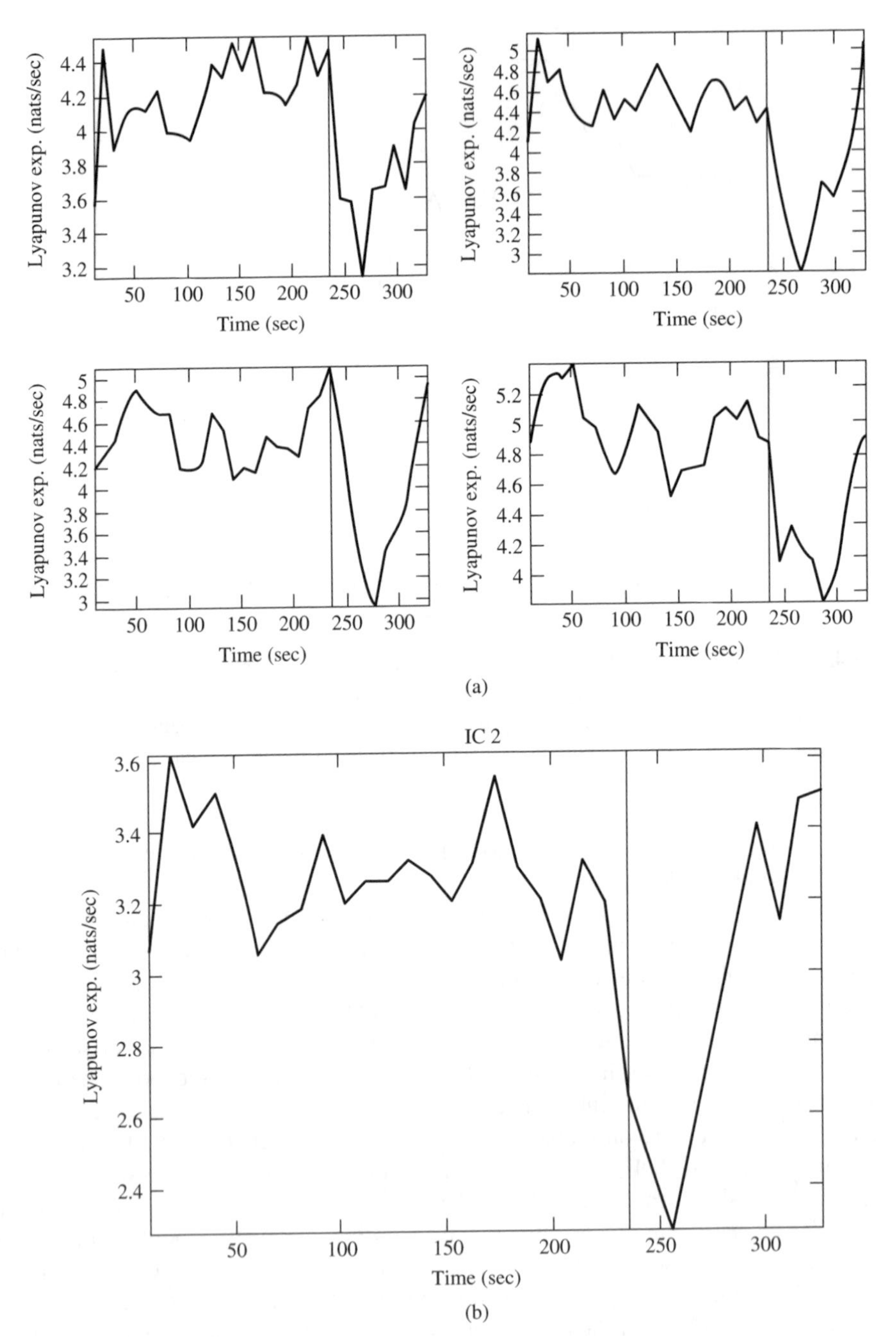

Figure 4.11 (a) Smoothed λ_1 evolution of four intracranial electrodes for a generalized seizure. The length of the recording is 337 seconds and records a generalized seizure starting at 236 seconds (marked by the vertical line). (b) Smoothed λ_1 evolution of the IC2 component estimated from the corresponding scalp EEG. The length of the recording is 337 seconds and records a generalized seizure at 236 seconds (marked by the vertical line)

buried in noise and artefacts. Incorporating a suitable preprocessing technique such as the BSS algorithm separates the seizure signal (long before the seizure) from the rest of the sources, noise, and artefacts within the brain. Therefore, the well-known traditional nonlinear methods for evaluation of the chaotic behaviour of the EEG signals can be applied. Using BSS, the prediction of seizure might be possible from scalp EEG data, even when this is not possible with intracranial EEG, especially when the epileptic activity is spread over a large portion of the brain, as in most cases. However, the results obtained from three sets with generalized seizure support the idea of unpredictability of this type of seizure, since they are preceded and followed by normal EEG activity.

The results obtained by the analysis of the scalp EEGs, although they are very promising, are subject to several limitations. The principal limitation is due to the length of the recordings. These recordings allowed a comparison to be made between the scalp and intracranial EEG, but they were of relatively short duration. Therefore, it is basically assumed that the epileptic component is active during the entire recording. Longer recordings are needed to examine the value of blind source separation better in the study of seizure predictability. For such recordings it cannot be assumed that the underlying sources are active during the entire recording and therefore the algorithm needs to detect the beginning and end of these activities. Furthermore, the algorithm employed to maintain the continuity fails in some cases where a segment of the scalp EEG is corrupted or the electrical activity is not correctly recorded. The number of corrupted segments increases for longer recordings and therefore a new methodology to maintain the continuity of the estimated sources for these particular segments should be combined with the overlap window approach. Another limitation arises from the fixed number of output signals selected for the BSS algorithm.

Generally, a seizure can be predicted if the signals are observed during a previous seizure and by analysing the interictal period. Otherwise, there remains a long way to go to be able to predict accurately the seizure from only the EEG. On the other hand, seizure is predictable if the EEG information is combined with other information, such as that gained from a video sequence of the patients, heart rate variability, and respiration.

4.4 Fusion of EEG–fMRI Data for Seizure Prediction

In the above sections it has been verified that some epileptic seizures may be predicted, but long-term recording is normally necessary. Processing of the EEGs using popular techniques such as BSS, however, is carried out on blocks of data. The processed blocks need to be aligned and connected to each other to provide the complete information. The inherent permutation problem of BSS can be solved by incorporating the functional MRI into the prediction system.

Clinical MRI has been of primary importance for visualization/detection of brain tumours, stroke, and multiple sclerosis. In 1990, Ogawa [85] showed that MRI can be sensitized to cerebral oxygenation, using deoxyhemoglobin as an endogenous susceptibility contrast agent. Using gradient-echo imaging, a form of MRI image encoding sensitive to local inhomogeneity of the static magnetic field, he demonstrated (for an animal) that the appearance of the blood vessels of the brain changed with blood oxygenation. In some later papers published by his group they presented the detection of human brain activations using this blood oxygenation level dependence (BOLD) [86,87]. It has now been established that by an increase in neuronal activity, local blood flow increases. The increase in perfusion, in excess of that needed to support the increased oxygen consumption due to neuronal activation, results in a local decrease in the concentration of deoxyhemoglobin. Since deoxyhemoglobin is paramagnetic, a reduction in its concentration results in an increase in the homogeneity of the static magnetic field, which yields an increase in the gradient-echo MRI signal. Although the BOLD fMRI does not measure brain activity directly, it relies on neurovascular coupling to encode the information about the brain function into detectible hemodynamic signals. It will clearly be useful to exploit this information in localization of seizure during the ictal period.

As will be stated in Chapter 5, brain sources can be localized within the brain. The seizure signals however, are primarily not known either in time or in space. Therefore, if the onset of seizure is detected within one ictal period, the location of the source may be estimated and the source originating from that location tracked. Using fMRI it is easy to detect the BOLD regions corresponding to the seizure sources. Therefore, in separation of the sources using BSS, it can be established that the source signals originating from those regions represent seizure signals. This process can be repeated for all the estimated source segments and thereby the permutation problem can be solved and the continuity of the signals maintained.

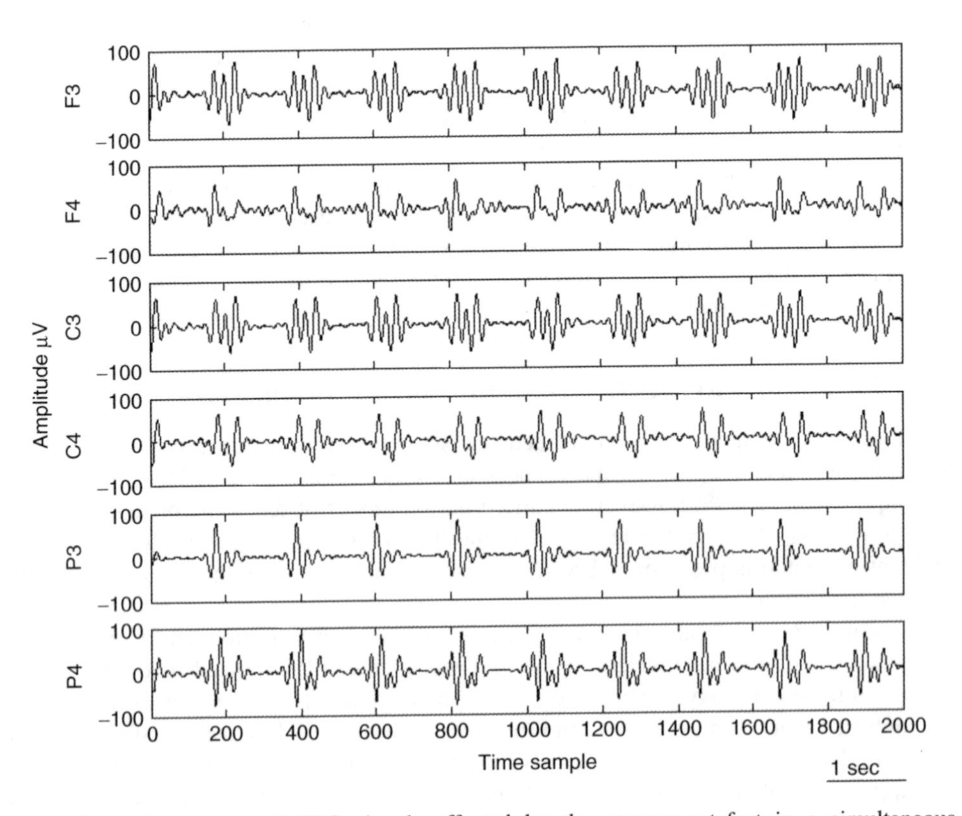

Figure 4.12 A segment of EEG signals affected by the scanner artefact in a simultaneous EEG–fMRI recording

Unfortunately, in a simultaneous EEG–fMRI recording the EEG signals are highly distorted by the fMRI effects. Figure 4.12 shows a multichannel EEG affected by fMRI. An effective preprocessing technique is then required to remove this artefact. Also, in a long-term recording it is difficult to keep the patient steady under the fMRI system. This also causes severe distortion of the fMRI data and new methods are required for its removal.

4.5 Summary and Conclusions

The detection of epileptic seizures using different techniques has been successful, particularly for the detection of adult seizures. The false alarm rate for the detection of seizure in the newborn is, however, still too high and therefore the design of a reliable newborn seizure detection system is still an open problem. On the other hand, although predictability of seizure from the EEGs (both scalp and intracranial) has been approved, more research is necessary to increase accuracy. Fusion of different neuroimaging modalities may indeed pave the path for more reliable seizure detection and prediction systems.

References

[1] Iasemidis, L. D., 'Epileptic seizure prediction and control', *IEEE Trans. Biomed. Engng.*, **50**, 2003, 549–558.

[2] Annegers, J. F., 'The epidemiology of epilepsy', in *The Treatment of Epilepsy*, Ed. E. Wyllie, Lea and Febiger, Philadelphia, Pennsylvania: 1993, pp. 157–164.

[3] Engel, Jr, J. and Pedley, T. A., *Epilepsy: A Comprehensive Text-book*, Lippinottc-Ravon, Philadelphias, Pennsylvania, 1997.

[4] Lehnertz, K., Mormann, F., Kreuz, T., Andrzejak, R. G., Rieke, C., Peter, D., and Elger, C. E., 'Seizure prediction by nonlinear EEG analysis', *IEEE Engng. Med. Biolog. Mag.*, **22**(1), 2003, 57–63.

[5] Peitgen, H., *Chaos and Fractals: New Frontiers of Science*, Springer-Verlag, New York, 2004.

[6] Niedermeyer, E., and Lopes da Silva, F., (Eds), *Electroencephalography, Basic Principles, Clinical Applications, and Retailed Fields*, 4th Edn, Lippincott, Williams and Wilkins, Philadelphia, Pennsylvania, 1999.

[7] Bravais, L. F., *Researchers sur les Symptoms et le Traitement de l'Épilesie Hémiplégique*, Thése de Paris, Paris, No. 118, 1827.

[8] Akay, M., *Nonlinear Biomedical Signal Processing*, Vol. II, *Dynamic Analysis and Modelling*, IEEE Press, New York, 2001.

[9] Chee, M. W. L., Morris, H. H., Antar, M. A., Van Ness, P. C., Dinner, D. S., Rehm, P., and Salanova, V., 'Presurgical evaluation of temporal lobe epilepsy using interictal temporal spikes and positron emission tomography', *Arch. Neurol.*, **50**, 1993, 45–48.

[10] Godoy, J., Luders, H. O., Dinner, D. S., Morris, H. H., Wyllie, E., and Murphy, D., 'Significance of sharp waves in routine EEGs after epilepsy surgery', *Epilepsia*, **33**, 1992, 513–288.

[11] Kanner, A. M., Morris, H. H., Luders, H. O., Dinner, D. S., Van Ness, P., and Wyllie, E., 'Usefulness of unilateral interictal sharp waves of temporal lobe origin in prolonged video EEG monitoring studies, *Epilepsia*, **34**, 1993, 884–889.

[12] Steinhoff, B. J., So, N. K., Lim, S., and Luders, H. O., 'Ictal scalp EEG in temporal lobe epilepsy with unitemporal versus bitemporal interictal epileptiform discharges', *Neurology*, 1995, 889–896.

[13] Ktonas, P. Y., 'Automated spike and sharp wave (SSW) detection', in *Methods of Analysis of Brain Electrical and Magnetic Signals, EEG Handbook*, Eds A. S. Gevins and A. Remond, Elsevier, Amsterdam, The Netherlands, 1987, Vol. 1, 211–241.

[14] Walter, D. O., Muller, H. F., and Jell, R. M., 'Semiautomatic quantification of sharpness of EEG phenomenon', *IEEE Trans. Biomed. Engng.*, **BME-20**, 1973, 53–54.

[15] Ktonas, P. Y., and Smith, J. R., 'Quantification of abnormal EEG spike characteristics', *Comput. Biol. Med.*, **4**, 1974, 157–163.

[16] Gotman, J., and Gloor, P., 'Automatic recognition and quantification of interictal epileptic activity in the human scalp EEG', *Electroencephalogr. Clin. Neurophysiol.*, **41**, 1976, 513–529.

[17] Davey, B. L. K., Fright, W. R., Caroll, G. J., and Jones, R. D., 'Expert system approach to detection of epileptiform activity in the EEG', *Med. Biol. Engng. Computing*, **27**, 1989, 365–370.

[18] Dingle, A. A., Jones, R. D., Caroll, G. J., and Fright, W. R., 'A multistage system to detect epileptiform activity in the EEG', *IEEE Trans. Biomed. Engng.*, **40**, 1993, 1260–1268.

[19] Glover, J. R. Jr., Ktonas, P. Y. Raghavan, N. Urunuela, J. M. Velamuri, S. S., and Reilly, E. L., 'A multichannel signal processor for the detection of epileptogenic sharp transients in the EEG', *IEEE Trans. Biomed. Engng.*, **33**, 1986, 1121–1128.

[20] Glover Jr., J. R., Raghavan, N. Ktonas, P. Y. and Frost, J. D., 'Context-based automated detection of epileptogenic sharp transients in the EEG: elimination of false positives', *IEEE Trans. Biomed. Engng.*, **36**, 1989, 519–527.

[21] Gotman, J., Ives, J. R., and Gloor, R., 'Automatic recognition of interictal epileptic activity in prolonged EEG recordings', *Electroencephalogr. Clin. Neurophysiol.*, **46**, 1979, 510–520.

[22] Gotman, J. and Wang, L. Y., 'State-dependent spike detection: concepts and preliminary results', *Electroencephalogr. Clin. Neurophysiol.*, **79**, 1991, 11–19.

[23] Gotman, J. and Wang, L. Y., 'State-dependent spike detection: validation', *Electroencephalogr. Clin. Neurophysiol.*, **83**, 1992, 12–18.

[24] Ozdamar, O., Yaylali, I., Jayakar, P., and Lopez, C. N., 'Multilevel neural network system for EEG spike detection', in *Computer-Based Medical Systems*, Proceedings of the 4th Annual IEEE Symposium, Eds I. N. Bankmann and J. E. Tsitlik, IEEE Computer Society, Washington DC, 1991.

[25] Webber, W. R. S., Litt, B., Lesser, R. P., Fisher, R. S., and Bankman, I., 'Automatic EEG spike detection: what should the computer imitate', *Electroencephalogr. Clin. Neurophysiol.*, **87**, 1993, 364–373.

[26] Webber, W. R. S., Litt, B., Wilson, K., and Lesser, R. P., 'Practical detection of epileptiform discharges (EDs) in the EEG using an artificial neural network: a comparison of raw and parameterized EEG data', *Electroencephalogr. Clin. Neurophysiol.*, **91**, 1994, 194–204.

[27] Wilson, S. B., Harner, R. N., Duffy, B. R., Tharp, B. R., Muwer, M. R., and Sperling, M. R., 'Spike detection. I. Correlation and reliability of human experts', *Electroencephalogr. Clin. Neurophysiol.*, **98**, 1996, 186–198.

[28] Kurth, C., Gilliam, F., and Steinhoff, B. J., 'EEG spike detection with a Kohonen feature map', *Ann. Biomed. Engng.*, **28**, 2000, 1362–1369.

[29] Kohonen, T., 'The self-organizing map', *Proc. IEEE*, **78**(9), 1990, 1464–1480.

[30] Kohonen, T., *The Self-Organizing Maps*, 2nd edn, Springer, New York, 1997.

[31] Nenadic, Z., and Burdick, J. W., 'Spike detection using the continuous wavelet transform', *IEEE Trans. Biomed. Engng.*, **52**(1), January 2005, 74–87.

[32] Boashash, B., Mesbah, M., and Colditz, P., 'Time frequency Analysis', Chapter 15, Article 15.5, Elsevier Science & Technology, UK, 2003, pp. 663–669.

[33] Subasi, A., 'Epileptic seizure detection using dynamic wavelet network', *Expert Systems with Applic.*, **29**, 2005, 343–355.

[34] Lutz, A., Lachaux, J-P., Martinerie, J., and Varela, F. J., 'Guiding the study of brain dynamics by using first-person data: synchrony patterns correlate with ongoing conscious states during a simple visual task', *Proc. Natl Acad. Sci.*, **99**(3), 2002, 1586–1591.

[35] Osorio, I., 'System for the prediction, rapid detection, warning, prevention, or control of changes in activity states in the brain of subject', US Patent 5995868, November 1999.

[36] Osorio, I., Frei, M. G., and Wilkinson, S. B., 'Real-time automated detection and quantitative analysis of seizures and short-term prediction of clinical onset', *Epilepsia*, **39**(6), 1998, 615–627.

[37] Bhavaraju, N. C., Frei, M. G., and Osorio, I., 'Analogue seizure detection and performance evaluation', *IEEE Trans. Biomed. Engng.*, **53**(2), February 2006, 238–245.

[38] Gonzalez-Vellon, B., Sanei, S., and Chambers, J., 'Support vector machines for seizure detection', in Proceedings of the ISSPIT, Darmstadt, Germany, 1993, pp. 126–129.

[39] Raad, A., Antoni, J., and Sidahmed, M., 'Indicators of cyclostationarity: proposal, statistical evaluation and application to diagnosis', in *Proceedings of the IEEE ICASSP*, Vol. VI, Hong Kong, 2003, 757–760.

[40] Acir, N., Oztura, I. Kuntalp, M., Baklan, B., and Guzelis, C., 'Automatic detection of epileptiform events in EEG by a three-stage procedure based on artificial neural network', *IEEE Trans. Biomed. Engng.*, **52**(1), January 2005, 30–40.

[41] Lopes da Silva, F. H. L., Blanes, W., Kalitzin, S. N., Parra, J., Suffczynski, P., and Velis, D. N., 'Dynamical diseases of brain systems: different routes to epileptic seizures', *IEEE Trans. Biomed. Engng.*, **50**(5), May 2003, 540–548.

[42] Volpe, J. J., *Neurology of the Newborn*, Saunders, Philadelphia, Pennsylvania, 1987.

[43] Liu, A., Hahn, J. S., Heldt, G. P., and Coen, R. W., 'Detection of neonatal seizures through computerized EEG analysis', *Electroencephalogr. Clin. Neurophysiol.*, **82**, 1992, 30–37.

[44] Shoker, L., Sanei, S., Wang, W., and Chambers, J., 'Removal of eye blinking artifact from EEG incorporating a new constrained BSS algorithm', *IEE J. Med. Biolog. Engng. and Computing*, **43**, 2004, 290–295.

[45] Celka, P., Boashash, B., and Colditz, P., 'Preprocessing and time–frequency analysis of newborn EEG seizures', *IEEE Engng. Med. Biol.*, September/October 2001, 30–39.

[46] Pfurtscheller, G., and Fischer, G., 'A new approach to spike detection using a combination of inverse and matched filter techniques', *Electroencephalogr. Clin. Neurophysiol.*, **44**, 1977, 243–247.

[47] Roessgen, M., Zoubir, A. M., and Boashash, B., 'Seizure detection of newborn EEG using a model-based approach', *IEEE Trans. Biomed. Engng.*, **45**(6), 1998, 673–685.

[48] Choudhuri, N., Ghosal, S., and Roy, A., 'Contiguity of the Whittle measure for a Gaussian time series', *Biometrika*, **91**(1), 2004, 211–218.

[49] Lopes da Silva, F. H., Hoeks, A., Smits, H., and Zetterberg, L. H., 'Model of brain rhythmic activity; the alpha rhythm of the thalamus', *Kybernetik*, **15**, 1974, 27–37.

[50] O'Toole, J., Mesbah, M., and Boashash, B., 'Neonatal EEG seizure detection using a time–frequency matched filter with a reduced template set', in *Proceedings of the 8th International Symposium on Signal Processing and Its Applications*, pp. August. Sydney, Australia, 2005, 215–218.

[51] Karayiannis, N. B. *et al.*, 'Detection of pseudosinusoidal epileptic seizure segments in the neonatal EEG by cascading a rule-based algorithm with a neural network', *IEEE Trans. Biomed. Engng.*, **53**(4), 2006, 633–641.

[52] Bishop, C. M., *Neural Networks for Pattern Recognition*, Oxford University Press, New York, 1995.

[53] Purushothaman, G., and Karayiannis, N. B., 'Quantum neural networks (QNNs): inherently fuzzy feedforward neural networks', *IEEE Trans. Neural Networks*, **8**(3), May 1997, 679–693.

[54] Takens, F., 'Detecting strange attractors in turbulence', in *Dynamical Systems and Turbulence*, Springer-Verlag, Berlin, 1981, pp. 366–381.

[55] Sauer, T., Yurke, J. A., and Casdagli, M., 'Embedology', *J. Statist. Phys.*, **65**(3/4), 1991, 579–616.

[56] Fraser, A. M., and Swinney, H. L., 'Independent coordinates for strange attractors from mutual information', *Phys. Rev. A*, **33**, 1986, 1134–1140.

[57] Kennel, M. B., Brown, R., and Abarbanel, H. D. I., 'Determining minimum embedding dimension using a geometrical construction', *Phys. Rev. A*, **45**, 1992, 3403–3411.

[58] Casdagli, M. C., 'Recurrence plots revisited', *Physica D*, **108**(1), 1997, 12–44.

[59] Kantz, H., and Schreiber, T., 'Dimension estimates and physiological data', *Chaos*, **5**(1), 1995, 143–154.

[60] Rosenstein, M. T., Collins, J. J., and Deluca, C. J., 'A practical method for calculating largest Lyapunov exponents from small data sets', *Physica D*, **65**, 1993, 117–134.

[61] Wolf, A., Swift, J. B., Swinney, H. L., and Vastano, J. A., 'Determining Lyapunov exponents from a time series', *Physica D*, **16**, 1985, 285–317.

[62] Kawabata, N., 'A nonstationary analysis of the electroencephalogram', *IEEE Trans. Biomed. Engng.*, **20**, 1973, 444–452.

[63] Iasemidis, L. D., Sackellares, J. C., Zaveri, H. P., and Willians, W. J., 'Phase space topography and the Lyapunov exponent of electrocorticograms in partial seizures', *Brain Topography*, **2**, 1990, 187–201.

[64] Iasemidis, L. D., D.-S. Shiau, J. C. Sackellares, Pardalos, P. M. and Prasad, A., 'A dynamical resetting of the human brain at epileptic seizures: application of nonlinear dynamics and global optimization techniques', *IEEE Trans. Biomed. Engng.*, **51**(3), 2004, 493–506.

[65] Lehnertz, K., and Elger, C. E., 'Spatio-temporal dynamics of the primary epileptogenic area in temporal lobe epilepsy characterized by neuronal complexity loss', *Electroencephalogr. Clin. Neurophysiol.*, **95**, 1995, 108–117.

[66] Lerner, D. E., 'Monitoring changing dynamics with correlation integrals: case study of an epileptic seizure', *Physica D*, **97**, 1996, 563–576.

[67] Osorio, I., Harrison, M. A. F., Lai, Y. C., and Frei, M. G., 'Observations on the application of the correlation dimension and correlation integral to the prediction of seizures', *J. Clin. Neurophysiol.*, **18**, 2001, 269–274.

[68] Van Quyen, M. L., Martinerie, J., Baulac, M., and Varela, F. J., 'Anticipating epileptic seizures in real time by a non-linear analysis of similarity between EEG recordings', *NeuroReport*, **10**, 1999, 2149–2155.

[69] Moser, H. R., Weber, B., Wieser, H. G., and Meier, P. F., 'Electroencephalogram in epilepsy: analysis and seizure prediction within the framework of Lyapunov theory', *Physica D*, **130**, 1999, 291–305.

[70] Litt, B., Estellera, R., Echauz, J., D'Alessandro, M., Shor, R., Henry, T., Pennell, P., Epstein, C., Bakay, R., Dichter, M., and Vachtsevanos, G., 'Epileptic seizures may begin hours in advance of clinical onset: a report of five patients', *Neuron*, **30**, 2001, 51–64.

[71] D'Alessandro, M., Esteller, R., Vachtsevanos, G., Hinson, A., Echauz, J., and Litt, B., 'Epileptic seizure prediction using hybrid feature selection over multiple intracranial eeg electrode contacts: A report of four patients', *IEEE Trans. Biomed. Engng.*, **50**, 2003, 603–615.

[72] Iasemidis, L. D., Principle, J. C., and Sackellares, J. C., 'Measurement and quantification of spatio-temporal dynamics of human epileptic seizures', in *Nonlinear Biomedical Signal Processing*, Ed. M. Akay, IEEE Press, New York, 2000, 296–318.

[73] Sackellares, J. C., Iasemidis, L. D., Shiau, D. S., Gilmore, R. L., and Roper, S. N., 'Epilepsy – when chaos fails', in *Chaos in the Brain?*, Eds K. Lehnertz and C. E. Elger, World Scientific, Singapore, 2000, pp. 112–133.

[74] Iasemidis, L. D., Shiau, D., Chaovalitwongse, W., Sackellares, J. C., Pardalos, P. M., Principe, J., Carney, P. R., Prasad, A., Veeramani, B., and Tsakalis, K., 'Adaptive epileptic seizure prediction system', *IEEE Trans. Biomed. Engng.*, **50**, 2003, 616–627.

[75] Hively, L. M., Protopopescu, V. A., and Gailey, P. C., 'Timely detection of dynamical change in scalp EEG signals', *Chaos*, **10**, 2000, 864–875.

[76] Hively, L. M., and Protopopescu, V. A., 'Channel-consistent forewarning of epileptic events from scalp EEG', *IEEE Trans. Biomed. Engng.*, **50**, 2003, 584–593.

[77] Iasemidis, L., Principe, J., Czaplewski, J., Gilmore, R., Roper, S., and Sackellares, J., 'Spatiotemporal transition to epileptic seizures: a nonlinear dynamical analysis of scalp and intracranial EEG recordings', in *Spatiotemporal Models in Biological and Artificial Systems*, Eds F. Silva, J. Principe, and L. Almeida, IOS Press, Amsterdam, 1997, pp. 81–88.

[78] Sackellares, J., Iasemidis, L., Shiau, D., Gilmore, R., and Roper, S., 'Detection of the preictal transition from scalp EEG recordings', *Epilepsia*, **40**(S7), 1999, 176.

[79] Shiau, D., Iasemidis, L., Suharitdamrong, W., Dance, L., Chaovalitwongse, W., Pardalos, P., Carney, P., and Sackellares, J., 'Detection of the preictal period by dynamical analysis of scalp EEG', *Epilepsia*, **44**(S9), 2003, 233–234.

[80] Takens, F., 'Detecting strange attractors in turbulence', in *Lectures Notes in Mathematics, Dynamical Systems and Turbulence, Warwick 1980*, Eds D. A. Rand and L. S. Young, Springer-Verlag, Berlin, 1981, pp. 366–381.

[81] Corsini, J., Shoker, L., Sanei, S., and Alarcon, G., 'Epileptic seizure predictability from scalp EEG incorporating constrained blind source separation', *IEEE Trans. Biomed. Engng.*, **53**(5), May 2006, 790–799.

[82] Fernandez, J., Alarcon, G., Binnie, C. D., and Polkey, C. E., 'Comparison of sphenoidal, foramen ovale and anterior temporal placements for detecting interictal epileptiform discharges in presurgical assessment for temporal lobe epilepsy', *Clin. Neurophysiol.*, **110**, 1999, 895–904.

[83] Nayak, D., Valentin, A., Alarcon, G., Seoane, J. J. G., Brunnhuber, F., Juler, J., Polkey, C. E., and Binnie, C. D., 'Characteristics of scalp electrical fields associated with deep medial temporal epileptiform discharges', *Clin. Neurophysiol.*, **115**, 2004, 1423–1435.

[84] Margerison, J. H., Binnie, C. D., and McCaul, I. R., 'Electroencephalographic signs employed in the location of ruptured intracranial arterial aneurysms', *Electroencephalogr. Clin. Neurophysiol.*, **28**, 1970, 296–306.

[85] Ogawa, S., Lee, T.-M., Nayak, A. S., and Glynn, P., 'oxygenation-sensitive contrast in magnetic resonance image of rodentbrain at high magnetic fields', *Magn. Resonance Med.*, **14**, 1990, 68–78.
[86] Huettel, S. A., Song, A. W., and McCarthy, G., *Functional Magnetic Resonance Imaging*, Sinauer, Sunderland, Massachusetts, 2004.
[87] Toga, A. W. and Mazziotta, J. C., *Brain Mapping: The Methods*, 2nd edn, Academic, San Diego, California, 2002.

5

EEG Source Localization

5.1 Introduction

The brain is divided into a large number of regions, each of which, when active, generates a local magnetic field or synaptic electric current. The brain activities can be considered to constitute signal sources which are either spontaneous and correspond to the normal rhythms of the brain, are a result of brain stimulation, or are related to physical movements. Localization of brain signal sources solely from EEGs has been an active area of research during the last two decades. Such source localization is necessary to study brain physiological, mental, pathological, and functional abnormalities, and even problems related to various body disabilities, and ultimately to specify the sources of abnormalities, such as tumours and epilepsy.

Radiological imaging modalities have also been widely used for this purpose. However, these techniques are unable to locate the abnormalities when they stem from mal functions. Moreover, they are costly and some may not be accessible by all patients at the time they needed are.

Recently, fMRI has been employed in the investigation of brain functional abnormalities such as epilepsy. From fMRI, it is possible to detect the effect of blood-oxygen-level dependence (BOLD) during metabolic changes, such as those caused by interictal seizure, in the form of white patches. The major drawbacks of fMRI, however, are its poor temporal resolution and its limitations in detecting the details of functional and mental activities. As a result, despite the cost limitation for this imaging modality, it has been reported that in 40–60 % of cases with interictal activity in EEG, fMRI cannot locate any focus or foci, especially when the recording is simultaneous with EEG.

Functional brain imaging and source localization based on scalp potentials require a solution to an ill-posed inverse problem with many possible solutions. Selection of a particular solution often requires *a priori* knowledge acquired from the overall physiology of the brain and the status of the subject.

Although in general localization of brain sources is a difficult task, there are some simple situations where the localization problem can be simplified and accomplished:

1. In places where the objective is to find the proximity of the actual source locations over the scalp. A simple method is just to attempt to somehow separate the sources

EEG Signal Processing S. Sanei and J. Chambers

using PCA or ICA, and backproject the sources of interest on to the electrodes and look at the scalp topography.

2. For some situations, where the aim is to localize a certain source of interest within the brain, e.g. a source of evoked potential (EP) or a source of movement-related potential (as widely used in the context of brain–computer interfacing). It is easy to fit a single dipole at various locations over a coarse spatial sampling of the source space and then to choose the location that produces the best match to the electrode signals (mixtures of the sources) as the focus of a spatially constrained but more finely sampled search. The major problem in this case is that the medium is not linear and this causes error, especially for sources deep inside the brain.

5.1.1 General Approaches to Source Localization

In order to localize multiple sources within the brain two general approaches have been proposed by researchers, namely:

(a) equivalent current dipole (ECD) and
(b) linear distributed (LD) approaches.

In the ECD approach the signals are assumed to be generated by a relatively small number of focal sources [1–4]. In the LD approaches all possible source locations are considered simultaneously [5–12].

In the inverse methods using the dipole source model the sources are considered as a number of discrete magnetic dipoles located in certain places in a three-dimensional space within the brain. The dipoles have fixed orientations and variable amplitudes.

On the other hand, in the current distributed-source reconstruction (CDR) methods, there is no need for any knowledge about the number of sources. Generally, this problem is considered as an underdetermined inverse problem. An L_p norm solution is the most popular regulation operator to solve this problem. This regularized method is based upon minimizing the cost function

$$\psi = \|\mathbf{L}\boldsymbol{x} - m\|_p + \lambda\|\mathbf{W}\boldsymbol{x}\|_p \tag{5.1}$$

where $\boldsymbol{x}$ is the vector of source currents, $\mathbf{L}$ is the lead field matrix, m is the EEG measurements, $\mathbf{W}$ is a diagonal location weighting matrix, λ is the regulation parameter, and $1 \leq p \leq 2$, the norm, is the measure in the complete normed vector (Banach) space [13]. A minimum L_p norm method refers to the above criterion when $\mathbf{W}$ is equal to the identity matrix.

5.1.2 Dipole Assumption

For a dipole at location $\mathbf{L}$, the magnetic field observed at electrode i at location $\mathbf{R}(i)$ is achieved as (Figure 5.1)

$$\mathbf{B}(i) = \frac{\mu}{4\pi}\frac{\mathbf{Q} \times (\mathbf{R}(i) - \mathbf{L})}{|\mathbf{R}(i) - \mathbf{L}|} \quad \text{for } i = 1, \ldots, n_{\mathrm{e}} \tag{5.2}$$

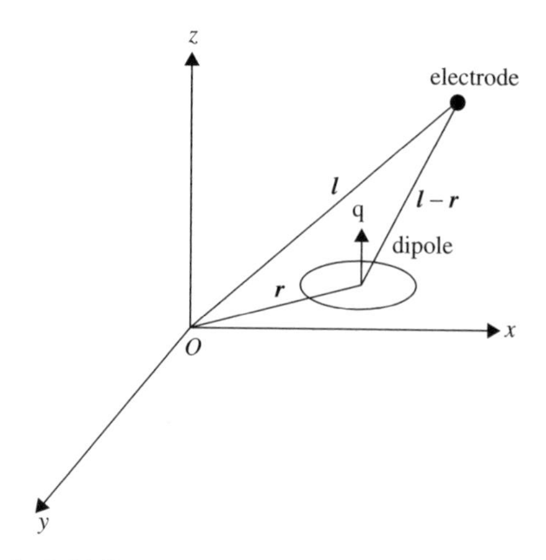

Figure 5.1 The magnetic field **B** at each electrode is calculated with respect to the moment of the dipole and the distance between the centre of the dipole and the electrode

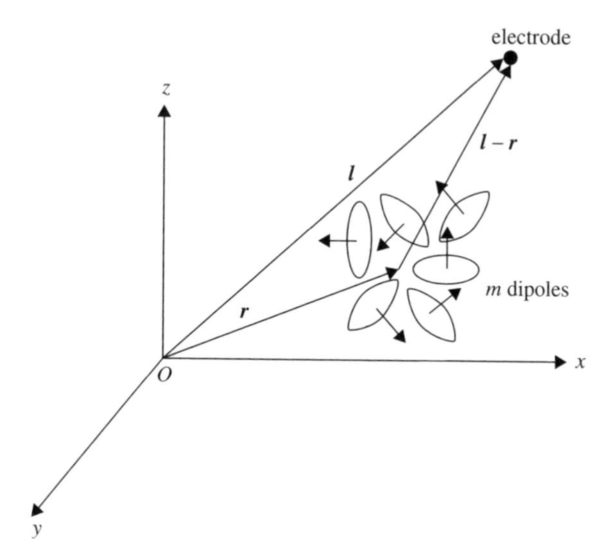

Figure 5.2 The magnetic field **B** at each electrode is calculated with respect to the accumulated moments of the m dipoles and the distance between the centre of the dipole volume and the electrode

where **Q** is the dipole moment, $|.|$ denotes absolute value, and $\times$ represents the outer vector product. This is frequently used as the model for magnetoencephalographic (MEG) data observed by magnetometers. This can be extended to the effect of a volume containing m dipoles at each one of the n_e electrodes (Figure 5.2).

In the m-dipole case the magnetic field at point j is obtained as

$$\mathbf{B}(i) = \frac{\mu_0}{4\pi} \sum_{j=1}^{m} \frac{\mathbf{Q}_j \times (\mathbf{R}(i) - \mathbf{L}_j)}{|\mathbf{R}(i) - \mathbf{L}_j|} \quad \text{for } i = 1, \ldots, n_e \tag{5.3}$$

where n_e is the number of electrodes and $\mathbf{L}_j$ represents the location of the jth dipole. The matrix $\mathbf{B}$ can be considered as $\mathbf{B} = [\mathrm{B}(1), \mathrm{B}(2), \ldots, \mathrm{B}(n_e)]$. On the other hand, the dipole moments can be factorized into the product of their unit orientation moments and strengths, i.e. $\mathbf{B} = \mathbf{GQ}$ (normalized with respect to $\mu_0/4\pi$), where $\mathbf{G} = [\boldsymbol{g}(1), \boldsymbol{g}(2), \ldots, \boldsymbol{g}(m)]$ is the propagating medium (mixing matrix) and $\mathbf{Q} = [\mathrm{Q}_1, \mathrm{Q}_2, \ldots \mathrm{Q}_m]^{\mathrm{T}}$. The vector $\boldsymbol{g}(i)$ has the dimension 1×3 (thus $\mathbf{G}$ is $m \times 3$). Therefore $\mathbf{B} = \mathbf{GMS}$. $\mathbf{GM}$ can be written as a function of location and orientation, such as $\mathbf{H}(\mathbf{L},\mathbf{M})$, and therefore $\mathbf{B} = \mathbf{H}(\mathbf{L}, \mathbf{M})\mathbf{S}$. The initial solution to this problem used a least-squares search that minimizes the difference between the estimated and the measured data:

$$J_{\mathrm{ls}} = \|\mathbf{X} - \mathbf{H}(\mathbf{L}, \mathbf{M})\mathbf{S}\|_{\mathrm{F}}^2 \tag{5.4}$$

where $\mathbf{X}$ is the magnetic (potential) field over the electrodes. The parameters to be estimated are location, dipole orientation, and magnitude for each dipole. This is subject to knowing the number of sources (dipoles). If too few dipoles are selected then resulting parameters are influenced by the missing dipoles. On the other hand, if too many dipoles are selected the accuracy will decrease since some of them are not valid brain sources. Also, the computation cost is high due a number of parameters being optimized simultaneously. One way to overcome this is by converting this problem into a projection minimization problem as

$$\mathbf{J}_{\mathrm{ls}} = \|\mathbf{X} - \mathbf{H}(\mathbf{L}, \mathbf{M})\mathbf{S}\|_{\mathrm{F}}^2 = \|\mathbf{P}_{\mathbf{H}}^{\perp}\mathbf{X}\|_{\mathrm{F}}^2 \tag{5.5}$$

The matrix $\mathbf{P}_{\mathbf{H}}^{\perp}$ projects the data on to the orthogonal complement of the column space of $\mathbf{H}(\mathbf{L},\mathbf{M})$. $\mathbf{X}$ can be reformed by a singular value decomposition (SVD), i.e. $\mathbf{X} = \mathbf{U\Sigma V}^{\mathrm{T}}$. Therefore

$$\mathbf{J}_{\mathrm{ls}} = \|\mathbf{P}_{\mathbf{H}}^{\perp}\mathbf{U\Sigma V}^{\mathrm{T}}\|_{\mathrm{F}}^2 \tag{5.6}$$

In this case orthogonal matrices preserve the Frobenius norm. The matrix $\mathbf{Z} = \mathbf{U\Sigma}$ is $m \times m$ unlike $\mathbf{X}$, which is $n_e \times T$, with T the number of samples. Generally, the rank of $\mathbf{X}$ satisfies rank($\mathbf{X}$) $\leq m$, $T >> m$, and $\mathbf{\Sigma}$ can have only m nonzero singular values. This means that the overall computation cost has been reduced by a large amount. The SVD can also be used to reduce the computation cost for $\mathbf{P}_{\mathbf{H}}^{\perp} = (\mathbf{I} - \mathbf{HH}^{\dagger})$. The pseudoinverse $\mathbf{H}^{\dagger}$ can be decomposed as $\mathbf{V}_{\mathbf{H}}\mathbf{\Sigma}_{\mathbf{H}}^{\dagger}\mathbf{U}_{\mathbf{H}}^{\mathrm{T}}$, where $\mathbf{H} = \mathbf{U}_{\mathbf{H}}\mathbf{\Sigma}_{\mathbf{H}}^{\dagger}\mathbf{V}_{\mathbf{H}}^{\mathrm{T}}$.

The dipole model, however, requires *a priori* knowledge about the number of sources, which is usually unknown.

Many experimental studies and clinical experiments have examined the developed source localization algorithms. Yao and Dewald [14] have evaluated different cortical source localization methods such as the moving dipole (MDP) method [15], the minimum L_p norm [16], and low-resolution tomography (LRT), as for LORETA (low-resolution electromagnetic tomography algorithm) [17] using simulated and experimental EEG data. In their study, only the scalp potentials have been taken into consideration.

In this study other source localization methods such as the cortical potential imaging method [18] and the three-dimensional resultant vector method [19] have not been included. These methods, however, follow similar steps to the minimum norm methods and dipole methods respectively.

5.2 Overview of the Traditional Approaches

5.2.1 ICA Method

In a simple approach by Spyrou *et al.* [20] ICA is used to separate the EEG sources. The correlation values of the estimated sources and mixtures are then used to build up the model of the mixing medium. The LS approach is then used to find the sources using the inverse of these correlations, i.e. $d_{ij} \approx 1/(C_{ij})^{0.5}$, where d_{ij} shows the distance between source i and electrode j, and C_{ij} shows the correlation of their signals. The method has been applied to separate and localize the sources of P3a and P3b subcomponents for five healthy subjects and five patients. A study of ERP components such as P300 and its constituent subcomponents is very important in analysis, diagnosing, and monitoring of mental and psychiatric disorders. Source location, amplitude, and latency of these components have to be quantified and used in the classification process. Figure 5.3 illustrates the results superimposed on two MRI templates for the above two cases. It can be seen that for healthy subjects the subcomponents are well apart whereas for schizophrenic patients they are geometrically mixed.

5.2.2 MUSIC Algorithm

Multiple signal classification (MUSIC) [21] has been used for localization of the magnetic dipoles within the brain [22–25] using EEG signals. In an early development of this algorithm a single-dipole model within a three-dimensional head volume is scanned and projections on to an estimated signal subspace are computed [26]. To locate the sources, the user must search the head volume for multiple local peaks in the projection metric. In an attempt to overcome the exhaustive search by the MUSIC algorithm a recursive MUSIC algorithm was developed [22]. Following this approach, the locations of the fixed,

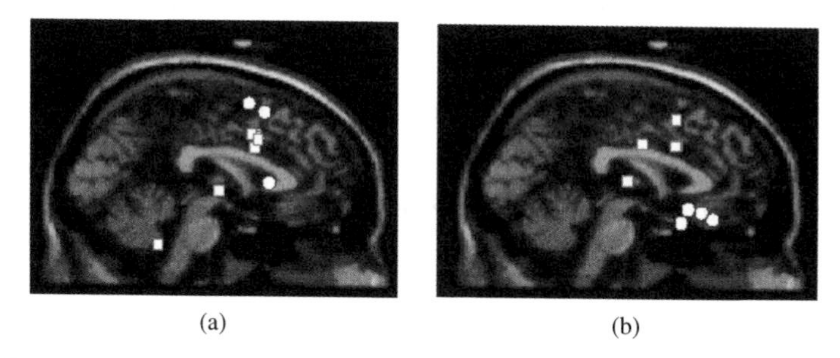

(a) (b)

Figure 5.3 Localization results for (a) the schizophrenic patients and (b) the normal subjects. The circles represent P3a and the squares represent P3b

rotating, or synchronous dipole sources are automatically extracted through a recursive use of subspace estimation. This approach tries to solve the problem of how to choose the locations that give the best projection on to the signal (EEG) space. In the absence of noise and using a perfect head and sensors model, the forward model for the source at the correct location projects entirely into the signal subspace. In practice, however, there are estimation errors due to noise. In finding the solution to the above problem there are two assumptions that may not be always true: first, the data are corrupted by additive spatially white noise and, second, the data are produced by a set of asynchronous dipolar sources. These assumptions are waved in the proposed recursive MUSIC algorithm [22].

The original MUSIC algorithm for the estimation of brain sources may be described as follows. Consider the head model for transferring the dipole field to the electrodes to be $\mathbf{A}(\boldsymbol{\rho}, \boldsymbol{\theta})$, where $\boldsymbol{\rho}$ and $\boldsymbol{\theta}$ are the dipole location and direction parameters respectively. The relationship between the observations (EEG), $\mathbf{X}$, the model $\mathbf{A}$, and the sources $\mathbf{S}$ is given as

$$\mathbf{X} = \mathbf{A}\mathbf{S}^{\mathrm{T}} + \mathbf{E} \tag{5.7}$$

where $\mathbf{E}$ is the noise matrix. The goal is to estimate the parameters $\{\boldsymbol{\rho}, \boldsymbol{\theta}, \mathbf{S}\}$, given the data set $\mathbf{X}$. The correlation matrix of $\mathbf{X}$, i.e. $\mathbf{R}_{\mathbf{x}}$, can be decomposed as

$$\mathbf{X}_{\mathrm{F}} = [\boldsymbol{\Phi}_{\mathrm{s}}, \boldsymbol{\Phi}_{\mathrm{e}}] \begin{bmatrix} \boldsymbol{\Lambda} + n_{\mathrm{e}}\sigma_{\mathrm{e}}^2\mathbf{I} & 0 \\ 0 & n_{\mathrm{e}}\sigma_{\mathrm{e}}^2\mathbf{I} \end{bmatrix} [\boldsymbol{\Phi}_{\mathrm{s}}, \boldsymbol{\Phi}_{\mathrm{e}}]^{\mathrm{T}} \tag{5.8}$$

or

$$\mathbf{R}_{\mathbf{X}} = \boldsymbol{\Phi}_{\mathrm{s}}\boldsymbol{\Lambda}_{\mathrm{s}}\boldsymbol{\Phi}_{\mathrm{s}}^{\mathrm{T}} + \boldsymbol{\Phi}_{\mathrm{e}}\boldsymbol{\Lambda}_{\mathrm{e}}\boldsymbol{\Phi}_{\mathrm{e}}^{\mathrm{T}} \tag{5.9}$$

where $\boldsymbol{\Lambda}_{\mathrm{s}} = \boldsymbol{\Lambda} + n_{\mathrm{e}}\sigma_{\mathrm{e}}^2\mathbf{I}$ is the $m \times m$ diagonal matrix combining both the model and noise eigenvalues and $\boldsymbol{\Lambda}_{\mathrm{e}} = n_{\mathrm{e}}\sigma_{\mathrm{e}}^2\mathbf{I}$ is the $(n_{\mathrm{e}} - m) \times (n_{\mathrm{e}} - m)$ diagonal matrix of noise-only eigenvalues. Therefore, the signal subspace span ($\boldsymbol{\Phi}_{\mathrm{s}}$) and noise-only subspace span ($\boldsymbol{\Phi}_{\mathrm{e}}$) are orthogonal. In practice, T samples of the data are used to estimate the above parameters, i.e.

$$\hat{\mathbf{R}}_{\mathbf{X}} = \mathbf{X}\mathbf{X}^{\mathrm{T}} = \hat{\boldsymbol{\Phi}}_{\mathrm{s}}\hat{\boldsymbol{\Lambda}}_{\mathrm{s}}\hat{\boldsymbol{\Phi}}_{\mathrm{s}}^{\mathrm{T}} + \hat{\boldsymbol{\Phi}}_{\mathrm{e}}\hat{\boldsymbol{\Lambda}}_{\mathrm{e}}\hat{\boldsymbol{\Phi}}_{\mathrm{e}}^{\mathrm{T}} \tag{5.10}$$

where the first m left singular vectors of the decomposition are designated as $\hat{\boldsymbol{\Phi}}_{\mathrm{s}}$ and the remaining eigenvectors as $\hat{\boldsymbol{\Phi}}_{\mathrm{e}}$. Accordingly, the diagonal matrix $\hat{\boldsymbol{\Lambda}}_{\mathrm{s}}$ contains the first m eigenvalues and $\hat{\boldsymbol{\Lambda}}_{\mathrm{e}}$ contains the remainder. To estimate the above parameters the general rule using least-squares fitting is

$$\{\hat{\boldsymbol{\rho}}, \hat{\boldsymbol{\theta}}, \hat{\mathbf{S}}\} = \arg \min_{\boldsymbol{\rho}, \boldsymbol{\theta}, \mathrm{S}} \|\mathbf{X} - \mathbf{A}(\boldsymbol{\rho}, \boldsymbol{\theta})\mathbf{S}^{\mathrm{T}}\|_{\mathrm{F}}^2 \tag{5.11}$$

where $\|.\|_F$ denotes the Frobenius norm. Optimal substitution [27] gives:

$$\{\hat{\boldsymbol{\rho}}, \hat{\boldsymbol{\theta}}\} = \arg \min_{\boldsymbol{\rho}, \boldsymbol{\theta}} \|\mathbf{X} - \mathbf{A}\mathbf{A}^{\dagger}\mathbf{X}\|_{\mathrm{F}}^2 \tag{5.12}$$

where $\mathbf{A}^{\dagger}$ is the Moore–Penrose pseudoinverse of $\mathbf{A}$ [28]. Given that the rank of $\mathbf{A}(\boldsymbol{\rho}, \boldsymbol{\theta})$ is m and the rank of $\hat{\boldsymbol{\Phi}}_{\mathrm{s}}$ is at least m, the smallest subspace correlation value,

$C_m = \text{subcorr}\{\mathbf{A}(\boldsymbol{\rho},\boldsymbol{\theta}), \hat{\boldsymbol{\Phi}}_s\}_m$, represents the minimum subspace correlation (maximum principal angle) between principal vectors in the column space of $\mathbf{A}(\boldsymbol{\rho},\boldsymbol{\theta})$ and the signal subspace $\hat{\boldsymbol{\Phi}}_s$. In MUSIC, the subspace correlation of any individual column of $\mathbf{A}(\boldsymbol{\rho},\boldsymbol{\theta})$ i.e. $\mathbf{a}(\boldsymbol{\rho}_i,\boldsymbol{\theta}_i)$ with the signal subspace must therefore equal or exceed this minimum subspace correlation:

$$C_i = \text{subcorr}\{\mathbf{a}(\boldsymbol{\rho}_i,\boldsymbol{\theta}_i), \hat{\boldsymbol{\Phi}}_s\} \geq C_m \tag{5.13}$$

$\hat{\boldsymbol{\Phi}}_s$ approaches $\boldsymbol{\Phi}_s$ when the number of sample increases or the SNR becomes higher. Then the minimum correlation approaches unity when the correct parameter set is identified, such that the m distinct sets of parameters $(\boldsymbol{\rho}_i,\boldsymbol{\theta}_i)$ have subspace correlations approaching unity. Therefore a search strategy is followed to find the $m$peaks of the metric [22]

$$\text{subcorr}^2\{\mathbf{a}(\boldsymbol{\rho},\boldsymbol{\theta}), \hat{\boldsymbol{\Phi}}_s\} = \frac{\mathbf{a}^T(\boldsymbol{\rho},\boldsymbol{\theta})\hat{\boldsymbol{\Phi}}_s\hat{\boldsymbol{\Phi}}_s^T\mathbf{a}(\boldsymbol{\rho},\boldsymbol{\theta})}{\|\mathbf{a}^T(\boldsymbol{\rho},\boldsymbol{\theta})\|_2^2} \tag{5.14}$$

where $\|\cdot\|_2^2$ denotes the squared Euclidean norm. For a perfect estimation of the signal subspace m global maxima equal to unity are found. This requires searching for both sets of parameters $\boldsymbol{\rho}$ and $\boldsymbol{\theta}$. In a quasilinear approach [22] it is assumed that $\mathbf{a}(\boldsymbol{\rho}_i,\boldsymbol{\theta}_i) = \mathbf{G}(\boldsymbol{\rho}_i\theta_i)$, where $\mathbf{G}(.)$ is called the gain matrix. Therefore, in the EEG source localization application, first the dipole parameters $\boldsymbol{\rho}_i$ that maximize the $\text{subcorr}\{\mathbf{G}(\boldsymbol{\rho}_i), \hat{\boldsymbol{\Phi}}_s\}$ are found and then the corresponding quasilinear $\boldsymbol{\theta}_i$ that maximize this subspace correlation are extracted. This avoids explicitly searching for these quasilinear parameters, reducing the overall complexity of the nonlinear search. Therefore, the overall localization of the EEG sources using classic MUSIC can be summarized in the following steps:

1. Decompose $\mathbf{X}$ or $\mathbf{XX}^T$ and select the rank of the signal subspace to obtain $\hat{\boldsymbol{\Phi}}_s$. Slightly overspecifying the rank has little effect on performance whereas underspecifying it can dramatically reduce the performance.
2. Form the gain matrix $\mathbf{G}$ at each point (node) of a dense grid of dipolar source locations and calculate the subspace correlations $\text{subcorr}\{\mathbf{G}(\boldsymbol{\rho}_i), \hat{\boldsymbol{\Phi}}_s\}$.
3. Find the peaks of the plot $\sqrt{1 - C_1^2}$, where C_1 is the maximum subspace correlation. Locate m or fewer peaks in the grid. At each peak, refine the search grid to improve the location accuracy and check the second subspace correlation.

A large second subspace correlation is an indication of a rotating dipole [22]. Unfortunately, there are often errors in estimating the signal subspace and the correlations are computed at only a finite set of grid points. A recursive MUSIC algorithm overcomes this problem by recursively building up the independent topography model. In this model the number of dipoles is initially considered as one and the search is carried out to locate a single dipole. A second dipole is then added and the dipole point that maximizes the second subspace correlation, C_2, is found. At this stage there is no need to recalculate C_1. The number of dipoles is increased and the new subspace correlations are computed. If m topographies comprise m_1 single-dipolar topographies and m_2 two-dipolar topographies, then the recursive MUSIC will first extract the m_1 single-dipolar models. At the

$(m_1 + 1)$th iteration, no single dipole location that correlates well with the subspace will be found. By increasing the number of dipole elements to two, the searches for both have to be carried out simultaneously such that the subspace correlation is maximized for C_{m_1+1}. The procedure continues to build the remaining m_2 two-dipolar topographies. After finding each pair of dipoles that maximizes the appropriate subspace correlation, the corresponding correlations are also calculated.

Another extension of MUSIC-based source localization for diversely polarized sources, namely recursively applied and projected (RAP) MUSIC [24], uses each successively located source to form an intermediate array gain matrix (similar to the recursive MUSIC algorithm) and projects both the array manifold and the signal subspace estimate into its orthogonal complement. In this case the subspace is reduced. Then the MUSIC projection to find the next source is performed. This method was initially applied to localization of the magnetoencephalogram (MEG) [24,25].

In a recent study by Xu *et al.* [29], another approach to EEG three-dimensional (3D) dipole source localization using a nonrecursive subspace algorithm, called FINES, has been proposed. The approach employs projections on to a subspace spanned by a small set of particular vectors in the estimated noise-only subspace, instead of the entire estimated noise-only subspace in the case of classic MUSIC. The subspace spanned by this vector set is, in the sense of the principal angle, closest to the subspace spanned by the array manifold associated with a particular brain region. By incorporating knowledge of the array manifold in identifying the FINES vector sets in the estimated noise-only subspace for different brain regions, this approach is claimed to be able to estimate sources with enhanced accuracy and spatial resolution, thus enhancing the capability of resolving closely spaced sources and reducing estimation errors. The simulation results show that, compared to classic MUSIC, FINES has a better resolvability of two closely spaced dipolar sources and also a better estimation accuracy of source locations. In comparison with RAP MUSIC, the performance of FINES is also better for the cases studied when the noise level is high and/or correlations among dipole sources exist [29].

A method for using a generic head model, in the form of an anatomical atlas, has also been proposed to produce EEG source localizations [30]. The atlas is fitted to the subject by a nonrigid warp using a set of surface landmarks. The warped atlas is used to compute a finite element model (FEM) of the forward mapping or lead-fields between neural current generators and the EEG electrodes. These lead-fields are used to localize current sources from the EEG data of the subject and the sources are then mapped back to the anatomical atlas. This approach provides a mechanism for comparing source localizations across subjects in an atlas-based coordinate system.

5.2.3 LORETA Algorithm

The low-resolution electromagnetic tomography algorithm (LORETA) for localization of brain sources has already been commercialized. In this method, the electrode potentials and matrix **X** are considered to be related as

$$\mathbf{X} = \mathbf{LS} \tag{5.15}$$

where **S** is the actual (current) source amplitudes (densities) and **L** is an $n_e \times 3m$ matrix representing the forward transmission coefficients from each source to the array of sensors.

$\mathbf{L}$ has also been referred to as the system response kernel or the lead-field matrix [31]. Each column of $\mathbf{L}$ contains the potentials observed at the electrodes when the source vector has unit amplitude at one location and orientation and is zero at all others. This requires the potentials to be measured linearly with respect to the source amplitudes based on the superposition principle. Generally, this is not true and therefore such assumption inherently creates some error. The fitted source amplitudes, $\mathbf{S}_n$, can be roughly estimated using an exhaustive search through the inverse least-squares (LS) solution, i.e.

$$\mathbf{S}_n = (\mathbf{L}^{\mathrm{T}}\mathbf{L})^{-1}\mathbf{L}^{\mathrm{T}}\mathbf{X} \tag{5.16}$$

$\mathbf{L}$ may be approximated by $3m$ simulations of current flow in the head, which requires a solution to the well-known Poisson equation [32]:

$$\nabla \cdot \sigma \nabla \mathbf{X} = -\rho \tag{5.17}$$

where σ is the conductivity of the head volume $(\Omega\,\mathrm{m})^{-1}$ and ρ is the source volume current density (A/m^3). A finite element method (FEM) or boundary element method (BEM) is used to solve this equation. In such models the geometrical information about the brain layers and their conductivities [33] have to be known. Unless some *a priori* knowledge can be used in the formulation, the analytic model is ill-posed and a unique solution is hard to achieve.

On the other hand, the number of sources, m, is typically much larger than the number of sensors, n_{e}, and the system in Equation (5.15) is underdetermined. Also, in the applications where more concentrated focal sources are to be estimated such methods fail. As will be seen later, using a minimum norm approach, a number of approaches choose the solution that satisfies some constraints, such as the smoothness of the inverse solution.

One approach is the minimum norm solution, which minimizes the norm of $\mathbf{S}$ under the constraint of the forward problem:

$$\min \|\mathbf{S}\|_2^2 \quad \text{subject to } \mathbf{X} = \mathbf{LS} \tag{5.18}$$

with a solution as

$$\mathbf{S} = \mathbf{L}^{\mathrm{T}}(\mathbf{L}\mathbf{L}^{\mathrm{T}})^{\dagger}\mathbf{X} \tag{5.19}$$

The motivation of the minimum norm solution is to create a sparse solution with zero contribution from most of the sources. This method has the serious drawback of poor localization performance in three-dimensional (3D) space. An extension to this method is the weighted minimum norm (WMN), which compensates for deep sources and hence performs better in 3D space. In this case the norms of the columns of $\mathbf{L}$ are normalized. Hence the constrained WMN is formulated as

$$\min \|\mathbf{WS}\|_2^2 \quad \text{subject to } \mathbf{X} = \mathbf{LS} \tag{5.20}$$

with a solution as

$$\mathbf{S} = \mathbf{W}^{-1}\mathbf{L}^{\mathrm{T}}(\mathbf{L}\mathbf{W}^{-1}\mathbf{L}^{\mathrm{T}})^{\dagger}\mathbf{X} \tag{5.21}$$

where $\mathbf{W}$ is a diagonal $3m \times 3m$ weighting matrix, which compensates for deep sources in the following way:

$$\mathbf{W} = \mathrm{diag}\left[\frac{1}{\|\mathbf{L}_1\|_2}, \frac{1}{\|\mathbf{L}_2\|_2}, \ldots, \frac{1}{\|\mathbf{L}_{3m}\|_2}\right] \tag{5.22}$$

where $\|\mathbf{L}_i\|_2$ denotes the Euclidean norm of the ith column of **L**, i.e. **W** corresponds to the inverse of the distances between the sources and electrodes.

In another similar approach a smoothing Laplacian operator is employed. This operator produces a spatially smooth solution agreeing with the physiological assumption mentioned earlier. The function of interest is then

$$\min \|\mathbf{BWS}\|_2^2, \quad \text{subject to } \mathbf{X} = \mathbf{LS} \tag{5.23}$$

where **B** is the Laplacian operator. This minimum norm approach produces a smooth topography in which the peaks representing the source locations are accurately located.

An FEM has been used to achieve a more anatomically realistic volume conductor model of the head, in an approach called the adaptive standardized LORETA/FOCUSS (focal underdetermined system solver) (ALF) [34]. It is claimed that using this application-specific method a number of different resolution solutions using different mesh intensities can be combined to achieve the localization of the sources with less computational complexities. Initially, FEM is used to approximate solutions to (5.23) with a realistic representation of the conductor volume based on magnetic resonance (MR) images of the human head. The dipolar sources are presented using the approximate Laplace method [35].

5.2.4 *FOCUSS Algorithm*

The FOCUSS [34] algorithm is a high-resolution iterative WMN method that uses the information from the previous iterations as

$$\min \|\mathbf{CS}\|_2^2 \quad \text{subject to } \mathbf{X} = \mathbf{LS} \tag{5.24}$$

where $\mathbf{C} = (\mathbf{Q}^{-1})^{\mathrm{T}}\mathbf{Q}^{-1}$ and $\mathbf{Q}_i = \mathbf{W}\mathbf{Q}_{i-1}[\mathrm{diag}(\mathbf{S}_{i-1}(1)\ldots\mathbf{S}_{i-1}(3m)]$ and the solution at iteration i becomes

$$\mathbf{S}_i = \mathbf{Q}_i\mathbf{Q}_i^{\mathrm{T}}\mathbf{L}^{\mathrm{T}}(\mathbf{L}\mathbf{Q}_i\mathbf{Q}_i^{\mathrm{T}}\mathbf{L}^{\mathrm{T}})^{\dagger}\mathbf{X} \tag{5.25}$$

The iterations will stop when there is no significant change in the estimation. The result of FOCUSS is highly dependent on the initialization of the algorithm. In practice, the algorithm converges close to the initialization point and may easily become stuck in some local minimum. A clever initialization of FOCUSS has been suggested to be the solution to LORETA [34].

5.2.5 *Standardized LORETA*

Another option, referred to as standardized LORETA (sLORETA), provides a unique solution to the inverse problem. It uses a different cost function, which is

$$\min \lfloor \|\mathbf{X} - \mathbf{LS}\|_2^2 + \lambda\|\mathbf{S}\|_2^2 \rfloor \tag{5.26}$$

Hence, sLORETA uses a zero-order Tikhonov–Phillips regularization [36,37], which provides a possible solution to the ill-posed inverse problems:

$$\mathbf{s}_i = \mathbf{L}_i^{\mathrm{T}}[\mathbf{L}_i\mathbf{L}_i^{\mathrm{T}} + \lambda_i\mathbf{I}]^{-1}\mathbf{X} = \mathbf{R}_i\mathbf{S} \tag{5.27}$$

where $\mathbf{s}_i$ indicates the candidate sources and $\mathbf{S}$ are the actual sources. $\mathbf{R}_i$ is the resolution matrix defined as

$$\mathbf{R}_i = \mathbf{L}_i^{\mathrm{T}}[\mathbf{L}_i\mathbf{L}_i^{\mathrm{T}} + \lambda_i\mathbf{I}]^{-1}\mathbf{L}_i \tag{5.28}$$

The reconstruction of multiple sources performed by the final iteration of sLORETA is used as an initialization for the combined ALF and weighted minimum norm (WMN or FOCUSS) algorithms [35]. The number of sources is reduced each time and Equation (5.19) is modified to

$$\mathbf{s}_i = \mathbf{W}_i\mathbf{W}_i^{\mathrm{T}}\mathbf{L}_f^{\mathrm{T}}[\mathbf{L}_f\mathbf{W}_i\mathbf{W}_i^{\mathrm{T}}\mathbf{L}_f^{\mathrm{T}} + \lambda\mathbf{I}]^{-1}\mathbf{X} \tag{5.29}$$

$\mathbf{L}_f$ indicates the final $n \times m$ lead-field returned by sLORETA. $\mathbf{W}_i$ is a diagonal $3m_f \times 3m_f$ matrix, which is recursively refined based on the current density estimated by the previous step:

$$\mathbf{W}_i = \mathrm{diag}[s_{i-1}(1), s_{i-1}(2), \ldots, s_{i-1}(3n_f)] \tag{5.30}$$

and the resolution matrix in (5.28) after each iteration changes to

$$\mathbf{R}_i = \mathbf{W}_i\mathbf{W}_i^{\mathrm{T}}\mathbf{L}_f^{\mathrm{T}}[\mathbf{L}_f\mathbf{W}_i\mathbf{W}_i^{\mathrm{T}}\mathbf{L}_f^{\mathrm{T}} + \lambda\mathbf{I}]^{-1}\mathbf{L}_f \tag{5.31}$$

Iterations are continued until the solution does not change significantly. In another approach by Liu *et al.* [38], called shrinking standard LORETA-FOCUSS (SSLOFO), sLORETA is used for initialization. It then uses the re-WMN of FOCUSS. During the process the localization results are further improved by involving the above standardization technique. However, FOCUSS normally creates an increasingly sparse solution during iteration. Therefore, it is better to eliminate the nodes with no source activities or recover those active nodes that might be discarded by mistake. The algorithm proposed in Reference [39] shrinks the source space after each iteration of FOCUSS, hence reducing the computational load [38]. For the algorithm not to get trapped in a local minimum a smoothing operation is performed. The overall SSLOFO is therefore summarized in the following steps [38,39]:

1. Estimate the current density $\hat{\mathbf{S}}_0$ using sLORETA.
2. Initialize the weighting matrix as $\mathbf{Q}_0 = \mathrm{diag}[\hat{\mathbf{S}}_0(1), \hat{\mathbf{S}}_0(2), \ldots, \hat{\mathbf{S}}_0(3m)]$.
3. Estimate the source power using standardized FOCUSS.
4. Retain the prominent nodes and their neighbouring nodes. Adjust the values on these nodes through smoothing.
5. Redefine the solution space to contain only the retained nodes, i.e. only the corresponding elements in $\mathbf{S}$ and the corresponding column in $\mathbf{L}$.
6. Update the weighting matrix.

7. Repeat steps 3 to 6 until a stopping condition is satisfied.
8. The final solution is the result of the last step before smoothing.

The stopping condition may be when a threshold is defined, or when there is no negligible change in the weights in further iterations.

5.2.6 Other Weighted Minimum Norm Solutions

In an LD approach by Phillips *et al.* [5] a weighted minimum norm (WMN) solution (or Tikhonov regularization) method [40] has been proposed. The solution has been regularized by imposing some anatomical and physiological information upon the overall cost function in the form of constraints. The squared error costs are weighted based on spatial and temporal properties. Information from, for example, hemodynamic measures of brain activity from other imaging modalities such as fMRI is used as constraints (or priors) together with the proposed cost function. In this approach it is assumed that the sources are sufficiently densely distributed and the sources are oriented orthogonal to the cortical sheet.

The instantaneous EEG source localization problem using a multivariate linear model and the observations, $\mathbf{X}$, as electrode potentials, is generally formulated on the basis of the observation model

$$\mathbf{X} = \Im(r, \mathbf{J}) + \mathbf{V} \tag{5.32}$$

where $\mathbf{X} = [\boldsymbol{x}(1), \boldsymbol{x}(2), \ldots, \boldsymbol{x}(T)]$ has dimension $n_e \times T$, T represents the length of the data in samples and n_e is the number of electrodes, $\boldsymbol{r}$ and $\mathbf{J} = [\boldsymbol{j}_1, \boldsymbol{j}_2, \ldots, \boldsymbol{j}_T]$ are respectively locations and moments of the sources, and $\mathbf{V}$ is the additive noise matrix. $\Im$ is the function linking the sources to the electrode potentials. In the calculation of $\Im$ a suitable three-layer head model is normally considered [41,42]. A structural MR image of the head may be segmented into three isotropic regions, namely brain, skull, and scalp, of the same conductivity [43] and used as the model. Most of these models consider the head as a sphere for simplicity.

However, in Reference [5] the EEG sources are modelled by a fixed and uniform three-dimensional grid of current dipoles spread within the entire brain volume. Also, the problem is an underdetermined linear problem as

$$\mathbf{X} = \mathbf{L}\mathbf{J} + \mathbf{V} \tag{5.33}$$

where $\mathbf{L}$ is the head-field matrix, which interrelates the dipoles to the electrode potentials. To achieve a unique solution for the above underdetermined equation some constraints have to be imposed. The proposed regularization method constrains the reconstructed source distribution by jointly minimizing a linear mixture of some weighted norm $\|\mathbf{H}\boldsymbol{j}\|_2$ of the current sources $\boldsymbol{j}$ and the main cost function of the inverse solution. Assuming the noise is Gaussian with a covariance matrix $\mathbf{C}_v$ then

$$\hat{\mathbf{J}} = \arg\min_{j} \left[\|\mathbf{C}_v^{-1/2}(\mathbf{L}\boldsymbol{j} - \boldsymbol{x})\|_2^2 + \lambda^2 \|\mathbf{H}\boldsymbol{j}\|_2^2 \right] \tag{5.34}$$

where the Lagrange multiplier λ has to be adjusted to make a balance between the main cost function and the constraint $\|\mathbf{H}\boldsymbol{j}\|_2$. The covariance matrix is scaled in such a way that trace($\mathbf{C}_v$) = rank($\mathbf{C}_v$) (recall that the trace of a matrix is the sum of its diagonal elements and the rank of a matrix refers to the number of independent columns (rows)). This can be stated as an overdetermined least-squares problem [18]. The solution to the minimization of Equation (5.34) for a given $\boldsymbol{\lambda}$ is in the form of

$$\hat{\mathbf{J}} = \mathbf{BX} \tag{5.35}$$

where

$$\mathbf{B} = [\mathbf{L}^{\mathrm{T}}\mathbf{C}_v^{-1}\mathbf{L} + \lambda^2(\mathbf{H}^{\mathrm{T}}\mathbf{H})]^{-1}\mathbf{L}^{\mathrm{T}}\mathbf{C}_v^{-1} = (\mathbf{H}^{\mathrm{T}}\mathbf{H})]^{-1}\mathbf{L}^{\mathrm{T}}[\mathbf{L}(\mathbf{H}^{\mathrm{T}}\mathbf{H})^{-1}\mathbf{L}^{\mathrm{T}} + \lambda^2\mathbf{C}_v]^{-1} \tag{5.36}$$

These equations describe the weighted minimum norm (WMN) solution to the localization problem. However, this is not complete unless a suitable spatial or temporal constraint is imposed. Theoretically, any number of constraints can be added to the main cost function in the same way and the hyperparameters such as Lagrange multipliers, $\boldsymbol{\lambda}$, can be calculated by expectation maximization [44]. However, more assumptions such as those about the covariance of the sources has to be implied in order to find effectively $\mathbf{L}$, which includes the information about the locations and the dipoles. One assumption, in the form of a constraint, can be based on the fact that $[\mathrm{diag}(\mathbf{L}^{\mathrm{T}}\mathbf{L})]^{-1}$, which is proportional to the covariance components, should be normalized. Another constraint is based on the spatial fMRI information which appears as the blood oxygenation level dependence (BOLD) when the sources are active. Evoked responses can also be used as temporal constraints.

5.2.7 Evaluation Indices

The accuracy of the inverse solution using the simulated EEG data has been evaluated by three indices: (1) the error distance (ED) i.e. the distance between the actual and estimated locations of the sources, (2) the undetected source number percentage (USP), and (3) the falsely detected source number percentage FSP. Obviously, these quantifications are based on the simulated models and data. For real EEG data it is hard to quantify and evaluate the results obtained by different inverse methods.

The ED between the estimated source locations, $\tilde{s}$, and the actual source locations, s, is defined as

$$\mathrm{ED} = \frac{1}{N_{\mathrm{d}}}\sum_{i=1}^{N_{\mathrm{d}}}\min_j(\|\tilde{s}_i - s_j\|) + \frac{1}{N_{\mathrm{ud}}}\sum_{j=1}^{N_{\mathrm{ud}}}\min_i(\|\tilde{s}_i - s_j\|_2) \tag{5.37}$$

where i and j are the indices of locations of the estimated and actual sources, and N_i andN_j are the total numbers of estimated and undetected sources. The USP and FSP are respectively defined as $\mathrm{USP} = N_{\mathrm{un}}/N_{\mathrm{real}} \times 100\,\%$ and $\mathrm{FSP} = N_{\mathrm{false}}/N_{\mathrm{estimated}} \times 100\,\%$ where N_{un}, N_{real}, N_{false}, and $N_{\mathrm{estimated}}$ are respectively the numbers of undetected, falsely detected, real, and estimated sources.

In practice three types of head volume conductor models can be used: a homogeneous sphere head volume conductor model, a boundary element method (BEM) model, or a finite element method (FEM) model. Since the FEM is computationally very intensive, the subject-specific BEM model, albeit an oversimplifying sphere head model, is currently used [45].

In terms of ED, USP, and FSP, LRT1 (i.e. $p = 1$) has been verified to give the best localization results. Use of the temporal properties of brain signals to improve the localization performance has also been attempted. An additional temporal constraint can be added assuming that for each location the change in the source amplitude with time is minimal. The constraint to be added is $\min \|\boldsymbol{s}(n) - \boldsymbol{s}(n-1)\|^2$, where n denotes the time index.

5.2.8 Joint ICA–LORETA Approach

In another study [46] Infomax ICA-based BSS has been implemented as a preprocessing scheme before the application of LORETA to localize the sources underlying the mismatch negativity (MMN). The MMN is an involuntary auditory ERP, which peaks at 100–200 ms when there is a violation of a regular pattern. This ERP appears to correspond to a primitive intelligence. MMN signals are mainly generated in the supramental cortex [47–51]. The LORETA analysis was performed with the scalp maps associated with selected ICA components to find the generators of these maps. Only values greater than 2.5 times the standard deviation of the standardized data (in the LORETA spatial resolution) were accepted as activations.

The inverse problem has also been tackled within a Bayesian framework [52]. In such methods some information about the prior probabilities are normally essential. Again the EEG generation model may be considered as

$$\boldsymbol{x}(n) = \mathbf{H}\boldsymbol{s}(n) + \boldsymbol{v}(n) \tag{5.38}$$

where $\boldsymbol{x}(n)$ is an $n_e \times 1$ vector containing the EEG sample values at time n, $\mathbf{H}$ is an $n_e \times m$ matrix representing the head medium model, $\boldsymbol{s}(n)$ are the $m \times 1$ vector sample values of the sources at time n and $\boldsymbol{v}(n)$ is the $n_e \times 1$ vector of noise samples at time n. The *a priori* information about the sources imposes some constraints on their locations and their temporal properties. The estimation may be performed using a maximum *a posteriori* (MAP) criterion in which the estimator tries to find $\boldsymbol{s}(n)$ that maximizes the probability distribution of $\boldsymbol{s}(n)$ given the measurements $\mathbf{x}(n)$. The estimator is denoted as

$$\hat{s}(n) = \max[p(\boldsymbol{s}(n)|\boldsymbol{x}(n))] \tag{5.39}$$

and following Bayes' rule, the posterior probability is

$$p(\boldsymbol{s}(n)|\boldsymbol{x}(n)) = p(\boldsymbol{x}(n)|\boldsymbol{s}(n))p(\boldsymbol{s}(n))/p(\boldsymbol{x}(n)) \tag{5.40}$$

where $p(\boldsymbol{x}(n)|\boldsymbol{s}(n))$ is the likelihood, $p(\boldsymbol{x}(n))$ is the marginal distribution of the measurements, or evidence and $p(\boldsymbol{s}(n))$ is the prior probability. The posterior can be written in terms of energy functions, i.e.

$$p(\boldsymbol{s}(n)|z(n)) = \frac{1}{z(n)}\exp[-U(\boldsymbol{s}(n))] \tag{5.41}$$

and $U(\boldsymbol{s}(n)) = (1-\lambda)U_1(\boldsymbol{s}(n)) + \lambda U_2(\boldsymbol{s}(n))$, where U_1 and U_2 correspond to the likelihood and the prior respectively, and $0 \leq \lambda \leq 1$. The prior may be separated into two functions, spatial priors U_s and temporal priors U_t. The spatial prior function can take into account the smoothness of the spatial variation of the sources. A cost function that determines the spatial smoothness is

$$\Phi(u) = \frac{u^2}{1 + (u/K)^2} \tag{5.42}$$

where K is the scaling factor that determines the required smoothness. Therefore the prior function for the spatial constraints can be written as

$$U_s(\mathbf{s}(n)) = \sum_{k=1}^{n_e} [\Phi_k^x(\nabla_x \mathbf{s}(n)|k) + \Phi_k^y(\nabla_y \mathbf{s}(n)|k)] \tag{5.43}$$

where the indices x and y correspond to horizontal and vertical gradients respectively. The temporal constraints are imposed by assuming that the projection of $\mathbf{s}(n)$ on to the space perpendicular to $\mathbf{s}(n-1)$ is small. Thus, the temporal prior function, as the second constraint, can be written as

$$U_t(\mathbf{s}(n)) = \|\mathbf{P}_{n-1}^{\perp} \mathbf{s}(n)\|^2 \tag{5.44}$$

where $\mathbf{P}_{n-1}^{\perp}$ is the projection on to the space perpendicular to $\mathbf{s}(n-1)$. Therefore the overall minimization criterion for estimation of $\mathbf{s}(n)$ will be

$$\hat{\mathbf{s}}(n) = \arg\min_{s} \left\{ \|\mathbf{x}(n) - \mathbf{H}\mathbf{s}(n)\|^2 + \alpha \sum_{k=1}^{n_e} [\Phi_k^x(\nabla_x \mathbf{s}(n)|k) + \Phi_k^y(\nabla_y \mathbf{s}(n)|k)] + \beta \|\mathbf{P}_{n-1}^{\perp} \mathbf{s}(n)\|^2 \right\} \tag{5.45}$$

where α and β are the penalty terms (regularization parameters).

According to the results of this study the independent components can be generated by one or more spatially separated sources. This confirms that each dipole is somehow associated with one dipole generator [53]. In addition, it is claimed that a specific brain structure can participate in different components, working simultaneously in different observations. The combination of ICA and LORETA exploits spatiotemporal dynamics of the brain as well as localization of the sources.

In Reference [38] four different inverse methods, namely WMN, sLORETA, FOCUSS, and SSLOFO, have been compared (based on a spherical head assumption and in the absence of noise). Figure 5.4 illustrates the results of these simulations.

5.2.9 *Partially Constrained BSS Method*

In a recent work [54] the locations of the known sources, such as some normal brain rhythms, have been used as a prior information in order to find the location of the abnormal or the other brain source signals using constrained BSS. The cost function of the BSS algorithm is constrained by this information and the known sources are iteratively

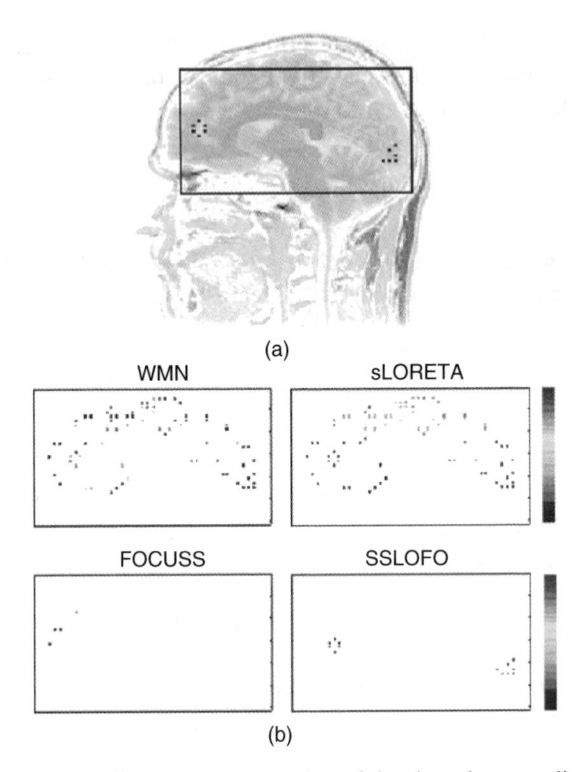

Figure 5.4 Brain source localization (reconstruction of the shaped source distribution in the realistic head model; (a) the positions of the simulated sources, (b) the results using the four inverse algorithms WMW, sLORETA, FOCUSS, and SSLOFO. (Adopted from Reference [38])

calculated. Consider that $\tilde{\mathbf{A}} = [\mathbf{A}_{\mathrm{k}} \vdots \mathbf{A}_{\mathrm{uk}}]$ is the mixing matrix including the geometrical information about the known, $\mathbf{A}_{\mathrm{k}}$, and unknown, $\mathbf{A}_{\mathrm{uk}}$, sources. $\mathbf{A}_{\mathrm{k}}$ is an $n_{\mathrm{e}} \times k$ matrix and $\mathbf{A}_{\mathrm{uk}}$ is an $n_{\mathrm{e}} \times (m - k)$ matrix. Given $\mathbf{A}_{\mathrm{k}}$, $\mathbf{A}_{\mathrm{uk}}$ may be estimated as follows:

$$\mathbf{A}_{\mathrm{uk}_{n+1}} = \mathbf{A}_{\mathrm{uk}_n} - \zeta \nabla_{\mathrm{A_{uk}}}(\mathbf{J}_{\mathrm{c}}) \tag{5.46}$$

where

$$\nabla_{\mathrm{A_{uk}}}(\mathbf{J}_{\mathrm{c}}) = 2([\mathbf{A}_{\mathrm{k}} \vdots \mathbf{A}_{\mathrm{uk}_n}] - \mathbf{R}_{n+1}\mathbf{W}_{n+1}^{-1}) \tag{5.47}$$

$$\mathbf{R}_{n+1} = \mathbf{R}_n - \gamma \nabla_{\mathbf{R}}(\mathbf{J}_{\mathrm{c}}) \tag{5.48}$$

and

$$\mathbf{J}_{\mathrm{c}} = \|\tilde{\mathbf{A}}_n - \mathbf{R}_{n+1}\mathbf{W}_{n+1}^{-1}\|_{\mathrm{F}}^2 \tag{5.49}$$

$$\nabla_{\mathbf{R}}(\mathbf{J}_{\mathrm{c}}) = 2\left[\mathbf{W}_{n+1}^{-1}\mathbf{A}_{\mathrm{k}} + \mathbf{R}_n\mathbf{W}_{n+1}^{-1}(\mathbf{W}_{n+1}^{-1})^{\mathrm{T}} - \mathbf{W}_{n+1}^{-1}\mathbf{A}_{\mathrm{uk}_n}\right] \tag{5.50}$$

with

$$\mathbf{W}_{n+1} = \mathbf{W}_n - \mu \nabla_{\mathbf{W}}\mathbf{J} \tag{5.51}$$

where $\mathbf{J}(\mathbf{W}) = \mathbf{J}_{\mathrm{m}}(\mathbf{W}) + \lambda\mathbf{J}_{\mathrm{c}}(\mathbf{W})$ and $\mathbf{J}_m$ is the main BSS cost function. The parameters μ, γ, and ζ are either set empirically or changed iteratively; they decrease when the convergence error decreases, and recall that $\mathbf{A}_{\mathrm{k}}$ is known and remains constant.

5.3 Determination of the Number of Sources

One of the major requirements for accurate separation and localization of brain sources is knowledge of the number of sources. The problem of detection of the number of independent (or uncorrelated) sources can be defined as analysing the structure of the covariance matrix of the observation matrix. This matrix can be expressed as $\mathbf{C} = \mathbf{C}_{\mathrm{sig}} + \mathbf{C}_{\mathrm{noise}}$, where $\mathbf{C}_{\mathrm{sig}}$ and $\mathbf{C}_{\mathrm{noise}}$ are the covariance of source signals and the covariance of noise respectively. PCA and SVD may perform well if the noise level is low. In this case the number of dominant eigenvalues represents the number of sources. In the case of white noise, the covariance matrix $\mathbf{C}_{\mathrm{noise}}$ can be expressed as $\mathbf{C}_{\mathrm{noise}} = \sigma_n^2\mathbf{I}$, where σ_{n}^2 is the noise variance and $\mathbf{I}$ is the identity matrix. In the case of coloured noise some similar methods can be implemented if the noise covariance is known apart from σ_{n}^2. The noise covariance matrix is a symmetric positive definite matrix $\mathbf{C}_{\mathrm{noise}} = \sigma_{\mathrm{n}}^2\mathbf{\Psi}$. Then, a nonsingular square matrix $\boldsymbol{\psi}$ $(m \times m)$ exists such that $\mathbf{\Psi} = \boldsymbol{\psi}\boldsymbol{\psi}^{\mathrm{T}}$ [55]. For both the white and coloured noise, cases the eigenvalues can be calculated from the observation covariance matrices, and then the eigenvalues can be analysed by the information theoretic criterion to estimate the number of independent sources [56].

However, it has been shown that this approach is suboptimal when the sources are temporally correlated [57]. Selection of an appropriate model for EEG analysis and source localization has been investigated by many researchers, and several criteria have been established to solve this problem. In most of these methods the amplitudes of the sources are tested to establish whether they are significantly larger than zero, in which case the sources are included in the model. Alternatively, the locations of the sources can be tested to determine whether they differ from each other significantly; in this case these sources should also be included in the model.

PCA and ICA may separate the signals into their uncorrelated and independent components respectively. By backprojecting the individual components to the scalp electrodes both of the above criteria may be tested. Practically, the number of distinct active regions within the backprojected information may denote the number of sources. The accuracy of the estimation also increases when the regions are clustered based on their frequency contents. However, due to the existence of noise with unknown distribution the accuracy of the results is still under question.

In Reference [58] the methods based on the residual variance (RV), the Akaike information criterion (AIC), the Bayesian information criterion (BIC), and the Wald tests on amplitudes (WA) and locations (WL) have been discussed. These methods have been later examined on MEG data [59] for both pure white error and coloured error cases. The same methods can be implemented for the EEG data as well. In this test the MEG data from $m$sensors and T samples are collected for each independent trial $j = 1, \ldots, n_{\mathrm{e}}$ in the $m \times T$ matrix $\mathbf{Y}_j = (\mathbf{y}_{1j}, \ldots, \mathbf{y}_{Tj})$, with $y_{ij} = (y_{1ij} \ldots, y_{mij})^{\mathrm{T}}$. Considering the average over trials as $\overline{\mathbf{Y}} = 1/n \sum_{j=1}^{\mathrm{n}} \mathbf{Y}_j$, the model for the averaged data can be given as

$$\overline{\mathbf{Y}} = \mathbf{GA} + \mathbf{E} \tag{5.52}$$

where **G** includes the sensor gains of the sources of unit amplitudes. Matrix **G** depends on the location and orientation parameters of the dipolar sources. Based on this model the tests for model selection are as follows.

The RV test defined as [58]:

$$\mathrm{RV} = 100\frac{\mathrm{tr}[(\overline{\mathbf{Y}} - \mathbf{GA})(\overline{\mathbf{Y}} - \mathbf{GA})^{\mathrm{T}}]}{\mathrm{tr}[\overline{\mathbf{Y}}\,\overline{\mathbf{Y}}]} \tag{5.53}$$

compares the squared residuals to the squared data for all sensors and samples simultaneously. The RV decreases as a function of the number of parameters, and therefore over-fits easily. The model is said to fit if the RV is below a certain threshold [60].

The AIC method penalizes the log-likelihood function for additional parameters required to describe the data. These parameters may somehow describe the sources [60]. The number of sources has been kept limited for this test since at some point any additional source hardly decreases the log-likelihood function, but increases the penalty. The AIC is defined as

$$\mathrm{AIC} = nmT\ln\left(\frac{\pi s^2}{n}\right) + \frac{1}{ns^2}\mathrm{tr}[(\overline{\mathbf{Y}} - \mathbf{GA})(\overline{\mathbf{Y}} - \mathbf{GA})'] + 2p \tag{5.54}$$

In this equation s^2 is the average of the diagonal elements of the spatial covariance matrix [58].

The BIC test resembles the AIC method but with more emphasis on the additional parameters. Therefore less over-fitting is expected when using BIC. This criterion is defined as [61]

$$\mathrm{BIC} = n_{\mathrm{e}}mT\ln\left(\frac{\pi s^2}{n_{\mathrm{e}}}\right) + \frac{1}{n_{\mathrm{e}}s^2}\mathrm{tr}[(\overline{\mathbf{Y}} - \mathbf{GA})(\overline{\mathbf{Y}} - \mathbf{GA})^{\mathrm{T}}] + p\ln(mT) \tag{5.55}$$

Similarly, the model with the minimum BIC is selected.

The Wald test is another important criterion, which gives the opportunity to test a hypothesis on a specific subset of the parameters [62]. Both amplitudes and locations of the sources can be tested using this criterion. If $\boldsymbol{r}$ is a q vector function of the source parameters (i.e. the amplitude and location), $\boldsymbol{r}_h$ the q vector of fixed hypothesized value of $\boldsymbol{r}$, $\mathbf{R}$ the $q \times k$ Jacobian matrix of $\boldsymbol{r}$ with respect to the k parameters, and $\mathbf{C}$ the $k \times k$ covariance matrix of source parameters, then the Wald test is defined as [63]

$$\mathbf{W} = \frac{1}{q}(r - \boldsymbol{r}_h)^{\mathrm{T}}(\mathbf{RC}^{-1}\mathbf{R}')^{-1}(\boldsymbol{r} - \boldsymbol{r}_h) \tag{5.56}$$

An advantage of using the WA technique in spatiotemporal analysis is the possibility of checking the univariate significance levels to determine at which samples the sources are active.

The tests carried out for two synthetic sources and different noise components [59] showed that the WL test has superior overall performance, and the AIC and WA perform well when the sources are close together.

These tests have been based on simulations of the sources and noise. It is also assumed that the locations and the orientations of the source dipoles are fixed and only the amplitudes change. For real EEG (or MEG) data, however, such information may be subject to change and is generally unknown. Therefore, up to now no robust criterion for estimation of the number of sources of the brain signals has been established.

5.4 Summary and Conclusions

Source localization, from only EEG signals, is an ill-posed optimization problem. This is mainly due to the fact that the number of sources is unknown. This number may change from time to time, especially when the objective is to investigate the EP or movement-related sources. Most of the proposed algorithms fall under one of the two methods of equivalent current dipole and linear distributed approaches. Some of the above methods such as sLORETA have been commercialized and reported to have a reasonable outcome for many applications. A hybrid system of different approaches seems to give better results. Localization may also be more accurate if the proposed cost functions can be constrained by some additional information stemming from clinical findings or from certain geometrical boundaries. Nonhomogeneity of the head medium is another major problem; comprehensive medical and physical experimental studies have to be carried out to find an accurate model of the head. Fusion of other modalities such as MEG or fMRI will indeed enhance the accuracy of the localization results. There are many potential applications for brain source localization such as for localization of the ERP signals [20], brain–computer interfacing [64], and seizure source localization [65].

References

[1] Miltner, W., Braun, C., Johnson, R. E., and Rutchkin, A. D. S., 'A test of brain electrical source analysis (BESA): a simulation study', *Electroencephalogr. Clin. Neurophysiol.*, **91**, 1994, 295–310.

[2] Scherg, M., and Ebersole, J. S., 'Brain source imaging of focal and multifocal epileptiform EEG activity', *Clin. Neurophysiol.*, **24**, 1994, 51–60.

[3] Scherg, M., Best, T., and Berg, P., 'Multiple source analysis of interictal spikes: goals, requirements, and clinical values', *J. Clin. Neurophysiol.*, **16**, 1999, 214–224.

[4] Aine, C., Huang, M., Stephen, J., and Christopher, R., 'Multistart algorithms for MEG empirical data analysis reliably characterize locations and time courses of multiple sources', *NeuroImage*, **12**, 2000, 159–179.

[5] Phillips, C., Rugg, M. D., and Friston, K. J., 'Systematic regularization of linear inverse solutions of the EEG source localization problem', *NeuroImage*', **17**, 2002, 287–301.

[6] Backus, G. E., and Gilbert, J. F., 'Uniqueness in the inversion of inaccurate gross earth data', *Phil. Trans. R. Soc.*, **266**, 1970, 123–192.

[7] Sarvas, J., 'Basic mathematical and electromagnetic concepts of the biomagnetic inverse problem', *Phys. Med. Biol.*, **32**, 1987, 11–22.

[8] Hamalainen, M. S., and Llmoniemi, R., 'Interpreting magnetic fields of the brain: minimum norm estimates', *Med. Biol. EMG Computing*, **32**, 1994, 35–42.

[9] Menendez, R. G., and Andino, S. G., 'Backus and Gilbert method for vector fields', *Human Brain Mapping*, **7**, 1999, 161–165.

[10] Pascual-Marqui, R. D., 'Review of methods for solving the EEG inverse problem', *Int. J. Bioelectromagnetism*, **1**, 1999, 75–86.

[11] Uutela, K., Hamalainen, M. S., and Somersalo, E., 'Visualization of magnetoencephalographic data using minimum current estimates', *NeuroImage*, **10**, 1999, 173–180.

[12] Phillips, C., Rugg, M. D., and Friston, K. J., 'Anatomically informed basis functions for EEG source localization: combining functional and anatomical constraints', *NeuroImage*, **16**, 2002, 678–695.

[13] Banach, S., *Théorie des Opérations Linéaires*, Vol. 1, Virtual Library of Science Mathematics. – Physical collection, Warsaw, 1932.

[14] Yao, J., and Dewald, J. P. A., 'Evaluation of different cortical source localization methods using simulated and experimental EEG data', *NeuroImage*, **25**, 2005, 369–382.

[15] Yetik, I. S., Nehorai, A., Hewine, J. D., and Murauchik, C. H., 'Distinguishing between moving and stationary sources using EEG/MEG measurements with an application to epilepsy', *IEEE Trans. Biomed. Engng*, **50**(3), 2005, 476–479.

[16] Xu, P., Tian, Y., Chen, H., and Yao, D., 'L_p noum iterative sparse solution for EEG source localization', *IEEE Trans. Biomed. Engng*, **54**(3), 2007, 400–409.

[17] Pascual-Marqui, R. D., Michel, C. M., and Lehmann, D., 'Low resolution electromagnetic tomography; a new method for localizing electrical activity in the brain', *Int. J. Psychophysiol.*, **18**, 1994, 49–65.

[18] He, B., Zhang, X., Lian, J., Sasaki, H., Wu, D., and Towle, V. L., 'Boundary element method-based cortical potential imaging of somatosensory evoked potentials using subjects' magnetic resonance images', *NeuroImage*, **16**, 2002, 564–576.

[19] Ricamato, A., Dhaher, Y., and Dewald, J., 'Estimation of active cortical current source regions using a vector representation scanning approach', *J. Clin. Neurophysiol.*, **20**, 2003, 326–344.

[20] Spyrou, L., Jing, M., Sanei, S., and Sumich, A., 'Separation and localization of P300 sources and their subcomponents using constrained blind source separation', *EURASIP J. of Appl. Signal Proc.*, Article ID 82912, 2007, 10 pages.

[21] Schmit, R. O., 'Multiple emitter location and signal parameter estimation', *IEEE Trans. Antennas Propagation*, **34**, 1986, 276–280; reprint of the original paper presented at RADC Spectrum Estimation Workshop, 1979.

[22] Mosher, J. C., and Leahy, R. M., 'Recursive music: a framework for EEG and MEG source localization', *IEEE Trans. Biomed. Engng*, **45**(11), 1998, 1342–1354.

[23] Mosher, J. C., Leahy, R. M., and Lewis, P. S., 'EEG and MEG: forward solutions for inverse methods', *IEEE Trans. Biomed. Engng*, **46**(3), 1999, 245–259.

[24] Mosher, J. C., and Leahy, R. M., 'Source localization using recursively applied and projected (RAP) MUSIC', *IEEE Trans. Biomed. Engng*, **47**(2), 1999, 332–340.

[25] Ermer, J. J., Mosher, J. C., Baillet, S., and Leahy, R. M., 'Rapidly recomputable EEG forward models for realizable head shapes', *J. Phys. Med. Biol.*, **46**, 2001, 1265–1281.

[26] Mosher, J. C., Lewis, P. S., and Leahy, R. M., 'Multiple dipole modelling and localization from spatio-temporal MEG data', *IEEE Trans. Biomed. Engng*, **39**, 1992, 541–557.

[27] Golub, G. H., and Pereyra, V., 'The differentiation of pseudo-inverses and nonlinear least squares problems whose variables separate', *SIAM J. Numer. Anal.*, **10**, 1973, 413–432.

[28] Golub, G. H., and Van, C. F. *Loan, Matrix Computations*, 2nd edn, John Hopkins University Press, Baltimore, Maryland, 1984.

[29] Xu, X.-L., Xu, B., and He, B., 'An alternative subspace approach to EEG dipole source localization', *Phys. Med. Biol.*, **49**, 2004, 327–343.

[30] Darvas, F., Ermer, J. J., Mosher, J. C., and Leahy, R. M., 'Generic head models for atlas-based EEG source analysis', *Human Brain Mapping*, **27**(2), 2006, 129–143.

[31] Buchner, H., Knoll, G., Fuchs, M., Rienacker, A., Beckmann, R., Wagner, M., Silny, J., and Pesch, J., 'Inverse localization of electric dipole current sources in finite element models of the human head', *Electroencephalogr. Clin. Neurophysiol.* **102**(4), 1997, 267–278.

[32] Steele, C. W., *Numerical Computation of Electric and Magnetic Fields*, Kluwer, Dordrecht, The Netherlands, 1996.

[33] Geddes, A., and Baker, L. E., 'The specific resistance of biological material – a compendium of data for the biomedical engineer and physiologist', *Med. Biol. Engng.*, **5**, 1967, 271–293.

[34] Schimpf, P. H., Liu, H., Ramon, C., and Haueisen, J., 'Efficient electromagnetic source imaging with adaptive standardized LORETA/FOCUSS', *IEEE Trans. Biomed. Engng.*, May 2005, **52**(5).

[35] Gorodnitsky, I. F., George, J. S., and Rao, B. D., 'Neuromagnetic source imaging with FOCUSS: a recursive weighted minimum norm algorithm', *J. Clin. Neurophysiol.*, **16**(3), 1999, 265–295.

[36] Pascual-Marqui, R. D., 'Standardized low-resolution brain electromagnetic tomography (sLORETA): technical details', *Method Findings Expl. Clin. Pharmacol.*, **24D**, 2002, 5–12.

[37] Hanson, P. C., *Rank-Efficient and Discrete Ill-Posed Problems*, SIAM, Philadelphia, Pennsylvania, 1998.

[38] Liu, H., Schimpf, P. H., Dong, G., Gao, X., Yang, F., and Gao, S., 'Standardized shrinking LORETA–FOCUSS (SSLOFO): a new algorithm for spatio-temporal EEG source reconstruction', *IEEE Trans. Biomed. Engng*, **52**(10), 2005, 1681–1691.

[39] Liu, H., Gao, X., Schimpf, P. H., Yang, F., and Gao, S., 'A recursive algorithm for the three-dimensional imaging of brain electric activity: shrinking LORETA–FOCUSS', *IEEE Trans. Biomed. Engng* **51**(10), 2004, 1794–1802.

[40] Tikhonov, A. N., and Arsenin, V. Y., *Solution of Ill Posed Problems*, John Wiley & Sons, Inc., New York, 1997.

[41] Ferguson, A. S., and Stronik, G., 'Factors affecting the accuracy of the boundary element method in the forward problem: I. Calculating surface potentials', '*IEEE Trans. Biomed. Engng*, **44**, 1997, 440–448.

[42] Buchner, H., Knoll, G., Fuchs, M., Rienacker, A., Beckmann, R., Wagner, M., Silny, J., and Pesch, J., 'Inverse localization of electric dipole current sources in finite element models of the human head', *Electroencephalogr. Clin. Neurophysiol.*, **102**(4), 1997, 267–278.

[43] Ashburner, J., and Friston, K. J., 'Multimodal image coregistration and partitioning – a unified framework', *NeuroImage*, **6**, 1997, 209–217.

[44] Vapnic, V., *Statistical Learning Theory*, John Wiley & Sons, Inc., New York, 1998.

[45] Fuchs, M., Drenckhahn, R., Wichmann, H. A., and Wager, M., 'An improved boundary element method for realistic volume-conductor modelling', *IEEE Trans. Biomed. Engng*, **45**, 1998, 980–977.

[46] Macro-Pallares, J., Grau, C., and Ruffini, G., 'Combined ICA–LORETA analysis of mismatch negativity', *NeuroImage*, **25**, 2005, 471–477.

[47] Alain, C., Woods, D. L., and Night, R. T., 'A distributed cortical network for auditory sensory memory in humans', *Brain Res.*, **812**, 1998, 23–27.

[48] Rosburg, T., Haueisen, J., and Kreitschmann-Andermahr, I., 'The dipole location shift within the auditory evoked neuromagnetic field components N100m and mismatch negativity (MMNm)', *Clin. Neurosci. Lett.*, **308**, 2004, 107–110.

[49] Jaaskelainen, I. P., Ahveninen, J., Bonmassar, G., and Dale, A. M., 'Human posterior auditory cortex gates novel sounds to consciousness', *Proc. Natl Acad. Sci.*, **101**, 2004, 6809–6814.

[50] Kircher, T. T. J., Rapp, A., Grodd, W., Buchkremer, G., Weiskopf, N., Lutzenberger, W., Ackermann, H., and Mathiak, K., 'Mismach negativity responses in schizophrenia: a combined fMRI and whole-head MEG study', *Am. J. Psychiatry*, **161**, 2004, 294–304.

[51] Muller, B. W., Juptner, M., Jentzen, W., and Muller, S. P., 'Cortical activation to auditory mismatch elicited by frequency deviant and complex novel sounds: a PET study', *NeuroImage*, **17**, 2002, 231–239.

[52] Serinagaoglu, Y., Brooks, D. H., and Macleod, R. S., 'Bayesian solutions and performance analysis in bioelectric inverse problems', *IEEE Trans. Biomed. Engng*, **52**(6), 2005, 1009–1020.

[53] Makeig, S., Debener, S., Onton, J., and Delorme, A., 'Mining event related brain dynamic trends', *Cognitive Sci.*, **134**, 2004, 9–21.

[54] Latif, M. A., Sanei, S., Chambers, J. A., and Shoker, L., 'Localization of abnormal EEG sources using blind source separation partially constrained by the locations of known sources', *IEEE Signal Proces. Let.*, **13**(3), 2006, 117–120.

[55] Bai, X., and He, B., 'Estimation of number of independent brain electric sources from the scalp EEGs', *IEEE Trans. Biomed.*, (This paper appears in: IEEE Transactions on Biomedical Engineering). **53**(10), 2006 1883–1892.

[56] Knösche, T., Brends, E., Jagers, H., and Peters, M., 'Determining the number of independent sources of the EEG: a simulation study on information criteria', *Brain Topography*, **11**, 1998, 111–124.

[57] Stoica, P., and Nehorai, A., 'MUSIC, maximum likelihood, and Cramer–Rao bound', *IEEE Trans. Acoust., Speech, and Signal Process.*, **37**(5), May 1989, 720–741.

[58] Waldorp, L. J., Huizenga, H. M., Nehorai, A., and Grasman, R. P., 'Model selection in electromagnetic source analysis with an application to VEFs', *IEEE Trans. Biomed. Engng*, **49**(10), March 2002, 1121–1129.

[59] Waldorp, L. J., Huizenga, H. M., Nehorai, A., and Grasman, R. P., 'Model selection in spatio-temporal electromagnetic source analysis', *IEEE Trans. Biomed. Engng*, **52**(3), March 2005, 414–420.

[60] Akaike, H., 'Information theory and an extension of the maximum likelihood principle', in: B. N. Petrov, F. Csaki (Eds.), Second International Symposium in Information Theory, Akademiai Kiado, Budapest, 1973, 267–281.

[61] Chow, G. C., 'A comparison of information and posterior probability criteria for model selection', *J. Econometrics*, **16**, 1981, 21–33.

[62] Huzenga, H. M., Heslenfeld, D. J., and Molennar, P. C. M., 'Optimal measurement conditions for spatiotemporal EEG/MEG source analysis', *Psychometrika*, **67**, 2002, 299–313.

[63] Seber, G. A. F., and Wild, C. J., *Nonlinear Regression*, John Wiley & Sons, Ltd, Toronto, Canada, 1989.

[64] Wentrup, M. G., Gramann, K., Wascher, E., and Buss, M., 'EEG source localization for brain–computer interfaces', in *Proceedings of the IEEE EMBS Conference*, 2005, Vienna, Austria, 128–131.

[65] Ding, L., Worrell, G. A., Lagerlund, T. D., and He, B., '3D source localization of interictal spikes in epilepsy patients with MRI lesions', *Phys. Med. Biol.*, **51**, 2006, 4047–4062.

6

Sleep EEG

Sleep is the state of natural rest observed in humans and animals, and even invertebrates such as the fruitfly *Drosophila*. It is an interesting and not perfectly known physiological phenomenon. The sleep state is one of the most important items of evidence for diagnosing mental disease. Sleep is characterized by a reduction in voluntary body movement, decreased reaction to external stimuli, an increased rate of anabolism (the synthesis of cell structures), and a decreased rate of catabolism (the breakdown of cell structures). Sleep is necessary and essential for the life of most creatures. The capability for arousal from sleep is a protective mechanism and also necessary for health and survival. Technically, sleep is different from unconsciousness [1].

Sleep is therefore defined as a state of unconsciousness from which a person can be aroused. In this state, the brain is relatively more responsive to internal stimuli than external stimuli. Sleep should be distinguished from coma. Coma is an unconscious state from which a person cannot be aroused.

Historically, sleep was thought to be a passive state. However, sleep is now known to be a dynamic process, and human brains are active during sleep. Sleep affects physical and mental health and the immune system.

States of brain activity during sleep and wakefulness result from different activating and inhibiting forces that are generated within the brain. Neurotransmitters (chemicals involved in nerve signalling) control whether some-one is asleep or awake by acting on nerve cells (neurons) in different parts of the brain. Neurons located in the brainstem actively cause sleep by inhibiting other parts of the brain that keep a person awake.

In humans, it has been demonstrated that the metabolic activity of the brain decreases significantly after 24 hours of sustained wakefulness. Sleep deprivation results in a decrease in body temperature, a decrease in immune system function as measured by a white blood cell count (the soldiers of the body), and a decrease in the release of growth hormone. Sleep deprivation can also cause increased heart rate variability [2].

Sleep is necessary for the brain to remain healthy. Sleep deprivation makes a person drowsy and unable to concentrate. It also leads to impairment of memory and physical performance and reduced ability to carry out mathematical calculations and other mental tasks. If sleep deprivation continues, hallucinations and mood swings may develop.

The release of growth hormone in children and young adults takes place during deep sleep. Most cells of the body show increased production and reduced breakdown of

EEG Signal Processing S. Sanei and J. Chambers

proteins during deep sleep. Sleep helps humans maintain optimal emotional and social functioning while we are awake by giving rest during sleep to the parts of the brain that control emotions and social interactions.

6.1 Stages of Sleep

Sleep is a dynamic process. Loomis provided the earliest detailed description of various stages of sleep in the mid-1930s, and in the early 1950s Aserinsky and Kleitman identified rapid eye movement (REM) sleep [1]. There are two distinct states that alternate in cycles and reflect differing levels of neuronal activity. Each state is characterized by a different type of EEG activity. Sleep consists of nonrapid eye movement (NREM) and REM sleep. NREM is further subdivided into four stages of I (drowsiness), II (light sleep), III (deep sleep), and IV (very deep sleep).

During the night the NREM and REM stages of sleep alternate. Stages I, II, III, and IV are followed by REM sleep. A complete sleep cycle, from the beginning of stage I to the end of REM sleep, usually takes about one and a half hours. However, generally, the ensuing sleep is relatively short and, for most practical purposes, a duration of 10–30 minutes suffices.

6.1.1 NREM Sleep

Stage I is the stage of drowsiness and very light sleep, and is considered a transition between wakefulness and sleep. During this stage, the muscles begin to relax. It occurs upon falling asleep and during brief arousal periods within sleep, and usually accounts for 5–10 % of the total sleep time. An individual can be easily awakened during this stage. Drowsiness shows marked age-determined changes. Hypnagogic rhythmical 4–6 cycles/s theta activity of late infancy and early childhood is a significant characteristic of such ages. Later in childhood and, in several cases, in the declining years of life, the drowsiness onset involves larger amounts of slow activity mixed with the posterior alpha rhythm [3]. In adults, however, the onset of drowsiness is characterized by gradual or brisk alpha dropout [3]. The slow activity increases as the drowsiness becomes deeper. Other findings show that in light drowsiness the P300 response increases in latency and decreases in amplitude [4], and the inter- and intrahemispheric EEG coherence alter [5]. Figure 6.1 shows a set of EEG signals recorded during the state of drowsiness. The seizure-type activity within the signal is very clear.

Deep drowsiness involves the appearance of vertex waves. Before the appearance of the first spindle trains, vertex waves occur (the transition from stage I to II). These sharp waves are also known as parietal humps [6]. The vertex wave is a compound potential, a small spike discharge of positive polarity followed by a large negative wave, which is a typical discharge wave. It may occur as an isolated event with larger amplitude than that of normal EEG. In aged individuals they may become small, inconspicuous, and hardly visible. Another signal feature for deep drowsiness is the positive occipital sharp transients (POST) of sleep.

Spindles (also called sigma activity), the trains of barbiturate-induced beta activity, occur independently at approximately 18–25 cycles/s, predominantly in the frontal lobe of the brain. They may be identified as a ‘group of rhythmic waves characterized by

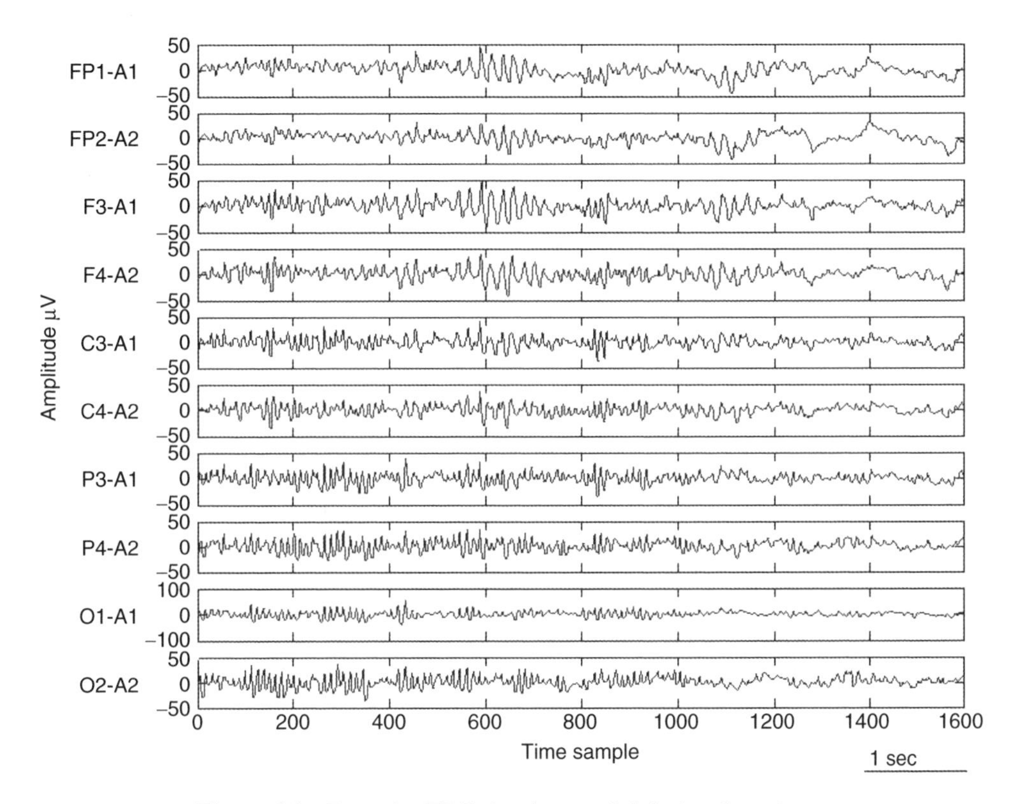

Figure 6.1 Examplar EEG signals recorded during drowsiness

progressively increasing, then gradually decreasing amplitude [3]. However, the use of middle electrodes shows a very definite maximum of the spindles over the vertex during the early stages of sleep.

Stage II of sleep occurs throughout the sleep period and represents 40–50 % of the total sleep time. During stage II, brain waves slow down with occasional bursts of rapid waves. Eye movement stops during this stage. Slow frequencies ranging from 0.7 to 4 cycles/s are usually predominant; their voltage is high, with a very prominent occipital peak in small children and gradually fall as age increases.

K-complexes appear in stage II and constitute a significant response to arousing stimuli. For topographical distribution over the brain, the K-complex shows a maximum over the vertex and has presence around the frontal midline [3]. For wave morphology, the K-complex consists of an initial sharp component, followed by a slow component that fuses with a superimposed fast component.

In stage III, delta waves begin to appear. They are interspersed with smaller, faster waves. Sleep spindles are still present at approximately 12–14 cycles/s but gradually disappear as the sleep becomes deeper.

In stage IV, delta waves are the primary waves recorded from the brain. Delta or slow wave sleep (SWS) usually is not seen during routine EEG [7]. However, it is seen during prolonged (>24 hours) EEG monitoring.

Stages III and IV are often distinguished from each other only by the percentage of delta activity. Together they represent up to 20 % of total sleep time. During stages III and IV all eye and muscle movement ceases. It is difficult to wake up someone during these two stages. If someone is awakened during deep sleep, he or she does not adjust immediately and often feels groggy and disoriented for several minutes after waking up. Generally, analysis of EEG morphology during stage IV has been of less interest.

6.1.2 REM Sleep

REM sleep including 20–25 % of the total sleep follows NREM sleep and occurs 4–5 times during a normal 8–9 hour sleep period. The first REM period of the night may be less than 10 minutes in duration, while the last period may exceed 60 minutes.

In an extremely sleepy individual, the duration of each bout of REM sleep is very short or it may even be absent. REM sleep is usually associated with dreaming. During REM sleep, the eyeballs move rapidly, the heart rate and breathing become rapid and irregular, blood pressure rises, and there is loss of muscle tone (paralysis), i.e. the muscles of the body are virtually paralysed. The brain is highly active during REM sleep, and the overall brain metabolism may be increased by as much as 20 %. The EEG activity recorded in the brain during REM sleep is similar to that recorded during wakefulness.

In a patient with REM sleep behaviour disorder (RBD), the paralysis is incomplete or absent, allowing the person to act out dreams, which can be vivid, intense, and violent. These dream-acting behaviours include talking, yelling, punching, kicking, sitting, jumping from the bed, arm flailing, and grabbing. Although the RBD may occur in association with different degenerative neurological conditions the main cause is still unknown.

Evaluation of REM sleep involves a long waiting period since the first phase of REM does not appear before 60–90 minutes after the start of sleep. The EEG in the REM stage shows low voltage activity with a slower rate of alpha.

6.2 The Influence of Circadian Rhythms

Biological variations that occur in the course of 24 hours are called circadian rhythms. Circadian rhythms are controlled by the biological clock of the body. Many bodily functions follow the biologic clock, but sleep and wakefulness comprise the most important circadian rhythm. The circadian sleep rhythm is one of several body rhythms modulated by the hypothalamus (a part of the brain as shown in Chapter 2).

Light directly affects the circadian sleep rhythm. Light is called *zeitgeber*, a German word meaning time-giver, because it sets the biological clock.

Body temperature cycles are also under the control of the hypothalamus. An increase in body temperature is seen during the course of the day and a decrease is observed during the night. The temperature peaks and troughs are thought to mirror the sleep rhythm. People who are alert late in the evening (i.e. evening types) have body temperature peaks late in the evening, while those who find themselves most alert early in the morning (i.e. morning types) have body temperature peaks early in the morning.

Melatonin (a chemical produced by the pineal gland in the brain and a hormone associated with sleep) has been implicated as a modulator of light entrainment. It is secreted

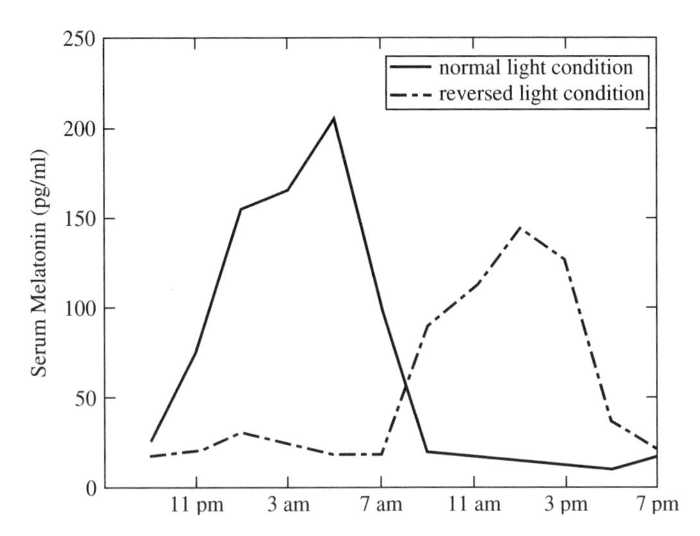

Figure 6.2 A typical concentration of melatonin in a healthy adult man. (Extracted from Reference [8])

maximally during the night. Prolactin, testosterone, and growth hormone also demonstrate circadian rhythms, with maximal secretion during the night. Figure 6.2 shows a typical concentration of melatonin in a healthy adult man.

Sleep and wakefulness are influenced by different neurotransmitters in the brain. Some substances can change the balance of these neurotransmitters and affect sleep and wakefulness. Caffeinated drinks (for example coffee) and medicines (for example diet pills) stimulate some parts of the brain and can cause difficulty in falling asleep. Many drugs prescribed for the treatment of depression suppress REM sleep.

People who smoke heavily often sleep very lightly and have reduced duration of REM sleep. Heavy smokers tend to wake up after three or four hours of sleep due to nicotine withdrawal.

Some people who have insomnia may use alcohol. Even though alcohol may help people to fall into a light sleep, it deprives them of REM sleep and the deeper and more restorative stages of sleep. Alcohol keeps them in the lighter stages of sleep from which they can be awakened easily. During REM sleep, some of the ability to regulate body temperature is lost. Therefore, abnormally hot or cold temperatures can disrupt our sleep. If REM sleep is disturbed, the normal sleep cycle progression is affected during the next sleeping time, and there is a possibility of slipping directly into REM sleep and going through long periods of REM sleep until the duration of REM sleep that is lost is caught up.

Generally, sleep disruption by any cause can be a reason for an increase in seizure frequency or severity. It can also have a negative effect on short-term memory, concentration, and mood. Seizure, itself, during the night can disrupt sleep and using any anticonvulsant drug may affect sleep in different ways. Both the frequency of seizure and the locality of seizure sources within the brain may change in different sleep stages and wakefulness.

6.3 Sleep Deprivation

Sleep deprivation is evaluated in terms of the tasks impaired and the average duration. In tasks requiring judgement, increasingly risky behaviours emerge as the total sleep duration is limited to five hours per night. The high cost of an action is seemingly ignored as the sleep-deprived person focuses on limited benefits. These findings can be explained by the fact that metabolism in the prefrontal and parietal associational areas of the brain decrease in individuals deprived of sleep for 24 hours. These areas of the brain are important for judgement, impulse control, attention, and visual association.

Sleep deprivation is a relative concept. Small amounts of sleep loss (for example one hour per night over many nights) produce subtle cognitive impairment, which may go unrecognized. More severe restriction of sleep for a week leads to profound cognitive deficits, which may also go unrecognized by the individual. If you feel drowsy during the day, fall asleep for very short periods of time (5 minutes or so), or regularly fall asleep immediately after lying down, you are probably sleep-deprived.

Many studies have made it clear that sleep deprivation is dangerous. With decreased sleep, higher-order cognitive tasks are impaired early and disproportionately. On tasks used for testing coordination, sleep-deprived people perform as poorly as or worse than people who are intoxicated. Total sleep duration of seven hours per night over one week has resulted in decreased speed in tasks of both simple reaction time and more demanding computer-generated mathematical problem solving. Total sleep duration of five hours per night over one week shows both a decrease in speed and the beginning of accuracy failure.

Using sleep deprivation for detection and diagnosis of some brain abnormalities has been reported by some researchers [9–11]. It consists of sleep loss for 24–26 hours. This was used by Klingler *et al.* [12] to detect the epileptic discharges that could otherwise be missed. Based on these studies it has also been concluded that sleep depravation is a genuine activation method [13]. Its efficacy in provoking abnormal EEG discharges is not due to drowsiness. Using the information in stage III of sleep, the focal and generalized seizure may be classified [14].

6.4 Psychological Effects

Sleep measurements and studies are often polygraphically oriented and use EEG in combination with a variety of other physiological parameters. EEG studies have documented abnormalities in sleep patterns in psychiatric patients with suicidal behaviour, including longer sleep latency, increased rapid eye movement (REM) time and increased phasic REM activity. Sabo *et al.* [15] compared sleep EEG characteristics of adult depressives with and without a history of suicide attempts and noted that those who attempted suicide had consistently more REM time and phasic activity in the second REM period but less delta wave counts in the fourth non-REM period. Another study [16] conducted at the same laboratory replicated the findings with psychotic patients. On the basis of two studies, the authors [16] suggested that the association between REM sleep and suicidality may cut across diagnostic boundaries and that sleep EEG changes may have a predictive value for future suicidal behaviour. REM sleep changes were later replicated by other studies in suicidal schizophrenia [17] and depression [18].

Three cross-sectional studies examined the relationship between sleep EEG and suicidality in depressed adolescents. Dahl *et al.* [19] compared sleep EEG between a depressed suicidal group, a depressed nonsuicidal group, and normal controls. Their results indicated that suicidal depressed patients had significantly prolonged sleep latency and increased REM phasic activity, with a trend for reduced REM latency compared to both nonsuicidal depressed and control groups. Goetz *et al.* [20] and McCracken *et al.* [21] replicated the finding of greater REM density among depressive suicidal adolescents.

Study of normal ageing and transient cognitive disorders in the elderly has also shown that the most frequent abnormality in the EEG of elderly subjects is slowing of alpha frequency whereas most healthy individuals maintain alpha activity within 9–11 Hz [22,23].

6.5 Detection and Monitoring of Brain Abnormalities During Sleep by EEG Analysis

EEG provides important and unique information about the sleeping brain. Polysomnography (PSG) has been the well-established method of sleep analysis and the main diagnostic tool in sleep medicine, which interprets the sleep signal macrostructure based on the criteria explained by Rechtschaffen and Kales (R&K) [24]. The spindles and slow-wave activities, arousals, and associated activities are detected from the EEG signals and monitored during sleep. The description of these activities relies on the division of the temporal domain signals into fixed segments of 20–30 seconds. For analysis and monitoring of sleep disorders the main stage is detection of the waveforms during different stages of sleep. A good example is diagnosis of sleep apnea syndrome (SAS), described in the following section.

6.5.1 Detection of the Rhythmic Waveforms and Spindles Incorporating Blind Source Separation

SAS with a high prevalence of approximately 2 % in women and 4 % in men between the ages of 30 to 60 years is the cause of many road accidents [16,25]. This syndrome is often treated by means of continuous positive airway pressure therapy or by surgery. An early diagnosis of the disease is important since adequate treatment can be provided. Diagnosis of this disease is normally by standard PSG techniques with overnight recordings of sleep stage, respiratory efforts, oronasal airflow, electrocardiographic findings, and oxyhemoglobin saturation parameters in an attended laboratory setting [17]. In order to facilitate recording of the EEGs during sleep with a small number of electrodes a method to best select the electrode positions and separate the ECG, EOG, and EMG has been proposed [18].

In this work, in order to investigate adequately the sleep signals and establish a night sleep profile, the electrophysiological activities manifested within the above signals have to be studied. Therefore a number of recording channels are needed and the signals have to be separately archived [26]. For the EEG the signals from C4-A1 or C3-A2 are used. For eye movement (EOG) two temporal electrodes located near each eye (E1 and E2), slightly moved relative to the median plain, are used so that horizontal and vertical eye movements relative to the same electrode A1 (or A2) located on the mastoids can be simultaneously observed. Muscular activity is obtained by processing two electrodes located on the chin.

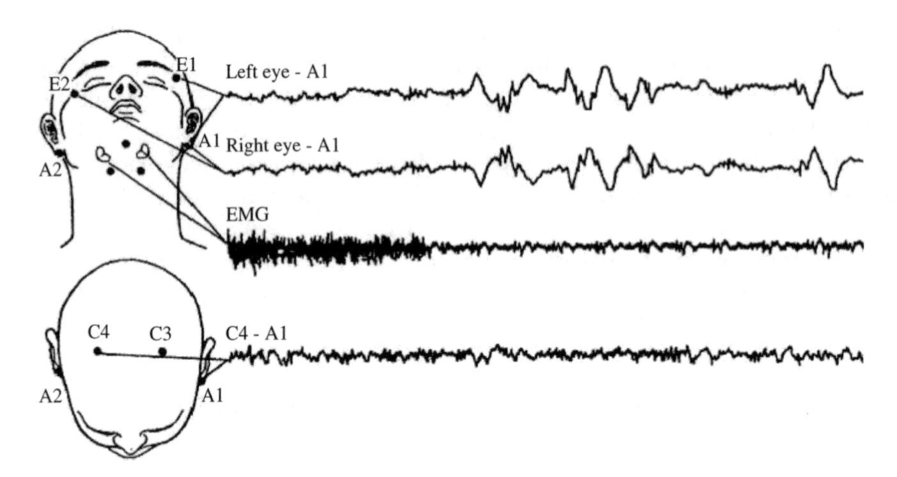

Figure 6.3 The configuration for sleep EEG recording proposed in Reference [18]. The position of the C3 and C4 electrodes is based on the conventional 10–20 EEG electrode positions. Reproduced with permission from Porée, F., Kachenoura, A., Gavrit, H., Morvan, C., Garrault, G., and Senhadji, L., 'Blind source separation for ambulatory sleep recording', *IEEE Trans. Information Technol. Biomed.*, **10**(2), 2006, 293–301. © 2006 IEEE

A reference electrode has also been used, which makes the total number of electrodes seven. Figure 6.3 shows the positions of the electrodes. These signals are then filtered to enhance the activity bands and reduce noise.

In order to separate the desired electrophysiological signals and obtain a night sleep profile a three-step approach has been followed. First, to extract the EMG signals a highpass filter is used to remove the frequency components above 30 Hz. The rest of the signals are lowpass-filtered to mitigate the effect of EMG. The remaining signals are then source-separated using blind source separation (BSS). The ICA algorithm introduced by Hyvärinen and Oja [27] has been used for this purpose. In this algorithm the columns of the unmixing matrix **W** are estimated as

$$w_i = E\left[z g'(w_i^{\mathrm{T}}\mathbf{z})\right] - E\left[z g''(w_i^{\mathrm{T}}z)\right] w_i \tag{6.1}$$

where g' and g'' are respectively the first and second derivatives of a nonquadratic function g, which approximates the negentropy, the index i refers to the ith column, and z is the whitened EEG signals. In practice, ensemble averaging is used instead of expectation. In each iteration, a symmetric orthogonalization of the unmixing matrix is performed by

$$\mathbf{W} \leftarrow (\mathbf{W}\mathbf{W}^{\mathrm{T}})^{-1/2}\mathbf{W} \tag{6.2}$$

until convergence is reached. The nonquadratic function has been chosen as $g(u) = \tanh(au)$ with $a<0$. The BSS algorithm is sought to separate the four desired signals of EEG, two EOGs and one ECG. In order to maintain the continuity of the estimated sources in the consecutive blocks of data the estimated independent components have been cross-correlated with the electrode signals, and those of consecutive signal segments most correlated with each particular electrode signal are considered to be the segments of the same source.

As a result of this work the alpha activity may not be seen consistently since the BSS system is generally underdetermined and therefore it cannot separate alpha activity from the other brain activities. However, the EMG complexes and REM are noticeable in the separated sources.

6.5.2 Application of Matching Pursuit

Some extensions to the R&K system using the conventional EEG signal processing methods have been proposed by Malinowska *et al.* [28]. These extensions include a finer timescale than the division into 20–30 second epochs, a measure of spindle intensity, and the differentiation of single and randomly evoked K-complexes in response to stimuli from spontaneous periodic ones. Figure 6.4 illustrates some typical spindles and K-complex waveforms.

The adaptive time–frequency (TF) approximation of signals using matching pursuit (MP) introduced initially by Mallat and Zhang [30] has been used in developing a method to investigate the above extensions [28]. MP has been reviewed in Chapter 2 of this book. The required dictionary of waveforms consists of Gabor functions mainly because these functions provide optimal joint TF localization [31]. Real-valued continuous time Gabor functions can be represented as

$$g_\gamma(t) = K(\gamma)\mathrm{e}^{-\pi((t-u)/s)^2}\cos\left[\omega(t-u)+\varphi\right] \tag{6.3}$$

where $K(\gamma)$ is such that the area under g_γ is equal to unity and the parameters of the Gabor function $\gamma = \{u, \omega, s\}$ provide a three-dimensional continuous space from which a finite dictionary must be chosen. In this application these parameters are drawn from uniform distributions over the signal range and correspond to the dictionary size. These parameters are fitted to the signal by the MP algorithm and often are directly used for analysis. Gabor functions represent a wide variety of EEG structures and are defined based on the amplitude and also the above parameters.

In this work the deep sleep stages (III and IV) are detected from the EEG signals based on the classical R&K criteria, derivation of the continuous description of slow-wave sleep

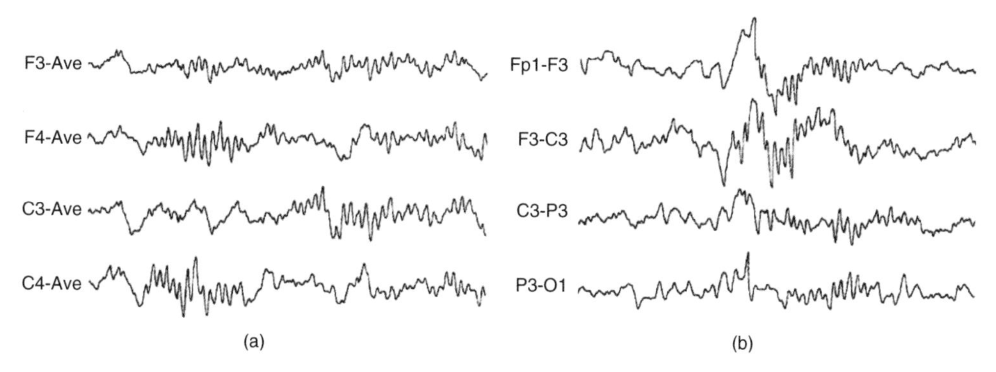

Figure 6.4 Typical waveforms for (a) spindles and (b) K-complexes. (Adopted from the website of Neural Networks Research Centre, Helsinki University of Technology [29])

Table 6.1 The time, frequency, and amplitudes of both SWA and sleep spindles [28]

	Time duration(s)	Frequency(Hz)	Minimum Amplitude (μV)
SWA	0.5–∞	0.5–4	$0.99 \times V_{EEG} + 28.18$
Sleep spindles	0.5–2.5	11–15	15

fully compatible with the R&K criteria has been attempted, a measure of spindling activity has been followed, and finally a procedure for detection of arousal has been presented. MP has been shown to separate well the various waveforms within the sleep EEGs. Assuming the slow wave activity (SWA) and sleep spindle to have the characteristics given in Table 6.1, they can be automatically detected by MP decomposition.

As the result of decomposition and examination of the components, if between 20 and 50 % of the duration of an epoch is occupied by SWA it corresponds to stage III and if above 50 % of the duration of an epoch is occupied by SWA it corresponds to stage IV. Therefore, stages III and IV of sleep can be recognized by applying the R&K criteria.

The first approach to the automatic detection of arousals was based upon the MP decomposition of only the C3-A2 single EEG channel and the standard deviation of EMG by implementing the rules established by the American Sleep Disorders Association (ASDA) [32]. The MP structure shows the frequency shifts within different frequency bands. Such shifts lasting three seconds or longer are related to arousal. To score a second arousal a minimum of 10 seconds of intervening sleep is necessary [28]. In Figure 6.5 the results of applying the MP algorithm using Gabor functions for the detection of both rhythms and transients can be viewed. Each blob in the TF energy map corresponds to one Gabor function. The 8–12 Hz alpha wave, sleep spindles (and one K-complex), and SWA of stages II and IV are presented in Figures 6.5 (a), (b), (c), and (d) respectively.

The sleep spindles exhibited inversely relate to the SWA [33]. The detected arousals decrease in relation to the amount of light NREM sleep, with particular concentration before the REM episodes [34].

The MP algorithm has also been extended to the differentiation of single, randomly evoked K-complexes in response to stimuli from spontaneous periodic ones [28]. The tools and algorithms developed for recognition and detection of the sleep stages can be applied to diagnosis of many sleep disorders such as apnea and the disturbances leading to arousal.

6.5.3 Detection of Normal Rhythms and Spindles using Higher Order Statistics

Long-term spectrum analysis has been widely used to detect and characterize sleep EEG waveforms [35,36]. However, these methods are unable to detect transient and isolated characteristic waves such as hump and K-complexes accurately.

In one approach, higher order statistics (HOS) of the time-domain signals together with the spectra of the EEG signals during the sleep have been utilized to characterize the dynamics of sleep spindles [37]. The spindles are considered as periodic oscillations with steady-state behaviour that can be modelled as a linear system with sinusoidal input or a nonlinear system with a limit cycle.

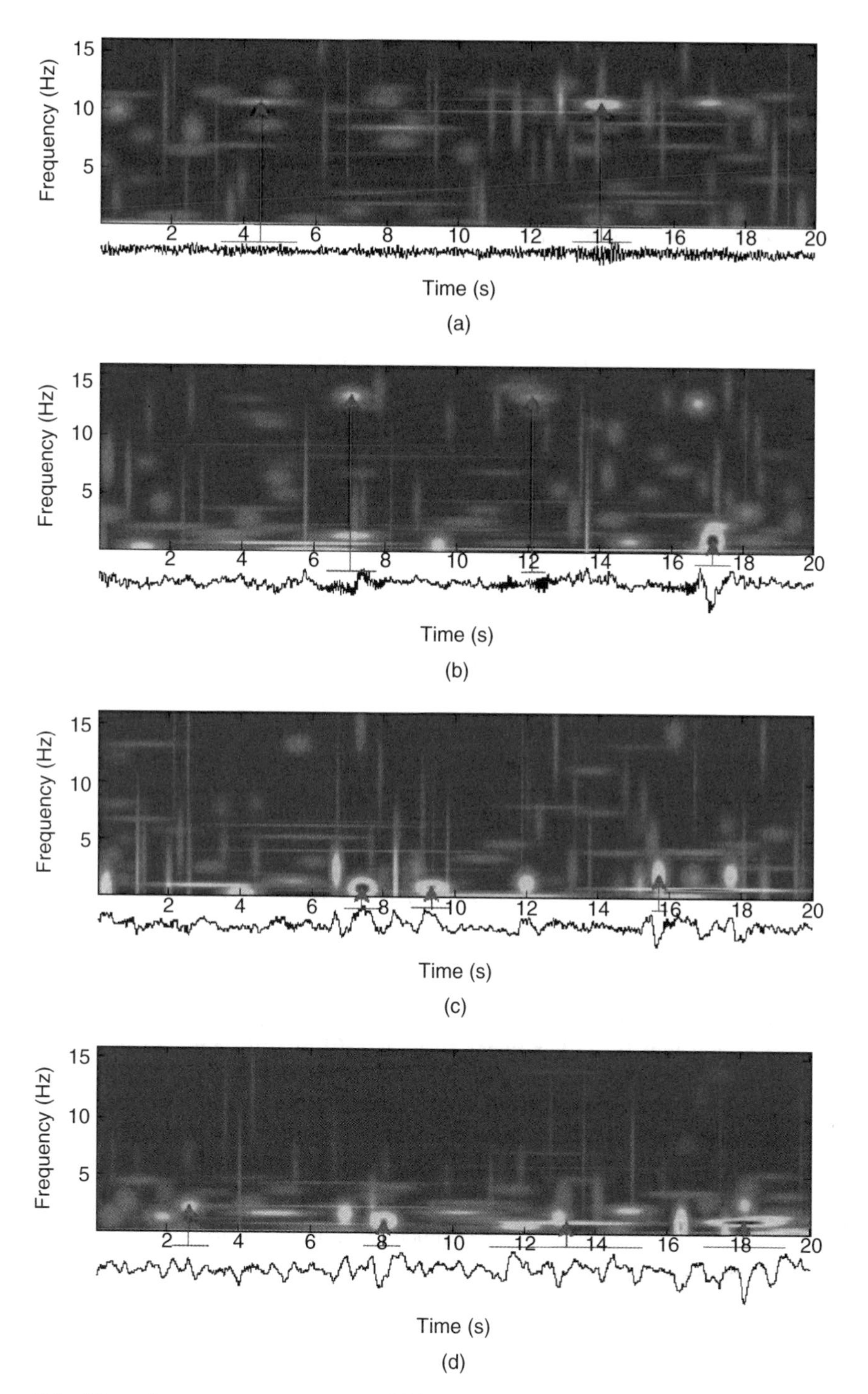

Figure 6.5 Time–frequency energy map of 20 s epochs of sleep EEG in different stages. The arrows point to the corresponding blobs for (a) awake alpha, (b) spindles and K-complex related to stage II, and (c) and (d) SWAs related to stages III and IV respectively. Reproduced with permission from Malinowska, U., Durka, P. J., Blinowska, K. J., Szelenberger, W., Wakarow, A. [28]

In this work, second-order and third-order correlations of the time-domain signals are combined to determine the stationarity of periodic spindle rhythms to detect transitions between multiple activities. The spectra (normalized spectrum and bispectrum) of the signals, on the other hand, describe frequency interactions associated with nonlinearities occurring in the EEGs.

The power spectrum of the stationary discrete signal, $x(n)$, is the power spectrum of its autocorrelation function, given by

$$P(\omega) = \sum_{n=-\infty}^{\infty} R(n)\mathrm{e}^{-\mathrm{j}n\omega} \cong \frac{1}{N}\sum_{i=1}^{N} X_i(\omega)X_i^*(\omega) \tag{6.4}$$

where $X_i(\omega)$ is the discrete Fourier transform of the ith segment of one EEG channel. Also, the bispectrum of data is defined as

$$\begin{aligned} B(\omega_1, \omega_2) &= \sum_{n_1=-\infty}^{\infty} \sum_{n_2=-\infty}^{\infty} x(n)x(n+n_1)x(n+n_2)e^{-j(n_1 w_1 + n_2 w_2)} \\ &\cong \frac{1}{N}\sum_{i=1}^{N} X_i(\omega_1)X_i(\omega_2)X_i(\omega_1+\omega_2) \end{aligned} \tag{6.5}$$

where N is the number of segments of each EEG channel. Using Equations (6.4) and (6.5), a normalized bispectrum (also referred to as bicoherence, second-order coherency, or bicoherency index) is defined as [38]

$$b^2(\omega_1, \omega_2) = \frac{|B(\omega_1, \omega_2)|^2}{P(\omega_1)P(\omega_2)P(\omega_1+\omega_2)} \tag{6.6}$$

which is an important tool for evaluating signal nonlinearities [39]. This measure (and the measure in Equation (6.5)) has been widely used for detection of coupled periodicities. Equation (6.6) acts as the discriminant of a linear process from a nonlinear one. For example, b^2 is constant for either linear systems [39] or fully coupled frequencies [40], and $b^2 = 0$ for either Gaussian signals or random phase relations where no quadratic coupling occurs. When the values of a normalized bispectrum vary above zero and below one ($0 < b^2 < 1$), then coupling of the frequencies occurs. The coherency value of one refers to quadratic interaction and an approximate zero value refers to either low or absent interactions [38].

In the hybrid method developed in Reference [37], however, to find the spindle periods a method similar to the average magnitude difference function (AMDF) algorithm, used for detection of the pitch frequency (period) of speech signals [40], has been applied to the short intervals of the EEG segments. The procedure has been applied to both second order and third order statistical measures as [37]:

$$D_n(k) = 1 - \frac{\gamma_n(k)}{\sigma_{\gamma_n}} \tag{6.7}$$

where

$$\gamma_n(k) = \sum_{m=-\infty}^{\infty} |x(n+m)w(m) - x(n+m-k)w(m-k)| \tag{6.8}$$

and $\sigma_{\gamma_n} = \sqrt{\sum_i \gamma_n^2(i)}$ is the normalization factor and $w(m)$ is the window function, and

$$Q_n(k) = 1 - \frac{\varphi_n(k)}{\sigma_{\varphi_n}} \tag{6.9}$$

where

$$\varphi_n(k) = \sum_{m=-\infty}^{\infty} |q(n+m)w(m) - q(n+m-k)w(m-k)|$$

and $q(n)$ is the inverse two-dimensional Fourier transform of the bispectrum and $\sigma_{\varphi_n} = \sqrt{\sum_i \varphi_n^2(i)}$. The measures are used together to estimate the periodicity of the spindles. For purely periodic activities these estimates are expected to give similar results. In this case Equation (6.9) manifests peaks (as in AMDF) where the first peak denotes the spindle frequency.

Based on this investigation, in summary it has been shown that:

1. Spindle activity may not uniformly dominate all regions of the brain.
2. During spindle activity frontal recordings still exhibit rich mixtures in frequency contents and coupling. On the other hand, a poor coupling may be observed at the posterior regions while showing dominant activity of the spindles.
3. It is concluded that spindle activity may be modelled using at least second-order nonlinearity.

6.5.4 Application of Neural Networks

Neural networks (NNs) can be used to classify different waveforms for recognition of various stages of sleep and also the types of mental illness. Neural networks have been widely used to analyse complicated systems without accurately modelling them in advance [41]. A number of typical waveforms from the sleep EEG can be used for training and classification. They include:

- Spindle
- Hump
- Alpha wave
- Hump train (although not present generally in the EEGs)
- Background wave

Each manifests itself differently in the time–frequency (TF) domain.

Time delay neural networks (TDNNs) may be used to detect the wavelets with roughly known positions on the time axis [42]. In such networks the deviation in location of the wavelet in time has to be small. For EEGs, however, a shift larger than the wavelet duration must be compensated since the occurrence times of the waveforms are not known.

In order to recognize the time-shifted pattern, another approach named as a sleep EEG recognition neural network (SRNN) has been proposed by Shimada *et al.* [43]. This NN has one input layer, two hidden layers, and one output layer. From the algorithmic point of view and the input–output connections, the SRNN and TDNN are very similar. As the main difference, in a TDNN each row of the second hidden layer is connected to a single cell of the output layer, while in an SRNN similar cells from both layers are connected together.

In order to use an SRNN the data are transformed into the TF domain. Instead of moving a sliding window over time, however, overlapped blocks of data are considered in this approach. Two-dimensional blocks with a horizontal axis of time and a vertical axis of frequency are considered as the inputs to the NN. Considering $y_{j,c}$ and $d_{j,c}$ to be the jth output neuron and the desired pattern for the input pattern c respectively, then

$$E = \frac{1}{2} \sum_{j=1}^{\substack{\text{Output}\\ \text{neurons}}} \sum_{c=1}^{\substack{\text{Input}\\ \text{patterns}}} (y_{j,c} - d_{j,c})^{1/2} \tag{6.10}$$

The learning rule therefore minimizes the following gradient:

$$\Delta w_{p,q} = \mu \frac{\partial E}{\partial w_{p,q}} \tag{6.11}$$

where $w_{p,q}$ are the link weights between neurons p and q and μ is the learning rate. In the learning phase the procedure [43] performs two passes, forward and backward, through the network. In the forward pass the inputs are applied and the outputs are computed. In the backward pass, the outputs are compared with the desired patterns and an error is calculated. The error is then backprojected through the network and the connection weight is changed by the gradient descent of the mean-squared error (MSE) as a function of weights. The final weights $w_{p,q}$ are obtained when the learning algorithm converges, however these are generally only locally optimal and various training runs with different initializations may be necessary. The weights are then used to classify a new waveform, i.e. to perform generalization. Further details of the performance of this scheme can be found in Reference [43].

6.5.5 Model-Based Analysis

Characterizing a physiological signal generation model for NREM has also been under study by several researchers [44–46]. These models describe how the depth of NREM sleep is related to the neuronal mechanism that generates slow waves. This mechanism is essentially feedback through closed loops in neuronal networks or through the interplay between ion currents in single cells. It is established that the depth of NREM sleep modulates the gain of the feedback loops [47]. According to this model, the sleep-related variations in the slow-wave power (SWP) result from variations in the feedback gain. Therefore, increasing the depth of sleep is related to an increasing gain in the neuronal feedback loops that generate the low-frequency EEG.

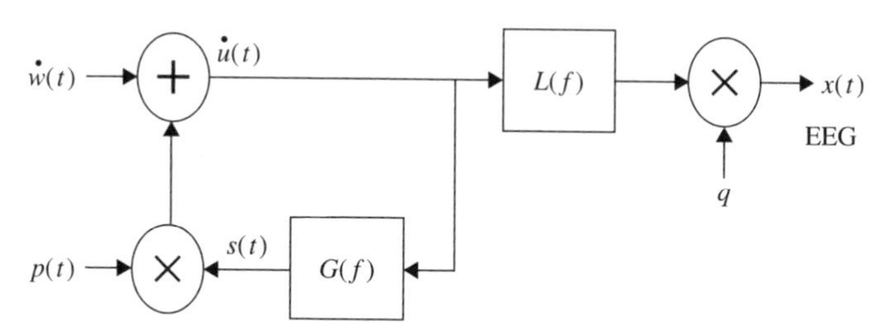

Figure 6.6 A model for neuronal slow-wave generation; $\dot{u}(t)$ is the EEG source within the brain, $G(f)$ denotes the frequency-selective feedback of the slow wave, $s(t)$, $p(t)$ is the feedback gain to be identified, $w(t)$ is the white noise, $L(f)$ is the lowpass filter representing the path from the source to electrodes, and q is the attenuation effect by the skull

In Reference [48] a model-based estimator of the slow-wave feedback gain has been proposed. The initial model depicted in Figure 6.6 is analogue and a discrete-time approximation of that has been built up. In the analogue model $G(f)$ is the complex frequency transfer function of a bandpass (resonance) filter as

$$G(f) = \frac{1}{1 + \mathrm{j}Y(f)} \tag{6.12}$$

in which $\mathrm{j} = \sqrt{-1}$ and

$$Y(f) = \frac{f_0}{B}\left(\frac{f}{f_0} - \frac{f_0}{f}\right) \tag{6.13}$$

where the resonance frequency f_0 and the bandwidth B are approximately 1 and 1.5 Hz respectively. The closed-loop equation can be written as

$$\dot{u}(t) = p(t)\,s(t) + \dot{w}(t) \tag{6.14}$$

where $\dot{u}(t) = \mathrm{d}u(t)/\mathrm{d}t$ and $\dot{w}(t) = \mathrm{d}w(t)/\mathrm{d}t$. It is seen that the transfer function depends on $p(t)$. For $p(t)$ constant,

$$U(f) = \frac{1}{1 - p\,G(f)} \tag{6.15}$$

and therefore

$$|U(f)|^2 = \frac{1 + Y^2(f)}{(1 - p)^2 + Y^2(f)} \tag{6.16}$$

In this case

$$S(f) = G(f)\,U(f) = \frac{G(f)}{1 - p\,G(f)} = \frac{1}{1 - p + \mathrm{j}Y(f)} \tag{6.17}$$

The output $x(t)$ is the lowpass copy of $\dot{u}(t)$ attenuated by the factor of q. For $p = 0$ there is no feedback and for $p = 1$ there is an infinite peak at $f = f_0$. The lowpass filter $L(f)$ is considered known with a cut-off frequency of approximately 1.8 Hz.

The feedback gain of the model $p(t)$ represents the sleep depth. Therefore, the objective would be to estimate the feedback gain. To do that, the observation $du(t)$ is sampled by a 50 Hz sampler (a sampling interval of $\Delta = 0.02$ s). Then, define $Du(k\Delta) = u(k\Delta + \Delta - u(k\Delta)$ over an interval $[0 \leq k\Delta < N\Delta - \Delta]$, with $N\Delta = T$. Also, it is considered that $p(t) = p$ is constant over the interval T. Equation (6.14) then becomes

$$Du(k\Delta) = p\, s(k\Delta)\, \Delta + Dw(k\Delta) \tag{6.18}$$

where $Dw(k\Delta) = w(k\Delta + \Delta) - w(k\Delta)$ is the increment of the standard continuous-time Wiener process $w(t)$. Assuming the initial state for the feedback filter $G(f)$ to be G_0 and $w(k\Delta)$ to have Gaussian distribution, the likelihood of $Du(k\Delta)$ can be factorised according to Bayes' rule as

$$\begin{aligned}
&P[Du(k\Delta) : 0 \leq k < N - 1 | G_0, p] \\
&\quad = \prod_{k=0}^{N-1} [Du(k\Delta) | [Du(m\Delta) : 0 \leq m < k], G_0, p] \\
&\quad = \prod_{k=0}^{N-1} [Du(k\Delta) | s(k\Delta), p] \\
&\quad = \prod_{k=0}^{N-1} \left(\frac{1}{\sqrt{2\pi\Delta}} \exp\left\{ -[Du(k\Delta) - p\, s(k\Delta)\, \Delta]^2 / 2\Delta \right\} \right) \\
&\quad = \frac{1}{\sqrt{2\pi\Delta}} \exp\left(\sum_{k=0}^{N-1} \left\{ -[Du(k\Delta) - p\, s(k\Delta)\, \Delta]^2 / 2\Delta \right\} \right)
\end{aligned} \tag{6.19}$$

To maximize this likelihood it is easy to conclude that the last term in the brackets has to be maximized. This gives [48]

$$\hat{p} = \frac{\sum_{k=0}^{N-1} [s(k\Delta)\, Du(k\Delta)]}{\sum_{k=0}^{N-1} s^2(k\Delta)\, \Delta} \tag{6.20}$$

Hence, $\hat{p}$ approximates the amount of slow wave and is often represented as the percentage of the slow wave. This completes the model and therefore the sleep EEG may now be constructed.

6.5.6 *Hybrid Methods*

Diagnosis of sleep disorders and other related abnormalities may not be complete unless other physiological symptoms are studied. These symptoms manifest themselves within other physiological and pneumological signals such as respiratory airflow, position of the patients, electromyogram (EMG) signal, hypnogram, level of SaO_2, abdominal effort, and thoracic effort, which may also be considered in the classification system.

A simple system to use the features extracted from the sleep signals in classification of the apnea stages and detection of the sleep apnea syndrome (SAS) has been suggested and used [9]. This system processes the signals in three phases. In the first phase the relevant characteristics of the signals are extracted and a segmentation based on significant time intervals of variable length is performed. The intermediate phase consists of assigning suitable labels to these intervals and combining these symbolic information sources with contextual information in order to build the necessary structures that will identify clinically significant events. Finally, all the relevant information is collected and a rule-based system is established to exploit the above data, provide the induction, and produce a set of conclusions.

In a rule-based system for detection of the SAS, two kinds of cerebral activities are detected and characterized from the EEG signals: rhythmic (alpha, beta, theta, and delta rhythms) and transitory (K-complexes and spindles) [49]. The magnitude and the change in the magnitude (evolution) are measured and the corresponding numerical values are classified together with the other features based on clinical and heuristic [10] criteria.

To complete this classifier the slow eye movement, very frequent during sleep stage I, and REM are measured using electroocclugram (EOG) signals. The distinction between the above two eye movements is based on the synchrony, amplitude, and slope of the EOG signals [9]. In another approach for sleep staging of the patients with obstructive sleep apnea the EEG features are used, classified, and compared with the results from cardiorespiratory features [11].

6.6 Concluding Remarks

Although the study of sleep EEG has opened a new path to investigate the psychology of a human being, there is still much more to discover from these signals. Various physiological and mental brain disorders manifest themselves differently in the sleep EEG signals. Different stages of sleep may be identified using simple established tools in signal processing. Detection and classification of mental diseases from the sleep EEG signals, however, requires more deep analysis of the data by developing and utilizing advanced digital signal processing techniques. This analysis becomes more challenging when other parameters such as age are involved. For example, in neonates many different types of complex waveforms may be observed for which the origin and causes are still unknown. On the other hand, there are some similarities between the normal rhythms within the sleep EEG signals and the EEGs of abnormal rhythms such as epileptic seizure and hyperventilation. An efficient algorithm (based on a sole EEG or combined with other physiological signals) should be able to differentiate between these signals.

References

[1] Steriade, M., 'Basic mechanisms of sleep generation', *Neurology*, **42** (Suppl. 6), 1992, 9–17.

[2] Kubicki, S., Scheuler, W., and Wittenbecher, H., 'Short-term sleep EEG recordings after partial sleep deprivation as a routine procedure in order to uncover epileptic phenomena: an evaluation of 719 EEG recordings', *Epilepsy Res. Suppl.*, **2**, 1991, 217–230.

[3] Niedermeyer, E., 'Sleep and EEG', Chapter 10, in *Electroencephalography Basic Principles, Clinical Applications, and Related Fields*, Eds E. Niedermeyer and F. Lopes da Silva, 4th edn, Lippincott, Williams and Wilkins, Philadelphia, Pennsylvania, 1999, 174–188.

[4] Koshino, Y., Nishio, M., Murata, T., Omori, M., Murata, I., Sakamoto, M., and Isaki, K., 'The influence of light drowsiness on the latency and amplitude of P300', *Clin. Electroencephalogr.*, **24**, 1993, 110–113.

[5] Wada, Y., Nanbu, Y., Koshino, Y., Shimada, Y., and Hashimoto, T., 'Inter- and intrahemispheric EEG coherenceduring light drowsiness', *Clin. Electroencephalogr*. **27**, 1996, 24–88.

[6] Niedermeyer, E., 'Maturation of EEG: development of walking and sleep patterns', Chapter. 11, in *Electroencephalography, Basic Principles, Clinical Applications, and Related Fields*, Eds E. Niedermeyer and F. Lopez da Silva, 4th edn, Lippincott, Williams and Wilkins, Philadelphia, Pennsylvania, 1999.

[7] Martin, C. R., and Marzec, M. L., 'Sleep scoring: 35 years of standardized sleep staging', *RT Mag.*, 2003.

[8] Brzezinski, A., *N. Engl. J. Med.*, **336**, 1997, 186–195.

[9] Cabrero-Canosa, M., Hernandez-Pereira, E., and Moret-Bonillo, V., 'Intelligent ignosis of sleep apnea syndrome', *IEEE Engng. in Med. Bio. Mag.*, 2004, 72–81.

[10] Karskadon, M. A., and Rechtschaffen, A., *Priniciples and Practice of Sleep Medicine*, Saunders, Philadelphia, Pennsylvania, 1989, 665–683.

[11] Redmond, S. J., and Heneghan, C., 'Cardiorespiratory-based sleep staging in subjects with obstructive sleep apnea', *IEEE Trans. Biomed. Engng.*, **51**(3), 2006.

[12] Klingler, D., Tragner, H., and Deisenhammer, E., 'The nature of the influence of sleep deprivation on the EEG', *Epilepsy Res. Suppl.*, **2**, 1991, 231–234.

[13] Jovanovic, U. J., 'General considerations of sleep and sleep deprivation', *Epilepsy Res. Suppl.*, **2**, 1991, 205–215.

[14] Naitoh, P., and Dement, W., 'Sleep deprivation in humans', in *Handbook of Electroencephalography and Clinical Neurophysiology*, 1976, pp. 146–151.

[15] Sabo, E., Reynolds, C. F., Kupfer, D. J., and Berman, S. R., 'Sleep, depression, and suicide', *Psychiat. Res.*, **36**(3), 1991, 265–277.

[16] Weitzenblum, E., and Racineux, J.-L., *Syndrome d'Apnées Obstructives du Sommeil*, 2nd edn, Masson, Paris, 2004.

[17] Man, G. C., and Kang, B. V., 'Validation of portable sleep apnea monitoring device', *Chest*, **108**(2), 1995, 388–393.

[18] Porée, F., Kachenoura, A., Gavrit, H., Morvan, C., Garrault, G., and Senhadji, L., 'Blind source separation for ambulatory sleep recording', *IEEE Trans. Information Technol. Biomed.*, **10**(2), 2006, 293–301.

[19] Dahl, R. E., and Puig-antich, J., 'EEG sleep in adolescents with major depression: the role of suicidality and inpatient status', *J. affective Disord.*, **19**(1), 1990, 63–75.

[20] Goetz, R. R., Puig-antich, J., Danl, R. E., Ryan, N. D., Asnis, G. M., Rabinovich, H., and Nelson, B., 'EEG sleep of young adults with major depression: a controlled study', *J. affective Disord.*, **22**(1-2), 1991, 91–100.

[21] McCracker, J. T., Poland, R. E., Lutchmansingh, P., and Edwards, C., 'Sleep electroencephalo-graphic abnormalities in adolescent depressives: effects of scopolamine', *Biol. Psychiat.*, **42**, 1997, 557–584.

[22] Van Swededn, B., Wauquier, A., and Niedermeyer, E., 'Normal ageing and transient cognitive disorders in elderly', Chapter 18, in *Electroencephalography Basic Principles, Clinical Applications, and Related Fields*, Eds E. Niedermeyer and F. Lopez da Silva, 4th edn, Lippincott, Williams and Wilkins, Philadelphia, Pennsylvania, 1999.

[23] Klass, D. W., and Brenner, R. P, 'Electroencephalography in the elderly', *J. Clin. Neurophysiol.*, **12**, 1995, 116–131.

[24] Rechtschaffen, A., and Kales, A. (Eds) *A Manual of Standardized Terminology and Scoring System for Sleep Stages in Human Subjects*, Series National Institute of Health Publications, No. 204, US Government Printing Office, Washington, DC, 1986.

[25] Young, T., Palta, M., Dempsey, J., Skatrud, J., Weber, S., and Badr, S., 'The occurrence of sleep-disordered breathing among middle-aged adults', *N. Engl. J, Med.*, **328**, 1993, 1230–1235.

[26] Benoit, O., and Goldenberg, F., *Exploration du Sommeil et de la Vigilance Chez l' Adulte*, Medicals Internationals, Cachan, France, 1997.

[27] Hyvärinen, A., and Oja, E., 'Fast fixed-point algorithm for independent component analysis', *Neural Comput.*, **9**, 1997, 1483–1492.

[28] Malinowska, U., Durka, P. J., Blinowska, K. J., Szelenberger, W., and Wakarow, A., 'Micro- and macrostructure of sleep EEG; a universal, adaptive time-frequency parametrization', *IEEE Engng. in Med. and Biolog. Mag.*, **25**(4), 2006, 26–31.

[29] http://www.cis.hut.fi/.

[30] Mallat, S., and Zhang, Z., 'Matching pursuit with time-frequency dictionaries', *IEEE Trans. Signal Process.*, **41**, 1993, 3397–3415.

[31] Mallat, S., *A Wavelet Tour of Signal Processing*, 2nd Ed, Academic, New York, 1999.

[32] American Sleep Disorder Association, 'EEG arousals: scoring rules and examples. A preliminary report from the Sleep Disorder Task Force of the American Sleep Disorder Association', *Sleep*, **15**(2), 1992, 174–184.

[33] Aeschbach, D., and Borb'ely, A. A., 'All-night dynamics of the human sleep EEG', *J. Sleep Res.*, **2**(2), 1993, 70–81.

[34] Terzano, M. G., Parrino, L., Rosa, A., Palomba, V., and Smerieri, A., 'CAP and arousals in the structural development of sleep: an integrative perspective', *Sleep Med.*, **3**(3), 2002, 221–229.

[35] Principe, J. C., and Smith, J. R., 'SAMICOS–a sleep analysing microcomputer system', *IEEE Trans. Biomed. Engng.*, **BME-33**, October 1986, 935–941.

[36] Principe, J. C., Gala, S. K., and Chang, T. G., 'Sleep staging automation based on the theory of evidence', *IEEE Trans. Biomed. Engng.*, **36**, May 1989, 503–509.

[37] Akgül, T., Sun, M., Sclabassi, R. J., and Çetin, A. E., 'Characterization of sleep spindles using higher order statistics and spectra', *IEEE Trans. Biomed. Engng.*, **47**(8), 2000, 997–1009.

[38] Nikias, C. L., and Petropulu, A., *Higher Order Spectra Analysis, A Nonlinear Signal Processing Framework*, Prentice-Hall, Englewood Cliffs, New Jersey, 1993.

[39] Rao, T. S., 'Bispectral analysis of nonlinear stationary time series', in *Handbook of Statistics*, Vol. 3, Eds D. R. Brillinger and P. R. Krishnaiah, Amsterdam, The Netherlands, 1993.

[40] Michel, O., and Flandrin, P., 'Higher order statistics for chaotic signal analysis', in *DSP Techniques and Applications*, Ed. E. T. Leondes, Academic, New York, 1995.

[41] Lippman, R. P., 'An introduction to computing with neural nets', *IEEE Acoust., Speech, Signal Process. Mag.*, **4**(2), April 1987, 4–22.

[42] Weibel, A., Hanazawa, T., Hinton, G., and Lang, K., 'Phoneme recognition using time-delay neural networks', *IEEE Trans. Acoust., Speech, Signal Process.*, **37**, March 1989, 328–339.

[43] Shimada, T., Shiina, T., and Saito, Y., 'Detection of characteristic waves of sleep EEG by neural network analysis', *IEEE Trans. Biomed. Engng.*, **47**(3), March 2000, 369–379.

[44] Kemp, B., Zwinderman, A. H., Tuk, B., Kamphuiesen, H. A. C., and Oberyé, J. J. L., 'Analysis of a sleep-dependent neuronal feedback loop: the slow-wave microcontinuity of the EEG', *IEEE Trans. Biomed. Engng.* **47**(9), 2000, 1185–1194.

[45] Kemp, B., 'NREM sleep-dependent neuronal feedback-slow-wave shape', *J. Sleep Res.*, **5**, 1996, 5106.

[46] Merica, H., and Fortune, R. D., 'A unique pattern of sleep structure is found to be identical at all cortical Sites: a neurobiological interpretation', *Cerebral Cortex*, **13**(10), 2003, 1044–1050.

[47] Steriade, M., McCormick, D. A., and Sejnowski, T. J., 'Thalamocortical oscillations in the sleeping and aroused brain', *Science*, **262**, 1993, 679–685.

[48] Kemp, B., Zwinderman, A. H., Tuk, B., Kamphuiesen, H. A. C., and Oberyé, J. J. L., 'Analysis of a sleep-dependent neuronal feedback loop: the slow-wave microcontinuity of the EEG', *IEEE Trans. Biomed. Engng.* **47**(9), 2000, 1185–1194.

[49] Steriade, M., Gloor, P., Llinas, R. R., Lopes da Silva, F. H., and Mesulam, M. -M., 'Basic mechanisms of cerebral rhythmic activities', *Electroencephalogr. Clin. Neurophysiol.*, **76**, 1990, 481–508.

7

Brain–Computer Interfacing

Brain–computer interfacing (BCI) (also called brain–machine interfacing (BMI)) is a challenging problem that forms part of a wider area of research, namely human–computer interfacing (HCI), which interlinks thought to action. BCI can potentially provide a link between the brain and the physical world without any physical contact. In BCI systems the user messages or commands do not depend on the normal output channels of the brain [1]. Therefore the main objectives of BCI are to manipulate the electrical signals generated by the neurons of the brain and generate the necessary signals to control some external systems. The most important application is to energize the paralysed organs or bypass the disabled parts of the human body. BCI systems may appear as the unique communication mode for people with severe neuromuscular disorders such as spinal cord injury, amyotrophic lateral sclerosis, stroke and cerebral palsy.

Approximately one hundred years after the discovery of the electrical activity of the brain the first BCI research was reported by Jacques Vidal [2,3] during the period 1973–7. In his research it was shown how brain signals could be used to build up a mental prosthesis. BCI has moved at a stunning pace since the first experimental demonstration in 1999 that ensembles of cortical neurons could directly control a robotic manipulator [4]. Since then there has been tremendous research in this area [5].

BCI addresses analysing, conceptualization, monitoring, measuring, and evaluating the complex neurophysiological behaviours detected and extracted by a set of electrodes over the scalp or from the electrodes implanted inside the brain. It is important that a BCI system be easy, effective, efficient, enjoyable to use, and user friendly. BCI is a multidisciplinary field of research since it deals with cognition, electronic sensors and circuits, machine learning, neurophysiology, psychology, sensor positioning, signal detection, signal processing, source localization, pattern recognition, clustering, and classification.

The main and prominent activities in BCI (especially noninvasive BCI) are:

1. The Berlin BCI (BBCI) group has followed the objective of transferring the effort of training from the human to machine since 2000. The major focus in their work is reducing the intersubject variability of BCI by minimizing the level of subject training. Some of their works have been reported in References [6] to [8].

EEG Signal Processing S. Sanei and J. Chambers

2. Wadsworth BCI research uses mainly the event-related desynchronization (ERD) of the mu rhythm for EEG classification of real or imaginary movements, achieved after training the subject [9,10].
3. The Graz BCI activity lead by Pfurtscheller has, as its core objective, the use of mu or beta rhythms for training and control. The expert users of their system are able to control a device based on the modulations of the precentral mu or beta rhythms of sensorimotor cortices in a similar way to Wadsworth BCI. However, while Wadsworth BCI directly presents the power modulations to the user, the Graz system for the first time also uses machine adaptation for controlling the BCI. They were also able to allow the grasping of the nonfunctional arm of a disabled patient by functional electrical stimulation (FES) of the arm controlled by EEG signals [11–15].
4. The Martigny BCI started with adaptive BCI in parallel with Berlin BCI. The researchers have proposed a neural network classifier based on linear discriminant analysis for classification of the static features [16]. In their approach three subjects are able to achieve 75 % correct classification by imagination of left- or right-hand movement or by relaxation with closed eyes in an asynchronous environment after a few days of training [17,18].
5. The thought translation device (TTD), which has been developed mainly for locked-in patients, enables the subjects to learn self-regulation of the slow cortical potentials at central scalp positions using EEG or electrocortiogram (ECOG). The subjects are able to generate binary decisions and hopefully provided a suitable communication channel to the outside world [19].

In this chapter the fundamental concepts and the requirement for the BCI design using EEG signals are reviewed. Development of BCI approaches within the last two decades are given next, and finally the well-established and latest advances in BCI are discussed.

7.1 State of the Art in BCI

The correspondence between EEG patterns and computer actions constitutes a machine-learning problem since the computer should learn how to recognize a given EEG pattern. As for other learning problems, in order to solve this problem, a training phase is necessary, in which the subject is asked to perform prescribed mental activities (MAs) and a computer algorithm is in charge of extracting the associated EEG patterns. After the training phase is finished the subject should be able to control the computer actions with his or her thoughts. This is the major goal for a BCI system.

In terms of signal acquisition, the BCI systems are classified into invasive (intracranial) and noninvasive. Noninvasive systems primarily exploit EEGs to control a computer cursor or a robotic arm. The techniques in developing such systems have been under development recently due to their hazardless nature and flexibility [1,19–27]. However, despite the advantage of not exposing the patient to the risks of brain surgery, EEG-based techniques provide limited information mainly because of the existence of system and physiological noise and the interfering undesired signals and artefacts. However, despite these shortcomings, EEG-based methods can detect modulations of brain activity that correlate with visual stimuli, gaze angle, voluntary intentions, and cognitive states [5]. These advantages led to development of several classes of EEG-based systems, which differ according to the cortical areas recorded, the extracted features of the EEGs, and the sensory modality providing feedback to subjects.

In noninvasive BCI approaches, although there may be different EEG electrode settings for certain BCI applications, an efficient BCI system exploits all the information content within the EEG signals. In all cases detection and separation of the control signals from the raw EEG signals is probably the first objective. The event-related source signals can be effectively clustered or separated if the corresponding control signals are well characterized. Given that these control sources are likely to be moving inside the brain, an exciting research area is also to localize and track these sources in real-time. The first step in developing an effective BCI paradigm is therefore to determine suitable control signals from the EEG. A suitable control signal has the following attributes: it can be precisely characterized for an individual, it can be readily modulated or translated to express the intention, and it can be detected and tracked consistently and reliably.

There are two main approaches towards BCI: one is based on event related potentials (ERPs) and another is based on the multiple sensor EEG activities recorded in the course of ordinary brain activity. The latter approach is more comprehensive and does not require any particular stimulus. However, as mentioned before, in many cases such as those where there is a direct connection from the electrodes to the mechanical systems, the number of recording channels, electrodes, is generally limited. In such cases EEG channel selection across various subjects has become another popular research area within the BCI community [28]. The general idea behind these approaches is based on a recursive channel elimination (RCE) criterion; channels that are well known to be important (from a physiological point of view) are consistently selected whereas task-irrelevant channels are disregarded. Nonnegative matrix factorization (NMF) has also been used to analyse neural activity and identify local spatiotemporal patterns of neural activity in the form of sparse basis vectors [29].

An ERP appears in response to some specific stimulus. The most widely used ERP evoked potential (EP) is the P300 signal, which can be auditory, visual, or somatosensory. It has a latency of approximately 300 ms and is elicited by rare or significant stimuli, and its amplitude is strongly related to the unpredictability of the stimulus; the more unforeseeable the stimulus, the higher the amplitude [30]. Another type of visual EP (VEP) is those for which the ERPs have a short latency, representing the exogenous response of the brain to a rapid visual stimulus. They are characterized by a negative peak around 100 ms (N1) followed by a positive peak around 200 ms (P2). The ERPs can provide control when the BCI produces the appropriate stimuli. Therefore the BCI approach based on ERP detection from the scalp EEG seems to be easy since the cortical activities can be easily measured noninvasively and in real-time. Also, an ERP-based BCI needs little training for a new subject to gain control of the system. However, the information achieved through ERP extraction and measurement is not accurate enough for extraction of movement related features and they have vast variability in different subjects with various brain abnormalities and disabilities. More importantly, the subject has to wait for the relevant stimulus presentation [31,32].

There are also two other approaches used for BCI. One is based on steady-state visual-evoked responses (SSVERs), which are natural responses for visual stimulations at specific frequencies. These responses are elicited by a visual stimulus that is modulated at a fixed frequency. The SSVERs are characterized by an increase in EEG activity around the stimulus frequency. With feedback training, subjects learn to voluntarily control their SSVER amplitude. Changes in the SSVER result in control actions occurring over fixed intervals of time [32,33]. The second approach is slow cortical potential shifts (SCPSs), which are shifts of cortical voltage lasting from a few hundred milliseconds up to several seconds. Subjects can learn to generate slow cortical amplitude shifts in an electrically positive or negative direction for binary control. This can be achieved if the subjects are provided with feedback on the evolution of their SCP and if they are positively reinforced for correct responses [34]. However, in both of the above methods the subject has to wait for the brain stimulus.

Generally, some BCI approaches rely on the ability of the subjects to develop control of their own brain activity using biofeedback [35–37] whereas others utilize classification algorithms that recognize EEG patterns related to particular voluntary intentions [38]. Initial attempts to enable the subjects to use the feedback from their own brain activity started in the 1960s. The system enables the subjects to gain voluntary control over brain rhythms. It has been claimed that after training with an EEG biofeedback, human subjects are able to detect their own alpha [39,40] and mu rhythms [41]. This was also tested on cats using their mu rhythms [42] and dogs [43] to control their hippocampal theta rhythm. Classification-based approaches have also been under research recently [38]. In a recent demonstration, subjects navigated through a virtual environment by imagining themselves walking [44]. These works paved the way for a vast amount of recent research developed by many workers in this area.

Beverina *et al.* [45] have used P300 and steady-state visual-evoked potentials (SSVEPs). They have classified the ERP feature patterns using support vector machines (SVMs). In their SSVEP approach they have used the signals from the occipital electrodes (O_z, O_2, PO_8). The stimulations have been considered random in time instants, and a visual feedback [35] has been suggested for training purposes. In normal cases it is possible through the feedback to make some subjects increment some brain wave activities. In another work, SSVEPs have been used in a visually elaborate immersive 3D game [46]. The SSVEP generated in response to phase-reversing checkboard patterns is used for the proposed BCI.

There are many attractions in using the normal EEGs (or spontaneous signals (SSs)) for BCI. A BCI system of this kind generates a control signal at given intervals of time based on the classification of EEG patterns resulting from particular mental activity (MA). In all types of BCI systems human factors such as boredom, fatigue, stress, or various illnesses are of great influence, and therefore motivating the subject often becomes very important.

Although most of the recent work in BCI research has been focused upon scalp EEGs, some work on invasive EEGs has also been reported. Invasive BCI approaches are based on recordings from ensembles of single brain cells or on the activity of multiple neurons. They rely on the physiological properties of individual cortical and subcortical neurons or combination of neurons that modulate their movement-related activities. These works started in the 1960s and 1970s through some experiments by Fetz and his coresearchers [47–52]. In these experiments, monkeys learnt how to control the activity of their cortical neurons voluntarily with the help of biofeedback, which indicated the firing rate of single neurons. A few years later, Schmidt [53] indicated that the voluntary motor commands could be extracted from raw cortical neural activity and used them to control a prosthetic device designed to restore motor functions in severely paralysed patients.

Most of the research on invasive BCI was carried out on monkeys. These works relied on single cortical site recordings either of local field potentials [54–57] or from small samples of neurons or multiple brain zones [58–60]. They are mostly recorded in the primary motor cortex [58,59], although some work has been undertaken on the signals recorded from the posterior parietal cortex [61]. Deep brain stimulators have recently been used to monitor patients with Parkinson's disease [62].

Philip Kennedy and his colleagues [63] have presented an impressive result from cortical implanted electrodes. In another work [4] the implanted electrodes were used inside the brain of a monkey to control a robot arm remotely. These works, however, require solutions to possible risk problems, advances in robust and reliable measurement technologies, and clinical competence. On the other hand, the disadvantage of using scalp recordings lies in the very low quality of the signals, due to attenuation of the electrical activity signals on their way to the electrodes.

In the following subsections a number of features used in BCI are explained. Initially, the changes in EEG before, during, and after the externally or internally paced events are observed. These events can be divided into two categories: in one type the ERPs included evoked potentials and in the second type event-related desynchronization (ERD) and synchronization (ERS). The main difference between the two types is that the ERP is a stimulus-locked or, more generally, a phase-locked reaction, while the ERD/ERS is a non-phase-locked response. In a finger movement process, for example, often discussed in BCI, premovement negativity prominent prior to movement onset, and postmovement beta oscillations occurring immediately after movement offset are respectively phase-locked (evoked) and non-phase-locked processes [64].

7.1.1 ERD and ERS

ERD and ERS can be considered as event-related potentials (which can also include evoked potentials). The cortical mu rhythm is an example of ERD. ERD is due to blocking of alpha activity just before and during the real or imagery movement.

ERD is measured in terms of the power of the peak in the alpha band to the bandwidth, that is

$$\text{ERS/ERD} = \frac{P(f,n) - P_{\text{ref}}(f)}{P_{\text{ref}}(f)} \tag{7.1}$$

where $P(f,n)$ is the value of a signal power at a given time–frequency point of an average power map and $P_{\text{ref}}(f)$ is an average power during some reference time calculated for frequency f. This represents the level of rhythmic activity within the alpha band just before or during the movement. Any attention dramatically attenuates the alpha rhythms, while an increase of task complexity or attention results in an increased magnitude of ERD.

Increased cellular excitability in thalamocortical systems results in a low amplitude desynchronized EEG. Therefore ERD may be due to an electrophysiological correlate of various activated cortical regions involved in processing sensory or cognitive information or production of a motor reaction. Involvement of more neurons increases the ERD magnitude. In the BCI context, explicit learning of a movement sequence, e.g. key pressing with different fingers, is accompanied by an enhancement of the ERD over the contralateral central regions. As the learning progresses and becomes more automatic the ERD decreases.

The cortical mu rhythm is of particular interest in BCI mainly because it can be modulated/translated through imaginary and can be monitored via a noninvasive technique. The overall alpha band may be divided into lower and higher alphas. Lower alpha ERD (6–10 Hz) is a response to any type of task and is topographically spread over almost all electrodes. Higher alpha ERD, restricted to parietooccipital areas, is found during visually presented stimulations.

The level of ERD is closely linked to semantic memory processes. Those with good memory show a larger ERD in the lower alpha band [65]. In an auditory memory task, the absence of ERD can be explained by the anatomical localization of the auditory cortex below the surface. Detection of auditory ERD from the EEGs is therefore normally difficult.

As related to BCI, voluntary movement also results in a circumscribed desynchronization in the upper alpha and lower beta bands, localized over sensorimotor regions [66]. ERD starts over the contralateral rolandic region and, during the movement, becomes bilaterally symmetrical with execution of movement. It is of interest that the time course of the contralateral mu desynchronization is almost identical to brisk and slow finger movements, starting about two seconds prior to movement onset. Generally, brisk and slow finger movements have different encoding processes. Brisk movement is preprogrammed and the afferents are delivered to the muscles as bursts. On the other hand, slow movement depends on the reafferent input from kinaesthetic receptors evoked by the movement itself.

Finger movement of the dominant hand is accompanied by a pronounced ERD in the ipsilateral side, whereas movement of the nondominant finger is preceded by a less lateralized ERD [66]. Circumscribed hand area mu ERD can be found in nearly every subject, whereas a foot area mu ERD is hardly localized close to the primary foot area between both hemispheres.

In another study [67] with subdural electrodes, it was discovered that mu rhythms are not only selectively blocked with arm and leg movements, but also with face movement. Using this method the electrodes are inserted under the skull and acquire the signals from over the cortex. These signals are called electrocorticograms (ECoGs). The ECoGs capture more detailed signals from smaller cortical areas than the conventional EEG-based systems. These signals also contain low-amplitude high-frequency gamma waves. Consequently, ECoG-based BCIs have better accuracy and require a shorter training time than those of EEGs [68].

In ERS, however, the amplitude enhancement is based on the cooperative or synchronized behaviour of a large number of neurons. In this case, the field potentials can be easily measured even using scalp electrodes. It is also interesting to note that approximately 85 % of cortical neurons are excitatory, with the other 15 % being inhibitatory.

7.1.2 Transient Beta Activity after the Movement

This activity, also called postmovement beta synchronization (PMBS), is another interesting robust event that starts during the movement and continues for about 600 ms [66]. It is found after finger or foot movement over both hemispheres without any significant bilateral coherence. The frequency band may vary from subject to subject; for finger movement the range is around 16–21 Hz [69] whereas for foot movement it is around 19–26 Hz [70]. The PMBS has similar amplitude for brisk and slow finger movements. This is interesting since brisk and slow movements involve different neural pathways. Moreover, this activity is significantly larger with hand movement as compared to finger movement [66]. Also, larger beta oscillations with wrist movement as compared to finger movement can be interpreted as the change of a larger population of motor cortex neurons from an increased neural discharge during the motor act to a state of cortical disfacilitation or cortical idling [66]. This means that movement of more fingers results in a larger beta wave. Beta activity is also important in the generation of a grasp signal, since it has less overlap with other frequency components [71].

7.1.3 Gamma Band Oscillations

Oscillation of neural activity (ERS) within the gamma band (35–45 Hz) has also been of interest recently. Such activity is very obvious after visual stimuli or just before the movement task. This may act as the carrier for the alpha and lower beta oscillations, and relate to binding of sensory information and sensorimotor integration. Gamma, together with other activities in the above bands, can be observed around the same time after performing a movement task. Gamma ERS manifests itself just before the movement, whereas beta ERS occurs immediately after the event.

7.1.4 Long Delta Activity

Rather than other known ERS and ERD activities within alpha, beta, and gamma bands a long delta oscillation starts immediately after the finger movement and lasts for a few seconds. Although this has not been reported often in the literature, it can be a prominent feature in distinguishing between movement and nonmovement states.

The main task in BCI is how to exploit the behaviour of the EEGs in the above frequency bands before, during, and after the imaginary movement, or after certain brain stimulation, in generation of the control signals. The following sections address this problem.

7.2 Major Problems in BCI

A simple BCI system setup is illustrated in Figure 7.1. As mentioned previously, the major problem in BCI is separating the control signals from the background EEG. Meanwhile, cortical connectivity, as an interesting identification of various task related brain activities, has to be studied and exploited. Detection and evaluation of various features in different domains will then provide the control signals. To begin, however, the EEG signals have to be preprocessed since the signals are naturally contaminated by various internal and external interferences. Data conditioning such as prewhitening may also be necessary before implementation of the source separation algorithms.

7.2.1 Preprocessing of the EEGs

In order to have an artefact-free EEG to extract the control signals, the EEGs have to be restored from the artefacts, such as eye-blinking, electrocardiograms (ECGs), and any other internal or external disturbing effects.

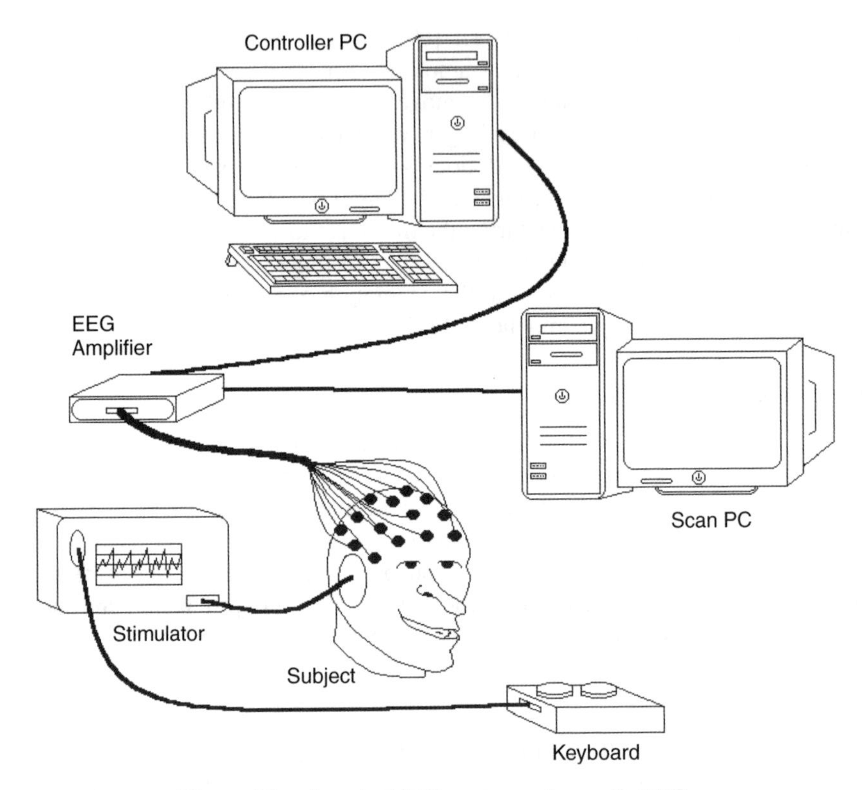

Figure 7.1 A typical BCI system using scalp EEGs

Eye-blinking artefacts are very clear in both frontal and occipital EEG recordings. ECGs, on the other hand, can be seen more over the occipital electrodes. Many attempts have been made by different researchers to remove these artefacts.

Most of the noise, external and even internal artefacts, such as ECGs, are filtered out by the hardware provided in new EEG machines. As probably the most dominant remaining artefact, interfering eye blinks (ocular artefact (OA)) generate a signal within EEGs that is on the order of ten times larger in amplitude than cortical signals, and can last between 200 and 400 ms.

There have been some works by researchers to remove OAs. Certain researchers have tried to estimate the propagation factors, as discussed in Reference [72], based on regression techniques in both the time and frequency domains. In this attempt there is a need for a reference electrooculogram (EOG) channel during the EEG recordings.

PCA and SVMs have also been utilized for this purpose [73]. These methods rely on the uncorrelatedness assumption of the EEGs and OAs. Adaptive filtering has also been utilized [74]. This approach has considered the EEG signals individually and therefore ignored the mutual information among the EEG channels. ICA has also been used in some approaches. In these works the EEG signals are separated into their constituent independent components (ICs) and the ICs are projected back to the EEGs using the estimated separating matrix after the artefact-related ICs are manually eliminated [75]. In

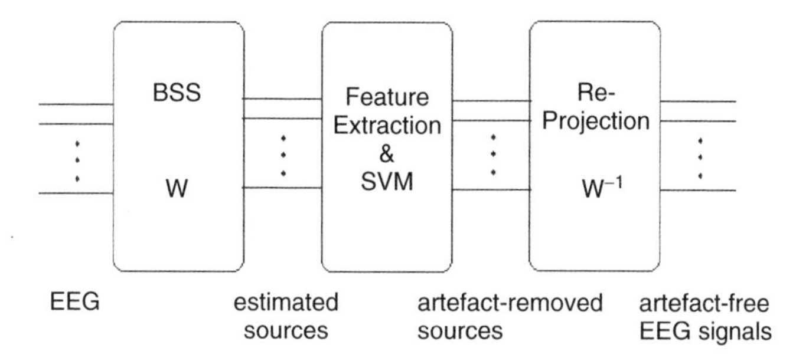

Figure 7.2 A hybrid BSS–SVM artefact removal system [77]

Reference [76] a BSS algorithm based on second order statistics separates the combined EEG and EOG signals into statistically independent sources. The separation is then repeated for a second time with the EOG channels inverted. The estimated ICs in both rounds are compared, and those ICs with different signs are removed. Although, due to the sign ambiguity of the BSS the results cannot be justified it is claimed that by using this method the artefacts are considerably mitigated. As noticed, there is also a need to separate EOG channels in this method.

In another recent attempt, an iterative SOBI-based BSS method followed by classification of the independent components using SVMs has been designed to effectively remove the EOG artefacts [77]. The method can also be easily extended to removal of the ECG artefacts. The proposed algorithm consists of BSS, automatic removal of the artefact ICs, and finally reprojection of the ICs to the scalp, providing artefact-free EEGs. This is depicted in Figure 7.2. Iterative SOBI as previously discussed has been effectively used to separate the ICs in the first stage. In the second stage only four features were carefully selected and used for classification of the normal brain rhythms from the EOGs. These features are as follows:

Feature I. A large ratio between the peak amplitude and the variance of a signal suggests that there is an unusual value in the data. This is a typical identifier for the eye blink because it causes a large deflection on the EEG trace. This is described mathematically as:

$$f_1 = \frac{\max(|\boldsymbol{u}(n)|)}{\sigma_u^2} \quad \text{for } n = 1, \ldots, N \tag{7.2}$$

where $\boldsymbol{u}(n)$ is one of the N ICs, max(.) is a scalar valued function that returns the maximum element in a vector, σ_u is the standard deviation of $\boldsymbol{u}(n)$, and $|\cdot|$ denotes absolute value. Normal EEG activity is tightly distributed about its mean. Therefore a low ratio is expected while the eye-blink signals manifest a large value.

Feature II. This is a measure of third order statistics of the data, skewness. This feature is defined as

$$f_2 = \left| \frac{E[\boldsymbol{u}^3(n)]}{\sigma_u^3} \right| \quad \text{for } n = 1, \ldots, N \tag{7.3}$$

for zero mean data. The EEG containing eye-blinks typically has a positive or negative skewness since the eye-blinking signal has a considerably larger value for this feature.

Feature III. The correlation between the ICs and the EEG signals from certain electrodes is significantly higher than those of other ICs. The electrodes with most contributed EOG are frontal electrodes F_{P1}, F_{P2}, F_3, and F_4 and occipital lobe electrodes O_1 and O_2 (in total six electrode signals). The reference dataset, i.e. the EEG from the aforementioned electrodes, is distinct from the training and test datasets. This will make the classification more robust by introducing a measure of the spatial location of the eye-blinking artefact. Therefore, the third feature can be an average of the correlations between the ICs and the signals from these six electrodes:

$$f_3 = \frac{1}{6}\sum_{i=1}^{6}\{|E[x_i^0(n)u(n+\tau)]|\} \quad \text{for } n = 1, \ldots, N \tag{7.4}$$

where $x_i^0(n)$ are eye-blinking reference signals and i indexes each of the aforementioned electrode locations. The value of this feature will be larger for ICs containing the eye-blinking artefact, since they will have a larger correlation for a particular value of τ, in contrast to ICs containing normal EEG activity.

Feature IV. The statistical distance between distributions of the ICs and the electrode signals that are more likely to contain EOG is used. This can be measured using the Kullback–Laibler (KL) distance, defined as

$$f_4 = \int_{-\infty}^{\infty} p(u(n)) \ln \frac{p(u(n))}{p(x_{\text{ref}})} \, \mathrm{d}u(n) \quad \text{for } n = 1, \ldots, N \tag{7.5}$$

where $p(.)$ denotes the PDF. When the IC contains OA the KL distance between its PDF and the PDF of the reference IC will be approximately zero, whereas the KL distance to the PDF of a normal EEG signal will be larger.

An SVM with an RBF nonlinear kernel is then used to classify the ICs based on the above features. Up to 99 % accuracy in detection of the EOG ICs has been reported [77].

After the artefact signals are marked, they will be set to zero and all the estimated sources are reprojected to the scalp electrodes to reconstruct the artefact-free EEG signals. Figure 7.3 shows the EEG signals containing OA and the restored signals after following the above procedure.

The same idea has been directly used for extraction of the movement related features [78] from the EEGs. In this work it is claimed that without any long-term training the decision of whether there is any movement for a certain finger or not can be achieved by BSS followed by a classifier. A combination of a modified genetic algorithm (GA) and an SVM classifier has been used to condition the selected features.

7.3 Multidimensional EEG Decomposition

All movement-related potentials are limited in duration and in frequency. In addition, each channel contains the spatial information of the EEG data. PCA and ICA have been

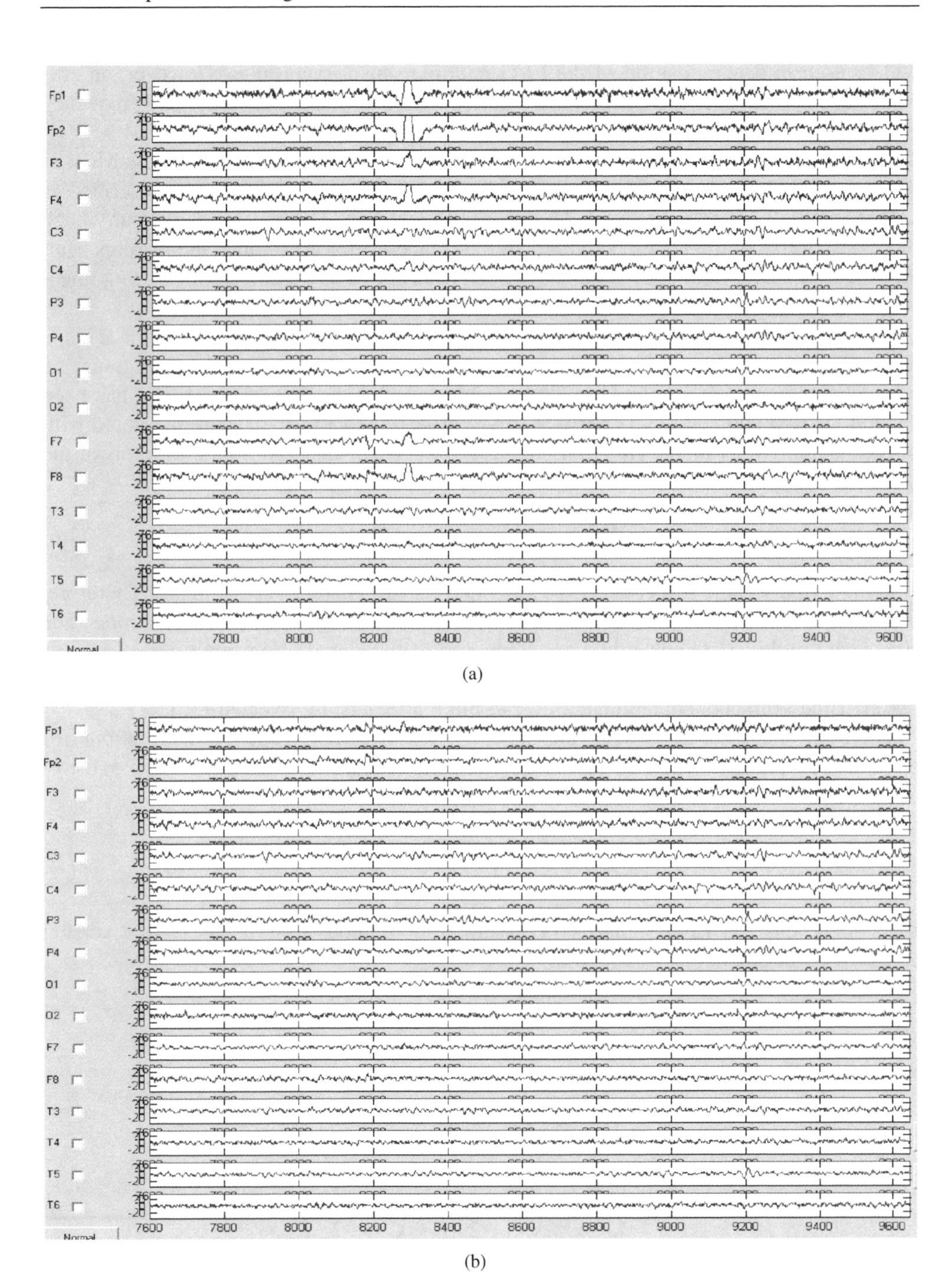

Figure 7.3 The EEG signals: (a) original EEG signals including eye-blinking artefacts mainly around the frontal electrodes and (b) the same set of EEGs after removing the artefact using the hybrid BSS–SVM system

widely used in decomposition of the EEG multiple sensor recordings. However, an efficient decomposition of the data requires incorporation of the space, time, and frequency dimensions.

Time–frequency (TF) analysis exploits variations in both time and frequency. Most of the brain signals are decomposable in the TF domain. This has been better described as sparsity of the EEG sources in the TF domain. In addition, TF domain features are much more descriptive of the neural activities. In Reference [79], for example, the features from the subject-specific frequency bands have been determined and then classified using linear discriminant analysis (LDA).

In a more general approach the spatial information is also taken into account. This is due to the fact that the majority of the events are localized in distinct brain regions. As a favourable approach, joint space–time–frequency classification of the EEGs has been studied for BCI applications [32,80]. In this approach the EEG signals are measured with reference to *digitally linked ears* (DLE). DLE voltage can easily be found in terms of the left and right earlobes as

$$V_{\mathrm{e}}^{\mathrm{DLE}} = V_{\mathrm{e}} - \tfrac{1}{2}(V_{A_1} + V_{A_2}) \tag{7.6}$$

where V_{A_1} and V_{A_2} are respectively the left and right earlobe reference voltages. Therefore, the multivariate EEG signals are composed of the DLE signals of each electrode. The signals are multivariate since they are composed of the signals from multiple sources. A decomposition of the multivariate signals into univariate classifications has been carried out after the segments contaminated by eye-blink artefacts are rejected [32].

There are many ways to write the general class of time–frequency distributions for classification purposes [81]. In the above work the characteristic function (CF) $M(\theta, \tau)$ as in

$$C(t, \omega) = \frac{1}{4\pi^2} \int_{\tau=-\infty}^{\infty} \int_{0}^{2\pi} M(\theta, \tau) \mathrm{e}^{-\mathrm{j}\theta t - \mathrm{j}\tau\omega} \, \mathrm{d}\theta \, \mathrm{d}\tau \tag{7.7}$$

for a single channel EEG signal, $x(t)$, assuming continuous time (a discretized version can be used in practice), is defined as

$$M(\theta, \tau) = \phi(\theta, \tau) A(\theta, \tau) \tag{7.8}$$

where

$$\begin{aligned} A(\theta, \tau) &= \int_{-\infty}^{\infty} x^*\left(u - \frac{1}{2}\tau\right) x\left(u + \frac{1}{2}\tau\right) \mathrm{e}^{\mathrm{j}\theta u} \, \mathrm{d}u \\ &= \int_{0}^{2\pi} \hat{X}^*\left(\omega + \frac{1}{2}\theta\right) \hat{X}\left(\omega - \frac{1}{2}\theta\right) \mathrm{e}^{\mathrm{j}\tau\omega} \, \mathrm{d}\omega \end{aligned} \tag{7.9}$$

and $\hat{X}(\omega)$ is the Fourier transform of $x(t)$, which has been used for classification. This is a representative of the joint time–frequency autocorrelation of $x(t)$. $\phi(\theta, \tau)$ is a kernel function that acts as a mask to enhance the regions in the TF domain so the signals to be classified are better discriminated. In Reference [32] a binary function has been suggested as the mask.

In the context of EEGs as multichannel data, a multivariate system can be developed. Accordingly, the multivariate ambiguity function (MAF) of such a system is defined as

$$\mathbf{MA}(\theta, \tau) = \int_{-\infty}^{\infty} \boldsymbol{x}\left(t + \frac{\tau}{2}\right)\boldsymbol{x}^{\mathrm{H}}\left(t - \frac{\tau}{2}\right)\mathrm{e}^{\mathrm{j}\theta t}\,\mathrm{d}t \tag{7.10}$$

where $(.)^{\mathrm{H}}$ denotes the conjugate transpose. This ambiguity function can also be written in a matrix form as

$$\mathrm{MA}(\theta, \tau) = \begin{bmatrix} a_{11} & \dots & a_{1N} \\ & \dots & \\ & \dots & \\ a_{N1} & \dots & a_{NN} \end{bmatrix} \tag{7.11}$$

where

$$a_{ij} = \int_{-\infty}^{\infty} x_j^*\left(t - \frac{\tau}{2}\right) x_i\left(t + \frac{\tau}{2}\right)\mathrm{e}^{\mathrm{j}\theta t}\,\mathrm{d}t \tag{7.12}$$

The diagonal terms are called autoambiguity functions and the off-diagonal terms are called cross-ambiguity functions. MAF can therefore be an indicator of the multivariate time–frequency–space autocorrelation of the corresponding multivariate system. The space dimension is taken into account by the cross ambiguity functions.

7.3.1 Space–Time–Frequency Method

The disjointness property of the sources in multidimensional space has been nicely exploited in developing a classifier based on the multidimensional features. As an example in Reference [82], the time–frequency transformation has been carried out for all the channels, representing the spatial information, and the atoms are masked and clustered and used for both reconstruction of the desired sources and classification of a finger movement signal based on the directionality of the motion of the sources. The block diagram in Figure 7.4 represents the approach.

The same conclusion can be obtained using multivariate autoregressive (MVAR) modelling and evaluation of the diagonal and off-diagonal terms. This will be discussed in a later section of this chapter.

The contrast of MAF is then enhanced using a multidimensional kernel function and the powers of cross-signals (cross-correlation spectra) are used for classification [32]. Using this method (as well as MVAR) the location of the event-related sources can be tracked and effectively used in BCI.

7.3.2 Parallel Factor Analysis

Parallel factor (PARAFAC) analysis, also called canonical decomposition (CANDECOMP), is another similar approach for analysis of brain signals in a multidimensional domain for BCI purposes. In this approach the events are considered sparse in the space–time–frequency domain and no assumption is made on either independency or uncorrelatedness of the

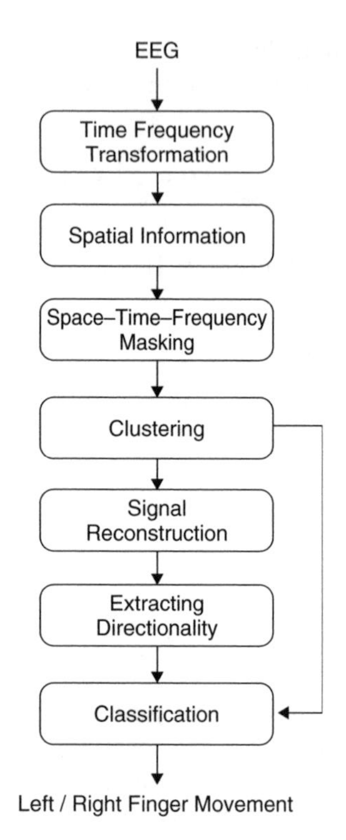

Figure 7.4 Classification of the finger movement sources using space–time–frequency decomposition

sources. Therefore, the main advantage of PARAFAC over PCA or ICA is that uniqueness is ensured under mild conditions, making it unnecessary to impose orthogonality or statistical independence constraints. Harshman [83] was the first researcher to suggest that PARAFAC be used for EEG decomposition. Harshman and Carol and Chang [84] independently proposed PARAFAC in 1970.

Möcks reinvented the model, naming it topographic component analysis, to analyse the ERP of channel × time × subjects [85]. The model was further developed by Field and Graupe [86]. Miwakeichi *et al.* eventually used PARAFAC to decompose the EEG data into its constituent space–time–frequency components [87]. In Reference [80] PARAFAC was used to decompose wavelet transformed event-related EEG given by the inter-trial phase coherence. Factor analysis can be expressed as

$$\mathbf{X} = \mathbf{A}\mathbf{S}^{\mathrm{T}} + \mathbf{E} \tag{7.13}$$

In this equation $\mathbf{A}$ is the $I \times F$ factor loading matrix, $\mathbf{S}$ the $J \times F$ factor score, $\mathbf{E}$ the $I \times J$ error, and F the number of factors. Similarly, PARAFAC of a three-way array $\mathbf{X}$ may be expressed by unfolding one modality to another, as

$$\mathbf{X} = \mathbf{A}(\mathbf{S}| \otimes |\mathbf{D})^{\mathrm{T}} + \mathbf{E} \tag{7.14}$$

where $\mathbf{D}$ is the $J \times F$ factor corresponding to the second modality. $\mathbf{S}| \otimes |\mathbf{D} = [\boldsymbol{s}_1 \otimes \boldsymbol{d}_1, \boldsymbol{s}_2 \otimes \boldsymbol{d}_2, \ldots, \boldsymbol{s}_F \otimes \boldsymbol{d}_F]$ is the Katri–Rao product [88]. In this formulation $\mathbf{X}$ and $\mathbf{E}$ are both $I \times JK$ matrices. Equivalently, the lth matrix corresponding to the jth slice of the second modality of the three way array can be expressed as

$$\mathbf{X} = \mathbf{A}\mathbf{D}_l\mathbf{S}^{\mathrm{T}} + \mathbf{E} \tag{7.15}$$

where $\mathbf{X}$ and $\mathbf{E}$ are $I \times l \times K$ matrices and $\mathbf{D}_j$ is an $F \times F$ diagonal matrix having the lth row of $\mathbf{D}$ along the diagonal. Alternating least squares (ALS) is the most common way to estimate the PARAFAC model. The cost function for estimating the matrices is denoted as

$$[\hat{\mathbf{A}}, \hat{\mathbf{S}}, \hat{\mathbf{D}}] = \arg \min_{\mathbf{A},\mathbf{S},\mathbf{D}} \|\mathbf{X} - \mathbf{A}(\mathbf{S}| \otimes |\mathbf{D})^{\mathrm{T}}\|_{\mathrm{F}}^2 \tag{7.16}$$

which corresponds to optimizing the likelihood of a Gaussian noise model. In each iteration, one parameter is estimated while the others are considered known, i.e. an alternating optimization. The algorithm can be initialized in several ways, i.e. by randomly defining all parameters and stopping when they have all converged. Figures 7.5 and 7.6 show

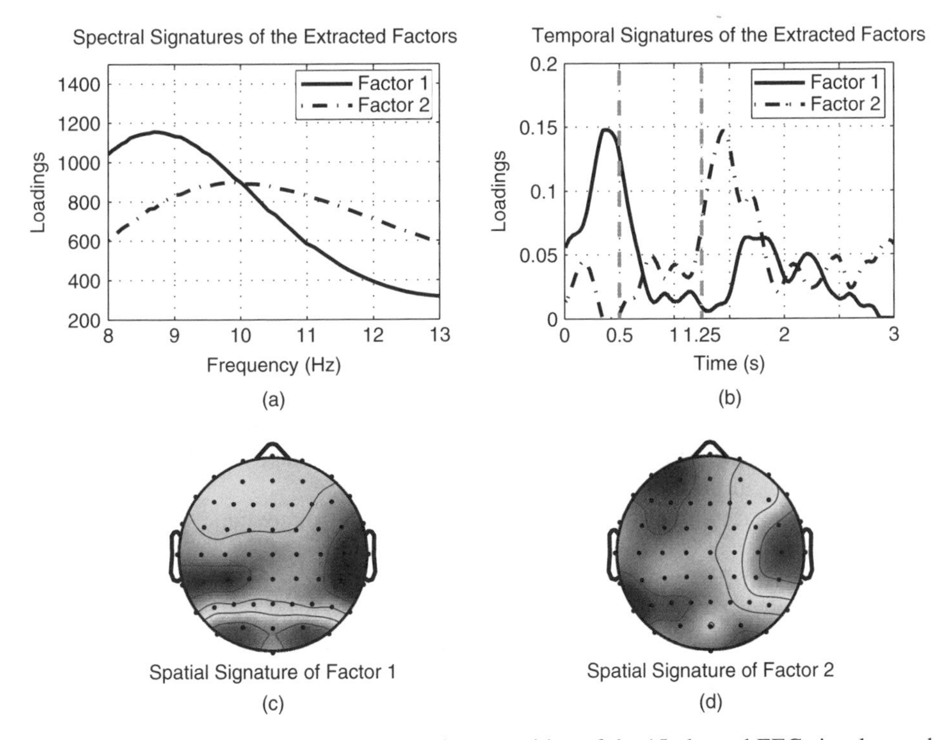

Figure 7.5 Sample space–time–frequency decomposition of the 15-channel EEG signal recorded during left index movement imagination. The factor demonstrated with the solid line indicates a clear ERD in the contralateral hemisphere: (a) spectral contents of the two identified factors, (b) temporal signatures of the factors, with the onset of preparation and execution cues shown in light and dark patches respectively, and (c) and (d) represent topographic mapping of the EEG for the two factors

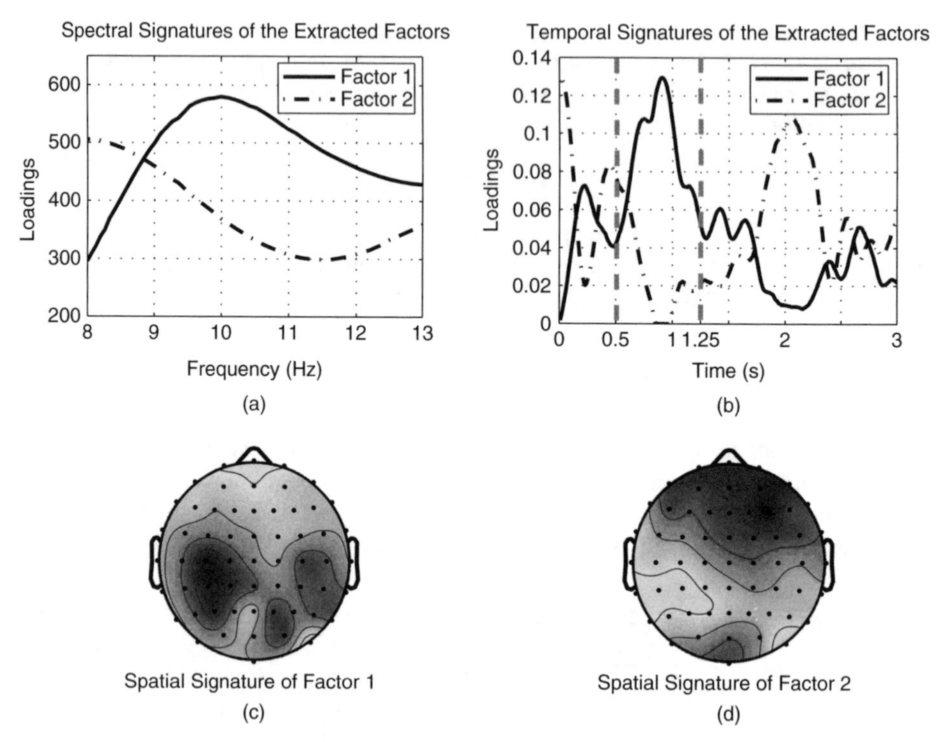

Figure 7.6 Sample space–time–frequency decomposition of the 15-channel EEG signal recorded during right index movement imagination. The factor demonstrated with the solid line indicates a clear ERD in the contralateral hemisphere: (a) spectral contents of the two identified factors, (b) temporal signatures of the factors, with the onset of preparation and execution cues shown in light and dark patches, respectively, and (c) and (d) show topographic mapping of the EEG for the two factors

the space–time–frequency decomposition of the 15 channel EEG signals recorded during left and right index finger movement imagination respectively. Spectral contents, temporal profiles of the two identified factors, and the topographic mapping of EEG for the two factors are shown in these images.

Accordingly, space–time–frequency features can be evaluated and used by a suitable classifier to distinguish between the left and right finger movements (or finger movement imagination) [89]. In an experiment the $I \times l \times K$ size $\mathbf{X}$ was formed by applying finite difference implementation of a spatial Laplacian filter [90] to 15 channels of the EEG and then transformed to the time–frequency domain using complex Morelet's wavelets $w(n, f_0)$ as

$$\mathrm{EEG}_{\mathrm{filtered}} = \mathrm{EEG}_i(n) - \frac{1}{4}\sum_{l \in N_i} \mathrm{EEG}_l \tag{7.17}$$

and

$$\mathbf{X}(n) = |w(n, f_0)\mathrm{EEG}_{\mathrm{filtered}}|^2 \tag{7.18}$$

The surface Laplacian filter may be considered as a spatial highpass filter.

7.4 Detection and Separation of ERP Signals

Utilization of the ERP signals provides another approach in BCI design. The ERP-based BCI systems often consider a small number of electrodes to study the movement-related potentials of certain body organs. However, in recent work multichannel EEGs have been used followed by an efficient means of the source separation algorithm in order to exploit the maximum amount of information within the recorded signals. Since the major problems in this context are related to detection, separation, and classification of the ERP signals a separate chapter (Chapter 3) has been dedicated to that. As stated previously, these systems are initiated by introducing certain stimulations of the brain. As soon as the movement-related ERP components are classified the system can be used in the same way as in the previous sections.

7.5 Source Localization and Tracking of the Moving Sources within the Brain

Source localization (discussed in Chapter 4) can be employed here to estimate the location of moving sources related to the finger movement. Application of the conventional dipole-fitting localization algorithms, however, is subject to having a preassumption about the number of sources. A simple source localizer may be designed using a feedback BSS system followed by an LS-based geometrical localization system. In this simple method BSS separates the EEG signals into their independent sources for a number of consecutive overlapping segments of the EEGs. For each segment the corresponding independent component (estimated source) is reprojected to the scalp using an inverse of the separating matrix. The resulting topographies are compared and the moving sources with maximum spatial and frequency correlations are selected. These sources are then localized using the LS method described in Chapter 4. Figure 7.7 represents the location of a number of steady sources and a moving source as a result of BSS and source localization for some synthetic sources.

7.6 Multivariant Autoregressive (MVAR) Modelling and Coherency Maps

The planning and the execution of voluntary movements are related to the premovement attenuation and postmovement increase in amplitude of alpha and beta rhythms in certain areas of the motor and sensory cortex [91,92]. It has also been found that during movement planning, two rhythmical components in the alpha frequency range, namely mu1 and mu2, play different functional roles. Differentiation of these two components may be achieved by using the matching pursuit (MP) algorithm [93] based on the signal energy in the two bands [94,95]. The MP algorithm refers to the decomposition of signals into basic waveforms from a very large and redundant dictionary of functions. MP has been utilized for many applications such as epileptic seizure detection, evaluation, and classification by many researchers.

Determination of the propagation of brain electrical activity, its direction, and the frequency content is of great importance. Directed transfer functions (DTFs) using a multivariate autoregressive (MVAR) model have been employed for this purpose [96]. In

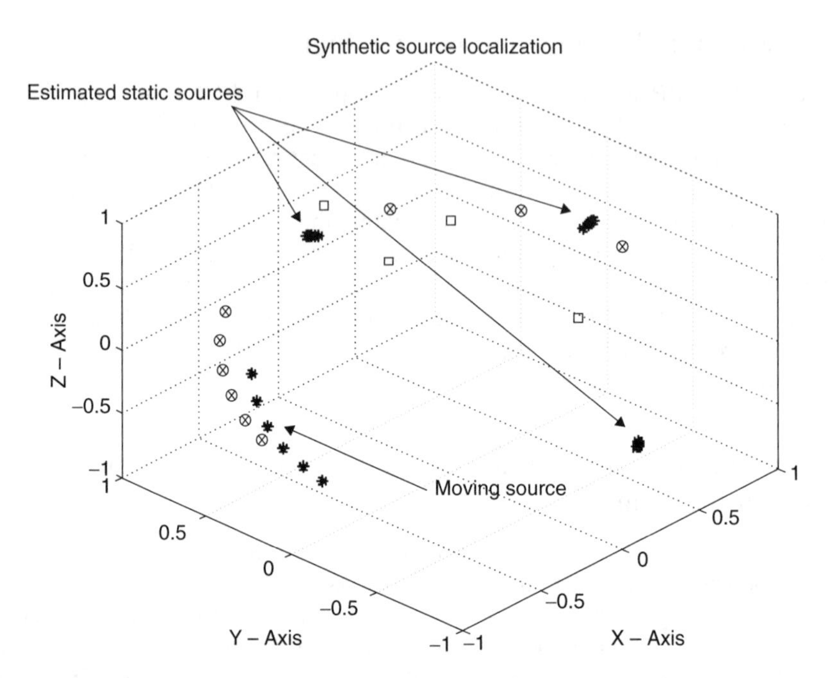

Figure 7.7 Detection and localization of moving sources using BSS followed by the LS-based geometrical localization. The circles denote the positions of the actual sources in different time instants and the strikes are the estimated positions of the sources for different time instants

this approach the signals from all EEG channels are treated as realizations of a multivariate stochastic process. A short-time DTF (SDTF) was also developed [95] for estimation and evaluation of AR coefficients for short time epochs of the EEGs.

MP and SDTF have been performed for analysis of the EEG activity during planning of self-paced finger movements. The results will be discussed with respect to representation of the features of cortical activities during voluntary action. The MP has been applied to decomposition of the EEGs to obtain reference-free data and averaged maps of power were constructed. ERP/ERS were calculated as described in Section 7.1.1.

The SDTF based on MVAR has been applied to a k-channel EEG. Similar to a one-dimensional AR, the MVAR model can be illustrated as

$$\mathbf{X}(n) = \sum_{i=1}^{p} \mathbf{A}_i \mathbf{X}(n-i) + \mathbf{E}(n) \tag{7.19}$$

or

$$\mathbf{E}(n) = -\sum_{i=0}^{p} \mathbf{A}_i \mathbf{X}(t-i) \quad \text{for } \mathbf{A}_0 = -1 \tag{7.20}$$

or in the frequency domain

$$\mathbf{X}(\omega) = \mathbf{A}^{-1}(\omega)\mathbf{E}(\omega) = \mathbf{H}(\omega)\mathbf{E}(\omega) \tag{7.21}$$

where $\mathbf{X}$ is the data vector, $\mathbf{E}$ is the vector of white noise samples, $\mathbf{A}_i$, $i = 1, \ldots, p$, are the model coefficient matrices, and p is the prediction order. Model order is normally selected at the point where the model error does not decrease considerably. A well-known criterion called Akaike information criterion (AIC) [97] has been widely used for this purpose. According to the AIC, the correct model order will be the value of p that makes the following criterion minimum:

$$\mathrm{AIC}(p) = 2\log[\det(\mathbf{V})] + \frac{2kp}{N} \tag{7.22}$$

where $\mathbf{V}$ is variance matrix of the model noise $\mathbf{E}(n)$, N is the data length for each channel, and k is the number of channels. $\mathbf{A}$ is determined using a correlation approach similar to that used by the Durbin algorithm. Then the DTF (or SDTF) describing the transition from channel i to channel j at frequency ω is defined as

$$\theta_{ji}^2(\omega) = |H_{ji}(\omega)|^2 \tag{7.23}$$

These transitions can be illustrated for different EEG channels for left and right hand finger movements, as depicted in Figures 7.8 and 7.9 respectively.

The direction of signal source movement is realized from the cross correlations between signals, which are computed for different time shifts in the procedure of correlation $\mathbf{R}(n)$ matrix estimation. These time shifts are translated into phase shifts by transformation to the frequency domain. The phase dependencies between channels are reflected in the transfer matrix. The DTF values express the direction of a signal component in a certain frequency (not the amount of delay) [98]. Analysis of the DTF values, however, will be difficult when the number of channels increases, resulting in an increase in the number of MVAR coefficients.

7.7 Estimation of Cortical Connectivity

The coherency of brain activities as described in the previous section may be presented from a different perspective, namely brain connectivity. This concept plays a central role in neuroscience. Temporal coherence between the activities of different brain areas are often defined as functional connectivity, whereas the effective connectivity is defined as the simplest brain circuit that would produce the same temporal relationship as observed experimentally between cortical regions [99]. A number of approaches have been proposed to estimate how different brain areas are working together during motor and cognitive tasks from the EEG and fMRI data [100–102].

Structural equation modelling (SEM) [103] has also been used to model such activities from high-resolution (both spatial and temporal) EEG data. Anatomical and physiological constraints have been exploited to change an underdetermined set of equations to a determined one.

The SEM consists of a set of linear structural equations containing observed variables and parameters defining causal relationships among the variables. The variables can be endogenous (i.e. independent from the other variables in the model) or exogenous (independent from the model itself).

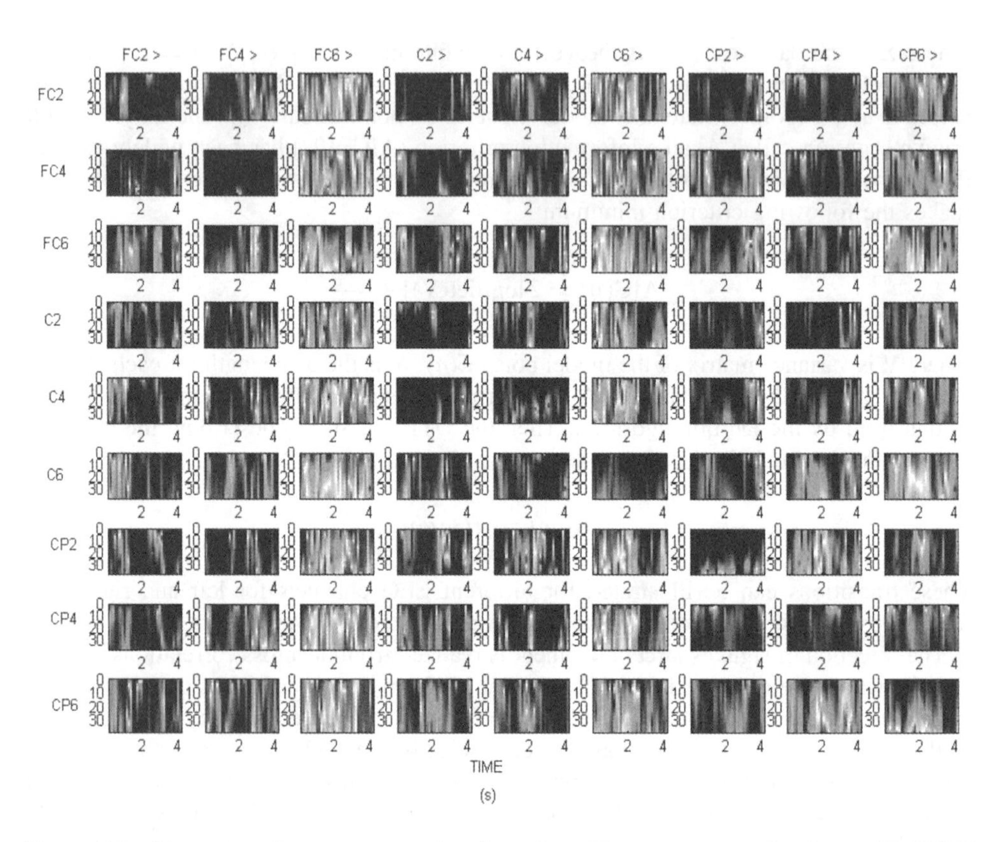

Figure 7.8 Illustration of source propagation from the coherency spectrum for the specified EEG channels for the left hand

Consider a set of variables (expressed as deviations from their means) with N observations. The SEM for these variables may be defined as

$$y = \mathbf{B}y + \mathbf{\Gamma}x + \xi \tag{7.24}$$

where y is an $m \times 1$ vector of dependent (endogenous) variables, x is an $n \times 1$ vector of independent (exogenous) variables, ξ is the $m \times 1$ vector of equation errors (random disturbances), and $\mathbf{B}$ $(m \times m)$ and $\mathbf{\Gamma}$ $(m \times n)$ are respectively the coefficient matrices of the endogenous and exogenous variables; ξ is assumed to be uncorrelated with the data and $\mathbf{B}$ to be a zero-diagonal matrix.

If z is a vector containing all the $p = m + n$ exogenous and endogenous variables in the following order

$$\mathbf{z}^{\mathrm{T}} = [x_1 \ldots x_n \, y_1 \ldots y_m] \tag{7.25}$$

then the observed covariances can be expressed as

$$\mathbf{\Sigma}_{\mathrm{obs}} = \frac{1}{N-1}\mathbf{Z}.\mathbf{Z}^{\mathrm{T}} \tag{7.26}$$

Figure 7.9 Illustration of source propagation from the coherency spectrum for the specified EEG channels for the right hand

where $\mathbf{Z}$ is the $p \times N$ matrix of the p observed variables for N observations. The covariance matrix implied by the model can be obtained as follows:

$$\boldsymbol{\Sigma}_{\text{mod}} = \begin{bmatrix} E[\boldsymbol{x}\boldsymbol{x}^{\mathrm{T}}] & E[\boldsymbol{x}\boldsymbol{y}^{\mathrm{T}}] \\ E[\boldsymbol{y}\boldsymbol{x}^{\mathrm{T}}] & E[\boldsymbol{y}\boldsymbol{y}^{\mathrm{T}}] \end{bmatrix} \tag{7.27}$$

where if $E[\boldsymbol{x}\boldsymbol{x}^{\mathrm{T}}] = \Phi$, then

$$E[\boldsymbol{x}\boldsymbol{y}^{\mathrm{T}}] = ((\mathbf{I} - \mathbf{B})^{-1}\boldsymbol{\Gamma}\boldsymbol{\Phi})^{\mathrm{T}} \tag{7.28}$$

$$E[\boldsymbol{y}\boldsymbol{x}^{\mathrm{T}}] = (\mathbf{I} - \mathbf{B})^{-1}\boldsymbol{\Gamma}\boldsymbol{\Phi} \tag{7.29}$$

and

$$E[\boldsymbol{y}\boldsymbol{y}^{\mathrm{T}}] = (\mathbf{I} - \mathbf{B})^{-1}(\boldsymbol{\Gamma}\boldsymbol{\Phi}\boldsymbol{\Gamma}^{\mathrm{T}} + \boldsymbol{\Psi})((\mathbf{I} - \mathbf{B})^{-1})^{\mathrm{T}} \tag{7.30}$$

where $\boldsymbol{\Psi} = E[\boldsymbol{\xi}\ \boldsymbol{\xi}^{\mathrm{T}}]$. With no constraints, the problem of the minimization of the differences between the observed covariances and those implied by the model is underdetermined mainly because the number of variables ($\mathbf{B}$, $\boldsymbol{\Gamma}$, $\boldsymbol{\Phi}$, and $\boldsymbol{\Psi}$) is greater than the number of equations $(m + n)(m + n + 1)/2$. The significance of SEM is that it eliminates

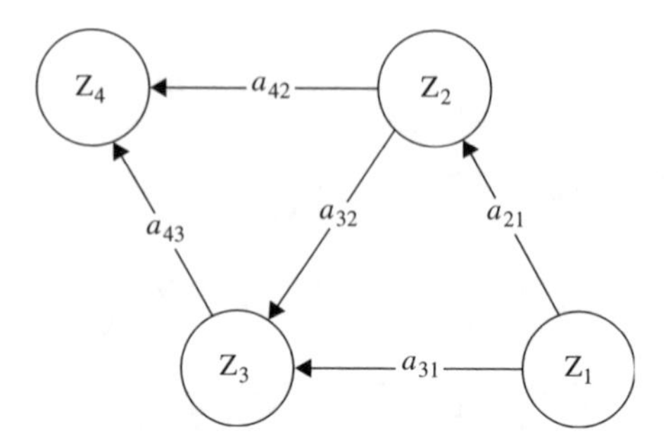

Figure 7.10 Connectivity pattern imposed in the generation of simulated signals. Values on the links represents the connection strength

some of the connections in the connectivity map based on some *a priori* anatomical and physiological (and possibly functional) information. For example, in Figure 7.10 if the connection a_{42} is not in the hypothesized model it may be set to zero.

Therefore, if r is the number of parameters to be estimated, then $r \leq (m+n)(m+n+1)/2$. The parameters are estimated by minimizing a function of the observed and implied covariances. The most widely used objective function for SEM is the maximum likelihood (ML) function [99]

$$F_{\mathrm{ML}} = \log|\boldsymbol{\Sigma}_{\mathrm{mod}}| + \mathrm{tr}(\boldsymbol{\Sigma}_{\mathrm{obs}}.\boldsymbol{\Sigma}_{\mathrm{mod}}^{-1}) - \log|\boldsymbol{\Sigma}_{\mathrm{obs}}| - p \tag{7.31}$$

where p is the number of observed variables (endogenous + exogenous).

7.8 Summary and Conclusions

BCI has been rapidly growing during the last three decades. A review of the ongoing research has been provided in this chapter. Static features measured in different EEG conventional frequency bands have been widely used in the classification of finger, arm, and leg movements. Dynamic features such as those characterizing the motion of movement-related sources have also been considered recently. Finally, estimation of the cortical connectivity patterns provides a new tool in evaluation of the directivity of brain signals and localization of the movement-related sources.

Although the advances in signal processing, especially in detection, separation, and classification of brain signals, have lead to very exciting results in BCI, as yet not all physiological, anatomical, and functional constraints have been taken into account. However, no robust solution to application of BCI for paralysed subjects exists. The effectiveness of the solution depends on the type and the level of subject disabilities. Moreover, there has not been any attempt to provide BCIs for subjects suffering mental disorders.

Often, EEG patterns change with time. Visual [35,36], auditory [37], and other types of feedback BCI systems [104] seem to provide more robust solutions in the presence of these changes. Development of on-line adaptive BCI systems, such as that in Reference [15], enhances the application of BCI in various areas. A complete feedback system, however, requires more thought and research to be undertaken.

Generally, to achieve a clinically useful BCI (invasive or noninvasive) system stable, low noise, and long recordings from multiple brain regions/electrodes are necessary. In addition, computationally efficient algorithms have to be developed in order to cope with the real-time applications. On the other hand, the subjects should learn how to use brain plasticity to incorporate prosthetic devices into the body representation. This will make the prosthetic feel like a natural part of the body of the subject, and thereby enhance the lifestyle of the subject, our ultimate aim!

References

[1] Wolpaw, J. R., Birbaumer, N., Heetderks, W. J., McFarland, D. J., Peckham, P. H., Schalk, G., Donchin, E., Quatrano, L. A., Robinson, C. J., and Vaughan, T. M., 'Brain–computer interface technology: a review of the first international meeting', *IEEE Trans. Rehabil. Engng*, **8**(2), 2000, 164–173.

[2] Vidal, J. J., 'Direct brain–computer communication', *Ann. Rev. Biophys. Bioengng*, **2**, 1973, 157–158.

[3] Vidal, J. J., 'Real-time detection of brain events in EEG', *IEEE Proc.*, **65**, 1977, 633–664.

[4] Chapin, J. K., Moxon, K. A., Markowitz, R. S., and Nicolelis, M. A., 'Real-time control of a robot arm using simultaneous recorded neurons in the motor cortex, *Nature Neurosci.*, **2**, 1999, 664–670.

[5] Lebedev, M. A., and Nicolelis, M. A. L., 'Brain–machine interfaces: past, present and future', *Trends in Neurosci.*, **29**(9), 2006.

[6] Blankertz, B., Dorhege, D., Krauledat, M., Müller, K.-R., Kunzmann, V., Losch, F., and Curio, G., 'The Berlin brain computer interface: EEG-based communication without subject training', *IEEE Trans. Neural System Rehabil. Engng*, **14**, 2006, 147–152.

[7] Blankertz, B., Dornheg, G., Schäfer, C., Krepki, R., Müller, K.-R., Kunzmann, V., Losch, F., and Curio, G., 'Boosting bit rates and error detection for the classification of fast-paced motor commands based on single-trial EEG analysis', *IEEE Trans. Neural System Rehabil. Engng.*, **11**, 2003, 127–131.

[8] Blankertz, B., Müller, K.-R., Curio, G., Vaughan, T. M., Schalk, G., Wolpaw, J. R., Schlogl, A., Neuper, C., Pfurtscheller, G., Hinterberger, T., Schroder, M., and Birbaumer, N., 'The BCI competition 2003: progress and perspectives in detection and discrimination of EEG single trials', *IEEE Trans. Biomed. Engng*, **51**, 2004, 1044–1051.

[9] Wolpaw, J. R., McFarland, D. J., and Vaughan, T. M., 'Brain–computer interface research at the Wadsworth Centre', *IEEE Trans. Neural System Rehabil. Engng*, **8**, 2000, 222–226.

[10] Wolpaw, J. R., and McFarland, D. J., 'Control of two-dimensional movement signal by a non-invasive brain–computer interface in human', *Natl Acad. Sci. (USA)*, **101**, 2003, 17849–17854.

[11] Peters, B. O., Pfurtscheller, G., and Flyvbjerg, H., 'Automatic differentiation of multichannel EEG signals', *IEEE Trans. Biomed. Engng*, **48**, 2001, 111–116.

[12] Müller-Putz, G. R., Neuper, C., Rupp, R., Keinrath, C., Gerner, H., and Pfurtscheller, G., 'Event-related beta EEG changes during wrist movements induced by functional electrical stimulation of forearm muscles in man', *Neurosci. Lett.*, **340**, 2003, 143–147.

[13] Müller-Putz, G. R., Scherer, R., Pfurtscheller, G., and Rupp, R., 'EEG-based neuroprosthesis control: a step towards clinical practice', *Neurosci. Lett.*, **382**, 2005, 169–174.

[14] Pfurtscheller, G., Muller, G. R., Pfurtscheller, J., Gerner, H. J., and Rupp, R., 'Thought-control of functional electrical stimulation to restore hand grasp in a patient with tetraplegia', *Neurosci. Lett.*, **351**, 2003, 33–36.

[15] Vidaurre, C., Schogl, A., Cabeza, R., Scherrer, R., and Phertscheller, G., 'A fully on-line adaptive BCI', *IEEE Trans. Biomed. Engng*, **53**(6), 2006, 1214–1219.

[16] Millan, J. D. R., Mourino, J., Franze, M., Cinotti, F., Heikkonen, J., and Babiloni, F., 'A local neural classifier for the recognition of EEG patterns associated to mental tasks', *IEEE Neural Network*, **13**, 2002, 678–686.

[17] Millan, J. D. R., and Mourino, J., 'Asynchronous BCI and local neural classifiers: an overview of the adaptive brain interface project', *IEEE Trans. Neural System Rehabil. Engng*, **11**, 2003, 1214–1219.

[18] Millan, J. D., Renkens, F., Mourino, J., and Gerstner, W., 'Noninvasive brain-actuated control of a mobile robot by human EEG', *IEEE Trans. Biomed. Engng*, **53**, 2004, 1214–1219.

[19] Birbaumer, N., Ghanayim, N., Hinterberger, T., Iversen, I., Kotchoubey, B., Kubler, A., Perelmouter, J., Taub, E., and Flor, H., 'A spelling device for the paralysed', *Nature*, **398**, 1999, 297–298.

[20] Wolpaw, J. R., McFarland, D., and Pfurtscheller, G., 'Brain computer interfaces for communication and control', *Clin. Neurophysiol.*, **113**(6), 2002, 767–791.

[21] Hinterberger, T., Veit, R., Wilhelm, B., Weiskopf, N., Vatine, J.-J., and Birbaumer, N., 'Neuronal mechanisms underlying control of a brain–computer interface', *Eur. J. Neurosci.*, **21**, 2005, 3169–3181.

[22] Kubler, A., Kotchoubey, B., Kaiser, J., Wolpaw, J. R., and Birbaumer, N., 'Brain–computer communication: unlocking the locked-in', *Psychol. Bull.*, **127**, 2001, 358–375.

[23] Kubler, A., Kotchoubey, B., Kaiser, J., Wolpaw, J. R., and Birbaumer, N., 'Brain–computer communication: self regulation of slow cortical potentials for verbal communication', *Arch. Phys. Med. Rehabil.*, **82**, 2001, 1533–1539.

[24] Obermaier, B., Muller, G. R., and Pfurtscheller, G., 'Virtual keyboard controlled by spontaneous EEG activity', *IEEE Trans. Neural System Rehabil. Engng*, **11**, 2003, 422–426.

[25] Obermaier, B., Muller, G. R., and Pfurtscheller, G., 'Information transfer rate in a five-classes brain–computer interface', *IEEE Trans. Neural System Rehabil. Engng*, **9**, 2001, 283–288.

[26] Wolpow, J. R., 'Brain computer interfaces (BCIs) for communication and control: a mini review', *Suppl. Clin. Neurophysiol.*, **57**, 2004, 607–613.

[27] Birbaumer, N., Weber, C., Neuper, C., Buch, E., Haagen, K., and Cohen, K., 'Brain–computer interface research: coming of age', *Clin. Neurophysiol.*, **117**, 2006, 479–483.

[28] Schröder, M. I., Lal, T. N., Hinterberger, T., Bogdan, M., Hill, N. J., Birbaumer, N., Rosenstiel, W., and Schölkopf, B., 'Robust EEG channel selection across subjects for brain–computer interfaces', *EURASIP J. Appl. Signal Proces.*, **19**, 2005, 3103–3112.

[29] Kim, S.-P., Rao, Y. N., Erdogmus, D., Sanchez, J. C., Nicolelis, M. A. L., and Principe, J. C., 'Determining patterns in neural activity for reaching movements using nonnegative matrix factorization', *EURASIP J. Appl. Signal Proces.* **19**, 2005, 3113–3121.

[30] Donchin, E., Spencer, K. M., and Wijesinghe, R., 'The mental prosthesis: assessing the speed of a P300-based brain–computer interface', *IEEE Trans. Rehabil. Engng*, **8**, 2000, 174–179.

[31] Bayliss, J. D., 'A flexible brain–computer interface', PhD Thesis, University of Rochester, New York, 2001.

[32] Molina, G. A., Ebrahimi, T., and Vesin, J.-M., 'Joint time–frequency–space classification of EEG in a brain computer interface application', *EURASIP J. on Appl. Signal Proces.*, **7**, 2003, 713–729.

[33] Middendorf, M., McMillan, G., Calhoun, G., and Jones, K. S., 'Brain–computer interfaces based on the steady-state visual-evoked response', *IEEE Trans. Rehabil. Engng*, **8**(2), 2000, 211–214.

[34] Kübler, A., Kotchubey, B., and Salzmann, H. P., 'Self regulation of slow cortical potentials in completely paralysed human patients', *Neurosci. Lett.*, **252**(3), 1998, 171–174.

[35] McFarland, D. J., McCane, L. M., and Wolpaw, J. R., 'EEG based communication and control: short-term role of feedback', *IEEE Trans. Rehabil. Engng*, **7**(1), 1998, 7–11.

[36] Neuper, C., Schlogl, A., and Phertscheller, G., 'Enhancement of left-right sensorimotor imagery', *J. Clin. Neurophysiol.*, **4**, 1999, 373–382.

[37] Hinterberger, T., Neumann, N., Pham, M., Kübler, A., Grether, A., Hofmayer, N., Wilhelm, B., Flor, H., and Birbaumer, Niels, 'A multimodal brain-based feedback and communication system', *Expl Brain Res.*, **154**(4), 2004, 521–526.

[38] Bayliss, J. D., and Ballard, D. H., 'A virtual reality testbed for brain–computer interface research', *IEEE Trans. Rehabil. Engng*, **8**, 2000, 188–190.

[39] Nowlis, D. P., and Kamiya, J., 'The control of electroencephalographic alpha rhythms through auditory feedback and the associated mental activity', *Psychophysiology*, **6**, 1970, 476–484.

[40] Plotkin, W. B., 'On the self-regulation of the occipital alpha rhythm: control strategies, states of consciousness, and the role of physiological feedback', *J. Expl Psychol. Gen.*, **105**(1), 1976, 66–99.

[41] Sterman, M. B., Macdonald, L. R., and Stone, R. K., 'Biofeedback training of the sensorimotor electroencephalogram rhythm in man: effects on epilepsy', *Epilepsia*, **15**, 1974, 395–416.

[42] Whyricka, W., and Sterman, M., 'Instrumental conditioning of sensorimotor cortex spindles in the walking cat', *Psychol. Behav.*, **3**, 1968, 703–707.

[43] Black, A. H., 'The direct control of neural processes by reward and punishment', *Am. Sci.*, **59**, 1971, 236–245.

[44] Pfurtscheller, G., Leeb, R., Keinrath, C., Friedman, D., Neuper, C., Guger, C., and Slater, M., 'Walking from thought', *Brain Res.*, **1071**, 2006, 145–152.

[45] Beverina, F., Palmas, G., Silvoni, S., Piccione, F., Camillo, S., and Giove, S., 'User adaptive BCIs: SSVEP and P300 based interfaces', *Psychol. J.*, **1**(4), 2003, 331–354.

[46] Lalor, E. C., Kelly, S. P., Finucane, C., Burke, R., Smith, R., Reilly, R. B., and McDarby, G., 'Steady-state VEP-based brain–computer interface control in an immersive 3D gaming environment', *EURASIP J. Appl. Signal Process.*, **19**, 2005, 3156–3164.

[47] Fetz, E. E., 'Operant conditioning of cortical unit activity', *Science*, **163**, 1969, 955–958.

[48] Fetz, E. E., 'Are movement parameters recognizably coded in activity of single neurons?', *Behav. Brain Sci.*, **15**, 1972, 679–690.

[49] Fetz, E. E., and Baker, M. A., 'Operantly conditioned patterns on precentral unit activity and correlated responses in adjacent cells and contralateral muscles', *J. Neurophysiol.*, **36**, 1973, 179–204.

[50] Fetz, E. E., and Finocchio, D. V., 'Operant conditioning of specific patterns of neural and muscular activity', *Science*, **174**, 1971, 431–435.

[51] Fetz, E. E., and Finocchio, D. V., 'Operant conditioning of isolated activity in specific muscles and precentral cells', *Brain Res.*, **40**, 1972, 19–23.

[52] Fetz, E. E., 'Correlations between activity of motor cortex cells and arm muscles during operantly conditioned response patterns', *Expl Brain Res.*, **23**, 1975, 217–240.

[53] Schmidt, E. M., 'Single neuron recording from motor cortex as a possible source of signals for control of external devices', *Ann. Biomed. Engng*, **8**, 1980, 339–349.

[54] Mehring, C., Rickert, J., Vaadia, E., Cardoso de Oliveira, S., Aertsen, A., and Rotter, S., 'Interface of hand movements from local field potentials in monkey motor cortex', *Natl Neurosci.*, **6**, 2003, 1253–1254.

[55] Rickert, J., Oliveira, S. C., Vaadia, E., Aertsen, A., Rotter, S., and Mehring, C., 'Encoding of movement direction in different frequency ranges of motor cortical local field potentials', *J. Neurosci.*, **25**, 2005, 8815–8824.

[56] Pezaran, B., Pezaris, J. S., Sahani, M., Mitra, P. P., and Andersen, R. A., 'Temporal structure in neuronal activity during working memory in macaque parietal cortex', *Natl Neurosci.*, **5**, 2002, 805–811.

[57] Scherberger, H., Jarvis, M. R., and Andersen, R. A., 'Cortical local field potential encodes movement intentions in the posterior parietal cortex', *Neuron.*, **46**, 2005, 347–354.

[58] Serruya, M. D., Hatsopoulos, N. G., Paninski, L., Fellows, M. R., and Donoghue, J. P., 'Instant neural control of a movement signal', *Nature*, **416**, 2002, 141–142.

[59] Taylor, D. M., Tillery, S. I., and Schwartz, A. B., 'Direct cortical control of 3D neuroprosthetic devices', *Science*, **296**, 2002, 1829–1832.

[60] Tillery, S. I., and Taylor, D. M., 'Signal acquisition and analysis for cortical control of neuroprosthetic', *Current Opinion Neurobiol.*, **14**, 2004, 758–762.

[61] Musallam, S., Corneil, B. D., Greger, B., Scherberger, H., and Andersen, R. A., 'Cognitive control signals for neural prosthetics', *Science*, **305**, 2004, 258–262.

[62] Patil, P. G., Carmena, J. M., Nicolelis, M. A., and Turner, D. A., 'Ensemble recordings of human subcortical neurons as a source of motor control signals for a brain-machine interface', *Neurosurgery*, **55**, 2004, 27–35.

[63] Kennedy, P. R., Bakay, R. A. E., Moore, M. M., Adams, K., and Goldwaithe, J., 'Direct control of a computer from the human central nervous system', *IEEE Trans. Rehabil. Engng*, **8**(2), 2000, 198–202.

[64] Pfurstcheller Jr, G., Stancak, A., and Neuper, C., 'Post-movement beta synchronization. A correlate of an idling motor area?, *Electroencephalogr. Clin. Neurophysiol.*, **98**, 1996, 281–293.

[65] Esch, W., Schimke, H., Doppelmayr, M., Ripper, B., Schwaiger, J., and Pfurtscheller, G., 'Event related desynchronization (ERD) and the Dm effect: does alpha desynchronization during encoding predict later recall performance?', *Int. J. Psychophysiol.*, **24**, 1996, 47–60.

[66] Pfurtscheller, G., 'EEG even-related desynchronization (ERD) and event-related synchronization (ERS)', Chapter 53, in *Electroencephalography, Basic Principles, Clinical Applications, and Related Fields*, Eds E. Niedermeyer and F. Lopes da Silva, Lippincott, Williams and Wilkins, Philadelphia, Pennsylvania, 1999, pp. 958–966.

[67] Arroyo, S., Lesser, R. P., Gordon, B., Uematsu, S., Jackson, D., and Webber, R., 'Functional significance of the mu rhythm of human cortex: an electrophysiological study with subdural electrodes', *Electroencephalogr. Clin. Neurophysiol.*, **87**, 1993, 76–87.

[68] Leuthardt, E. C., Schalk, G., Wolpaw, J. R., Ojemann, J. G., and Moran, D. W., 'A brain–computer interface using electrocorticographic signals in humans', *J. Neural Engng*, **1**, 2004, 63–71.

[69] Pfurtscheller, G., Stancak Jr, A., and Edlinger, G., 'On the existence of different types of central beta rhythms below 30 Hz', *Electroencephalogr. Clin. Neurophysiol.*, **102**, 1997, 316–325.

[70] Neuper, C., and Pfurtscheller, G., 'Post movement synchronization of beta rhythms in the EEG over the cortical foot area in man', *Neurosci. Lett.*, **216**, 1996, 17–20.

[71] Pfurtscheller, G., Müller-Putz, G. R., Pfurtscheller, J., and Rupp, R., 'EEG-based asynchronous BCI controls functional electrical stimulation in a tetraplegic patient', *EURASIP J. App. Signal Proces.*, **19**, 2005, 3152–3155.

[72] Gratton, G., 'Dealing with artefacts: the EOG contamination of the event-related brain potentials over the scalp', *Electroencephalogr. Clin. Neurophysiol.*, **27**, 1969, 546.

[73] Lins, O. G., Picton, T. W., Berg, P., and Scherg, M., 'Ocular artefacts in EEG and event-related potentials, i: scalp topography', *Brain Topography*, **6**, 1993, 51–63.

[74] Celka, P., Boshash, B., and Colditz, P., 'Preprocessing and time–frequency analysis of new born EEG seizures', *IEEE Engng Med. Biol. Mag.*, **20**, 2001, 30–39.

[75] Jung, T. P., Humphies, C., and Lee, T. W., 'Extended ICA removes artefacts from electroencephalographic recordings', *Adv. Neural Inf. Proces. Systems*, **10**, 1998, 894–900.

[76] Joyce, C. A., Gorodnitsky, I., and Kautas, M., 'Automatic removal of eye movement and blink artefacts from EEG data using blind component separation', *Psychophysiology*, **41**, 2004, 313–325.

[77] Shoker, L., Sanei, S., and Chambers, J., 'Artefact removal from electro-encephalograms using a hybrid BSS–SVM algorithm', *IEEE Signal Process. Lett.*, **12**(10), October 2005.

[78] Peterson, D. A., Knight, J. N., Kirby, M. J., Anderson, C. W., and Thaut, M. H., 'Feature selection and blind source separation in an EEG-based brain–computer interface', *EURASIP J. Appl. Signal Process.*, **19**, 2005, 3128–3140.

[79] Coyle, D., Prasad, G., and McGinnity, T. M., 'A time–frequency approach to feature extraction for a brain–computer interface with a comparative analysis of performance measures', *EURASIP J. Appl. Signal Process.*, **19**, 2005, 3141–3151.

[80] Mørup, M., Hansen, L. K., Herrmann, C. S., Parnas, J., and Arnfred, S. M., 'Parallel factor analysis as an exploratory tool for wavelet transformed event-related EEG', *NeuroImage*, **29**(3), 2006, 938–947.

[81] Cohen, L., *Time Frequency Analysis*, Prentice-Hall Signal Processing Series, Prentice-Hall, Upper Saddle River, New Jersey, 1995.

[82] Shoker, L., Nazarpour, K., Sanei, S., and Sumich, A., 'A novel space–time–frequency masking approach for quantification of EEG source propagation, with application to brain computer interfacing', in Proceedings of EUSIPCO, Florence, Italy, 2006.

[83] Harshman, R. A., 'Foundation of the PARAFAC: models and conditions for an "explanatory" multimodal factor analysis', UCLA Work, Paper Phon. **16**, 1970, pp. 1–84.

[84] Carol, J. D., and Chang, J., 'Analysis of individual differences in multidimensional scaling via an N-way generalization of "Eckart–Young" decomposition', *Psychometrika*, **35**, 1970, 283–319.

[85] Möcks, J., 'Decomposing event-related potentials: a new topographic components model', *Biol. Psychol.*, **26**, 1988, 199–215.

[86] Field, A. S., and Graupe, D., 'Topographic component (parallel factor) analysis of multichannel evoked potentials: practical issues in trilinear spatiotemporal decomposition', *Brain Topogr.*, **3**, 1991, 407–423.

[87] Miwakeichi, F., Martinez-Montes, E., Valdes-Sosa, P. A., Nishiyama, N., Mizuhara, H., and Yamaguchi, Y., 'Decomposing EEG data into space–time–frequency components using parallel factor analysis', *NeuroImage*, **22**, 2004, 1035–1045.

[88] Bro, R., 'Multi-way analysis in the food industry: models, algorithms and applications', PhD Thesis, University of Amsterdam, and Royal Veterinary and Agricultural University, 1997.

[89] Nazarpour, K., Shoker, L., Sanei, S., and Chambers, J., 'Parallel space–time–frequency decomposition of EEG signals for brain computer interfacing', in Proceedings of the European Signal Processing Conference, Italy, 2006.

[90] Hjorth, B., 'An on-line transformation of EEG scalp potentials into orthogonal source derivations', *Electroencephalogr. Clin. Neurophysiol.*, **39**(5), 1975, 526–530.
[91] Pfurtscheller, G., and Neuper, C., 'Event-related synchronization of mu rhythm in the EEG over the cortical hand area in man', *Neurosci. Lett.*, **174**, 1994, 93–96.
[92] Ginter Jr, J., Blinowska, K. J., Kaminski, M., and Durka, P. J., 'Phase and amplitude analysis in time–frequency space – application to voluntary finger movement', *J. Neurosci. Methods*, **110**, 2001, 113–124.
[93] Mallat, S., and Zhang, Z., 'Matching pursuits with time–frequency dictionaries', *IEEE Trans. Signal Process.*, **41**(12), 1993, 3397–3415.
[94] Durka, P. J., Ircha, D., and Blinowska, K. J., 'Stochastic time–frequency dictionaries for matching pursuit', *IEEE Trans. Signal Process.*, **49**(3), 2001, 507–510.
[95] Durka, P. J., Ircha, D., Neuper, C., and Pfurtscheller, G., 'Time–frequency microstructure of event-related EEG desynchronization and synchronization', *Med. Biolog. Engng and Computing*, **39**(3), 2001, 315–321.
[96] Kaminski, M. J., and Blinowska, K. J., 'A new method of the description of the information flow in the structures', *Biol. Cybernetics*, **65**, 1991, 203–210.
[97] Bozdogan, H., 'Model-selection and Akaike's information criterion (AIC): the general theory and its analytical extensions', *Psychometrika*, **52**, 1987, 345–370.
[98] Ginter Jr, J., Bilinowska, K. J., Kaminski, M., and Durka, P. J., 'Phase and amplitude analysis in time–frequency–space – application to voluntary finger movement', *J. Neurosci. Methods*, **110**, 2001, 113–124.
[99] Astolfi, L., Cincotti, F., Babiloni, C., Carducci, F., Basilisco, A., Rossini, P. M., Salinari, S., Mattia, D., Cerutti, S., Dayan, D. B., Ding, L., Ying Ni, Bin He, and Babiloni, F., 'Estimation of the cortical connectivity by high resolution EEG and structural equation modelling: Simulations and application to finger tapping data', *IEEE Trans. Biomed. Engng*, **52**(5), 2005, 757–767.
[100] Gerloff, C., Richard, J., Hardley, J., Schulman, A. E., Honda, M., and Hallett, M., 'Functional coupling and regional activation of human cortical motor areas during simple, internally paced and externally paced finger movement', *Brain*, **121**, 1998, 1513–1531.
[101] Urbano, A., Babiloni, C., Onorati, P., and Babiloni, F., 'Dynamic functional coupling of high resolution EEG potentials related to unilateral internally triggered one-digit movement', *Electroencephalogr. Clin. Neurophysiol.*, **106**(6), 1998, 477–487.
[102] Jancke, L., Loose, R., Lutz, K., Specht, K., and Shah, N. J., 'Cortical activations during paced finger-tapping applying visual and auditory pacing stimuli', *Cognitive Brain Res.*, **10**, 2000, 51–56.
[103] Bollen, K. A., *Structural Equations with Latent Variable*, John Wiley & Sons, Inc., New York, 1989.
[104] Kauhanen, L., Palomaki, T., Jylamki, P., Aloise, F., Nuttin, M., and Millan, J. del R., 'Haptic feedback compared with visual feedback for BCI', in Proceedings of the 3rd International BCI Workshop and Training Course, 2006, 66–67.

Index

EEG Signal Processing S. Sanei and J. Chambers